D1797003

MEDICAL
INTELLIGENCE
UNIT

Marfan Syndrome:
A Primer for Clinicians and Scientists

Peter N. Robinson, M.D.

Institute of Medical Genetics
Charité University Hospital
Humboldt University
Berlin, Germany

Maurice Godfrey, Ph.D.

Department of Pediatrics and Center
for Human Molecular Genetics
Munroe-Meyer Institute
University of Nebraska Medical Center
Omaha, Nebraska, U.S.A.

Springer Science+Business Media, LLC

MARFAN SYNDROME:
A PRIMER FOR CLINICIANS AND SCIENTISTS
Medical Intelligence Unit

Copyright ©2004 Springer Science+Business Media New York
Originally published by Kluwer Academic Publishers in 2004
Softcover reprint of the hardcover 1st edition 2004

ISBN 978-1-4613-4757-6   ISBN 978-1-4419-9013-6 (eBook)
DOI 10.1007/978-1-4419-9013-6

Marfan Syndrome: A Primer for Clinicians and Scientists, edited by Peter N. Robinson and Maurice Godfrey, Landes / Kluwer dual imprint / Landes series: Medical Intelligence Unit

While the authors, editors and publisher believe that drug selection and dosage and the specifications and usage of equipment and devices, as set forth in this book, are in accord with current recommendations and practice at the time of publication, they make no warranty, expressed or implied, with respect to material described in this book. In view of the ongoing research, equipment development, changes in governmental regulations and the rapid accumulation of information relating to the biomedical sciences, the reader is urged to carefully review and evaluate the information provided herein.

Library of Congress Cataloging-in-Publication Data

Marfan syndrome : a primer for clinicians and scientists / [edited by] Peter N. Robinson, Maurice Godfrey.
    p. ; cm. -- (Medical intelligence unit)
  Includes bibliographic references and index.

  1. Marfan syndrome. I. Robinson, Peter N. (Peter Nicholas), 1963- II. Godfrey, Maurice. III. Series: Medical intelligence unit (Unnumbered : 2003)
  [DNLM: 1. Marfan Syndrome. WD 375 M326 2004]
  RC580.M37M37 2004
  616.7'73--dc22
                    2004015623

CONTENTS

EDITORS

Peter N. Robinson, M.D.
Institute of Medical Genetics
Charité University Hospital
Humboldt University
Berlin, Germany
peter.robinson@charite.de
Chapter 1

Maurice Godfrey, Ph.D.
Department of Pediatrics and Center
for Human Molecular Genetics
Munroe-Meyer Institute
University of Nebraska Medical Center
Omaha, Nebraska, U.S.A.
mgodfrey@unmc.edu
Chapter 10

CONTRIBUTORS

Emilio Arteaga-Solis
Department of Pediatrics
Mount Sinai School of Medicine
New York University
New York, New York, U.S.A.
Chapter 16

Clair Baldock
Wellcome Trust Centre
  for Cell-Matrix Research
Schools of Medicine
  and Biological Sciences
University of Manchester
Manchester, United Kingdom
Chapter 12

Frank Barthel
German Heart Institute Berlin
Berlin, Germany
Chapter 6

Boris Bätge
Klinikum Neustadt
Neustadt, Germany
Chapter 11

Catherine Boileau
INSERM
Clinique Maurice Lamy
Hôpital Necker-Enfants Malades
Paris, France
E-mail: boileau@necker.fr
Chapter 8

Duke E. Cameron
Division of Cardiac Surgery
  and the Dana and Albert "Cubby"
  Broccoli Center for Aortic Diseases
Johns Hopkins Medical Insitution
Baltimore, Maryland, U.S.A.
E-mail: dcameron@jhmi.edu
Chapter 5

Anne H. Child
Department of Cardiological Sciences
St George's Hospital Medical School
London, United Kingdom
E-mail: achild@sghms.ac.uk
Chapter 1

Gwenaëlle Collod-Béroud
Laboratoire de Génétique Moléculaire
Institut Universitaire de Recherche
  Clinique
Montpellier, France
E-mail: gwenaelle.beroud@igh.cnrs.fr
Chapter 8

Paul Coucke
Ghent University Hospital
Department Medical Genetics
Gent, Belgium
Chapter 7

Anne De Paepe
Ghent University Hospital
Department Medical Genetics
Gent, Belgium
E-mail : Anne.DePaepe@UGent.be
Chapter 7

Harry Dietz
Departments of Pediatrics, Medicine
  and Molecular Biology and Genetics
Howard Hughes Medical Institute
John Hopkins University School
  of Medicine
Baltimore, Maryland, U.S.A.
Chapter 16

Mark A. Gibson
Department of Pathology
University of Adelaide
Adelaide, South Australia, Australia
E-mail: mark.gibson@adelaide.edu.au
Chapter 13

Vincent L. Gott
Division of Cardiac Surgery
  and the Dana and Albert "Cubby"
  Broccoli Center for Aortic Diseases
Johns Hopkins Medical Insitution
Baltimore, Maryland, U.S.A.
E-mail: vgott@csurg.jhmi.jhu.edu
Chapter 5

Penny A. Handford
Division of Molecular
  and Cellular Biochemistry
Department of Biochemistry
University of Oxford
Oxford, United Kingdom
Chapter 15

Sumera N. Hasham
University of Texas Medical School
  at Houston
Houston, Texas, U.S.A.
Chapter 9

J. Louise Haston
Wellcome Trust Centre
  for Cell-Matrix Research
Schools of Medicine
  and Biological Sciences
University of Manchester
Manchester, United Kingdom
Chapter 12

Roland Hetzer
German Heart Institute Berlin
Berlin, Germany
Chapter 6

Cay M. Kielty
Wellcome Trust Centre
  for Cell-Matrix Research
Schools of Medicine
  and Biological Sciences
University of Manchester
Manchester, United Kingdom
Chapter 12

Bart Loeys
Ghent University Hospital
Department Medical Genetics
Gent, Belgium
Chapter 7

Victor A. McKusick
Institute of Medical Genetics
Johns Hopkins University School
  of Medicine
Baltimore, Maryland, U.S.A.
Introduction

Dianna M. Milewicz
University of Texas Medical School
  at Houston
Houston, Texas, U.S.A.
Dianna.M.Milewicz@uth.tmc.edu
Chapter 9

Luitgard Neumann
Institute of Human Genetics
Charité University Hospital
Humboldt University
Berlin, Germany
Chapter 1

Reinhard Pregla
German Heart Institute Berlin
Berlin, Germany
Chapter 6

Francesco Ramirez
Laboratory of Genetics
  and Organogenesis
Hospital for Special Surgery
  at Weill Medical College
  of Cornell University
New York, New York, U.S.A.
E-mail: ramirezf@hss.edu
Chapter 16

Dieter P. Reinhardt
Universität zu Lübeck
Institut für Medizinische
  Molekularbiologie
Lübeck, Germany
E-mail: reinhardt@molbio.uni-luebeck.de
Chapter 11

Thomas Rosenberg
Gordon Norrie Centre
  for Genetic Eye Diseases
National Eye Clinic
  for the Visually Impaired
Hellerup, Denmark
E-mail: roseeye@visaid.dk
Chapter 3

Maike Rybczynski
Clinic of Internal Medicine III
University Hospital
  Hamburg-Eppendorf
Hamburg, Germany
Chapter 4

Lynn Y. Sakai
Shriners Hospital for Children
Oregon Health and Science University
Portland, Oregon, U.S.A.
E-mail: LYS@SHCC.org
Chapter 14

Michael J. Sherratt
Wellcome Trust Centre
  for Cell-Matrix Research
Schools of Medicine
  and Biological Sciences
University of Manchester
Manchester, United Kingdom
Chapter 12

Michael Shindle
Department of Orthopaedic Surgery
The Johns Hopkins Hospital
Baltimore, Maryland, U.S.A.
Chapter 2

C. Adrian Shuttleworth
Wellcome Trust Centre
  for Cell-Matrix Research
Schools of Medicine
  and Biological Sciences
University of Manchester
Manchester, United Kingdom
Chapter 12

Paul Sponseller
Department of Orthopaedic Surgery
The Johns Hopkins Hospital
Baltimore, Maryland, U.S.A.
E-mail: psponse@jhmi.edu
Chapter 2

Kerstin Tiedemann
Universität zu Lübeck
Institut für Medizinische
　Molekularbiologie
Lübeck, Germany
Chapter 11

Yskert von Kodolitsch
Clinic of Internal Medicine III
University Hospital
　Hamburg-Eppendorf
Hamburg, Germany
E-mail: kodolitsch@uke.uni-hamburg.de
Chapter 4

Tim J. Wess
Centre for Extracellular Matrix Biology
Department of Biological Sciences
University of Stirling
Stirling, United Kingdom
Chapter 12

Pat Whiteman
Division of Molecular
　and Cellular Biochemistry
Department of Biochemistry
University of Oxford
Oxford, United Kingdom
Chapter 15

PREFACE

The Marfan syndrome is a common autosomal dominant hereditary disorder of connective tissue with prominent manifestations in the eye, skeleton, and cardiovascular system. The compilation of the clinical and pathological aspects of the Marfan syndrome began some 50 years ago by Professor Victor A. McKusick. In the intervening years his updates and, more recently, those of his disciples or students of his disciples, provided periodic refreshers of the "state of the art" in the Marfan syndrome. These updates were most often published in the genetics or cardiovascular literature or in books devoted to heritable connective tissue disorders. In the present volume we present a compendium dedicated to the Marfan syndrome and its related disorders and their associated genes and proteins.

The book is organized into four main sections covering the most important clinical aspects of the Marfan syndrome, mutation analysis in Marfan syndrome and related disorders, the molecular biology of fibrillin-1 and fibrillin-rich microfibrils, and the pathophysiology of Marfan syndrome.

Diagnosis and Treatment of the Marfan Syndrome

The diagnosis of Marfan syndrome is easy to make in individuals with a range of classic features in multiple organ systems. However, the extreme clinical variability of this syndrome, the age-dependent nature of many of the manifestations of Marfan syndrome together with the phenotypic overlap with other disorders and indeed with the healthy general population can present clinicians with difficulties. Careful and individualized phenotypic assessment is the cornerstone of the diagnostic process. The first chapter by Child, Neumann, and Robinson therefore presents an overview of diagnosis and treatment of Marfan syndrome.

Clinical manifestations of the skeletal system are the most obvious features of MFS in many individuals and may be the findings that first suggest the diagnosis. Sponseller and Shindle discuss issues surrounding the diagnosis and treatment of MFS-associated skeletal anomalies in Chapter 2. Dural ectasia (a widening of the dural sac) is rare in the normal population but has a high prevalence in persons with MFS. For this reason, dural ectasia is included as a major criterion in the Gent nosology. Sponseller and Shindle present guidelines for the diagnosis and evaluation of dural ectasia by CT or MRT imaging.

The eye is clinically involved in a majority of persons with MFS. Bilateral ectopia lentis represents a major criterion of the Gent nosology, but a series of other manifestations such as high myopia, glaucoma, and retinal detachment are not uncommon. Thomas Rosenberg offers an overview of MFS-associated ocular pathology, clinical symptoms, ophthalmologic differential diagnosis and treatment of ocular manifestations in Chapter 3.

Von Kodolitsch and Rybczynski provide an overview of the natural history of Marfan syndrome with a focus on cardiovascular manifestations as well as a comprehensive review of cardiovascular issues associated with Marfan syndrome. The author discusses current recommendations for monitoring and treating the cardiovascular manifestations of this disorder.

In the last several decades, there has been remarkable progress in the surgical treatment of aortic dilatation and dissection beginning with the introduction of the Bentall composite graft procedure. These developments have had a major positive impact on the life expectancy of persons with MFS. Cameron and Gott discuss indications for surgery of the ascending aorta, surgical techniques and long-term results in Chapter 5.

While aortic dilatation and dissection are the primary limiting factor for the life expectancy of untreated MFS patients, lesions in the atrioventricular valves, the myocardium and distal aortic segments are not uncommon in persons with MFS and can have significant clinical consequences that require careful clinical attention. In Chapter 6, Hetzer, Pregla, and Barthel review these aspects of MFS based upon their experience with a large group of MFS patients followed over a period of ten years.

Mutation Analysis in Fibrillinopathies and Related Disorders

Dietz and coworkers discovered the first mutation in *FBN1* two unrelated individuals with MFS in 1991. *FBN1* mutation analysis has proved to be surprisingly difficult, although significant advances have been made in recent years. In Chapter 7, De Paepe, Loeys, and Coucke provide an overview of *FBN1* mutation analysis in Marfan syndrome. Most *FBN1* mutations discovered to date have been unique to one individual or family, and mutations have been found in almost all of the 65 exons of *FBN1*. Therefore, it is quite useful to be able to compare an identified sequence change in *FBN1* with previously identified mutations. In Chapter 8, Collod-Béroud and Boileau present the Marfan Mutation database, which is a very useful resource for gathering information on *FBN1* mutations. The database can be used to determine if a given sequence change has been identified previously, or as the basis for studies on the distribution of individual mutation types, and for genotype-phenotype correlation studies.

Ascending aortic dilatation and dissection is a major manifestation of Marfan syndrome. In rare cases, *FBN1* mutations have been associated with isolated aneurysms of the ascending aorta, and there are other genetic disorders associated with thoracic aortic aneurysm and dissection (TAAD). Hasham and Milewicz provide a discussion of these disorders in Chapter 9.

Chapter 10 by Maurice Godfrey offers an overview of congenital contractural arachnodactyly (CCA), a disorder with phenotypic similarities to MFS that is caused by mutations in the gene for fibrillin-2 (*FBN2*). Affected individuals often have a Marfanoid appearance in addition to features not generally seen in MFS such as congenital contractures and crumpled ears. The distinct clinical phenotype presumably is a reflection of functional differences between fibrillin-1 and fibrillin-2, although the precise molecular basis thereof is just beginning to be unraveled.

The Biology of Fibrillin-Rich Microfibrils

The first two sections of this book reviewed the clinical phenotypes; molecular bases; diagnostic options and outcomes; and treatment modalities of the Marfan syndrome and related heritable connective tissue disorders. In this section, we begin to explore the gene products that make up the extracellular matrix microfibrils.

Tiedemann, Bätge, and Reinhardt begin by describing the assembly of these microfibrils. They provide a step by step description of the process from fibrillin monomers to tissue microfibrils. They carefully review their and others' studies of the roles of processing, cross-links, and interactions with other glycoproteins in microfibril assembly. They also discuss the affects of fibrillin mutations in the Marfan syndrome on microfibrillar assembly. Finally, they outline the tasks that remain to truly understand microfibrillar assembly with the ultimate goal towards novel therapeutics.

Microfibrils are structural matrix macromolecules. They assemble to form tissues with specific physical properties. Kielty and her co-authors discuss these physical properties. They begin by describing the tissue organization of microfibrils followed by analysis of the biomechanical properties of microfibrils.

It is important to keep in mind that microfibrils are composed of more than fibrillin alone. In the final chapter of this section, Mark Gibson provides a compendium of the numerous components of elastin-associated microfibrils.

The Pathophysiology of Marfan Syndrome

The first two sections summarized the phenotypes and genotypes that are observed in the Marfan syndrome and related disorders. The previous section reviewed the biology of fibrillin and microfibrils. The final three chapters review the pathophysiology of fibrillin mutations.

Lynn Sakai describes the possible mechanisms that lead from mutations in *FBN1* to fragmentation and loss fibrillin-1 containing microfibrils to the Marfan syndrome. This chapter also describes a preliminary blood test for fibrillin-1 fragments in blood. The author notes that control blood samples have low levels of fibrillin-1 fragments. Do people with the Marfan syndrome have significantly higher levels? Do levels of fibrillin-1 fragments in the blood of people with the Marfan syndrome increase with disease severity and progression? The potential promise of these analyses awaits ongoing investigations. The chapter also delineates some of the other extracellular elements that together with microfibrils provide normal tissue homeostasis.

Whiteman and Handford explore the structure of fibrillin using peptides of combinations of the two principal domain types in fibrillin. They review their elegant x-ray crystallographic studies that provide a glimpse into the structure of fibrillin. They use this "structural knowledge" to begin the dissection of the structural alterations caused by mutations in fibrillin. These data provide a framework of how fibrillin mutations may lead to pathogenesis of the Marfan syndrome.

Readers are cautioned that the domain identification nomenclature that Whiteman and Handford use is not the same as that used to identify the individual exons in the chapters that discuss fibrillin mutations.

In the final chapter in this book Arteaga-Solis, Dietz, and Ramirez outline the phenotypes generated by targeting *FBN1* and *FBN2* in murine models. They describe three different *FBN1* mutations and the pathogenetic insights that have been gained from these models. They also describe the syndactyly that results in *FBN2* null mice, a phenotype that is not observed in the known human fibrillin-2 disease, congenital contractural arachnodactyly. They present their evidence regarding the interactions of signaling molecules or cytokines and the extracellular matrix. The authors end by speculating on the mechanisms of pathology in fibrillinopathies and the potential for novel treatment modalities.

Peter N. Robinson and Maurice Godfrey
Berlin, May 2004

The editors are very grateful for the timely cooperation and comprehensive reviews of the authors and the patient, professional, and encouraging support of the publishers Dr. Ron Landes and Cynthia Conomos.

PNR thanks his three niñas Elisabeth, Sophie, and Anna for their love and patience.

MG thanks Matilde, Maximilian, Alessandro, and Guillermo for their constant love and support on both sides of Atlantic.

Finally, the editors extend a special and personal thanks to Professor McKusick, singularly recognized as the father of genetic medicine and a modern Vesalius for Medical Genetics. No review of the Marfan syndrome would be complete without Professor McKusick's historical perspective. We are honored and delighted to begin this book with Victor A. McKusick's semiautobiographic account.

Introduction

Historical Introduction
The Marfan Syndrome: From Clinical Delineation to Mutational Characterization, a Semiautobiographic Account

Victor A. McKusick

In 1876, E. Williams,[1] an ophthalmologist in Cincinnati, Ohio, described ectopia lentis in a brother and sister who were exceptionally tall and had been loosejointed from birth. Although there is a Williams syndrome that has aortic manifestations (supravalvar aortic stenosis), the name Williams was never associated with the disorder we now call Marfan syndrome. The reason is clear: Williams was geographically removed from the leading medical centers and published in the Transactions of the American Ophthalmological Society; surely his report attracted little attention and the non-ocular features were not emphasized.

The case report[2] that brought the disorder to attention was provided by a prominent Parisian professor of pediatrics, Antoine Bernard-Jean Marfan (1858-1942), who did much to establish pediatrics as a specialty in France and elsewhere. He was the author of widely read textbooks and monographs on pediatric topics and was editor of *Le Nourrisson* for a great many years. In addition to the syndrome under discussion here, his name is often attached to "Marfan's law" (that immunity to pulmonary phthisis is conferred by the healing of a local tuberculous lesion) and Marfan's subxiphoid approach for aspirating fluid from the pericardial sac.[3] (Please pardon my use of the possessive form of the eponym in these two instances!)

Pictures of Marfan (Fig. 2A, 1960)[4] show him as a goateed gentleman confirming the report to me by Edwards A. Park, Johns Hopkins Professor of Pediatrics, who visited Marfan's clinic in the 1920s. Dr. Park told me that Marfan was the only man he ever knew who could play Mephistopheles without special makeup!

The patient Gabrielle P. was 5 1/2 years old at the time of Marfan's original report in 1896,[5] and 11 1/2 years old at the time of the follow-up report by Méry and Babonneix,[6] who called the condition "hyperchondroplasie". By the time of the latter report, the method introduced by Wilhelm Röntgen in the same year as Marfan's report had found its way into clinical practice and was used to study Gabrielle's scoliosis. Gabrielle had fibrous contractures of the fingers, but no ocular or cardiac abnormalities were noted. The illustrations of the hands and feet in Marfan's original report (Fig. 2B, 1960)[4] are consistent with the Marfan syndrome, but are also consistent with congenital contracture arachnodactyly, which is the diagnosis proposed by Beals and Hecht in 1971.[7]

Marfan called Gabrielle's condition dolichostenomelia (long, thin extremities). He used the simile "spider legs" (in French); arachnodactyly was first used by Achard in 1902[8] but there is some question also about whether that patient had Marfan syndrome as now defined.

Salle in 1912[9] reported necropsy observations in the case of a 2 1/2 -month-old infant who may well have had the socalled neonatal form of Marfan syndrome with striking changes in the

Marfan Syndrome: A Primer for Clinicians and Scientists, edited by Peter N. Robinson and Maurice Godfrey. ©2004 Eurekah.com and Kluwer Academic / Plenum Publishers.

mitral valve. In 1914, Boerger[10] first clearly related ectopia lentis to the other manifestations, 36 years after the report of Williams.[1]

Publishing in a German ophthalmologic journal in 1931, Weve[11] of Utrecht first clearly demonstrated the heritable nature of the syndrome that was designated arachnodactyly at the time and its transmission as a dominant trait. Furthermore, he conceived of this syndrome as a disorder of mesenchymal tissue and accordingly designated it *dystrophia mesodermalis congenita, typus Marfanis*.

The major cardiovascular complications, namely aortic dilatation and dissection of the aorta, were clearly described in 1943 by R.W. Baer, Helen Taussig and Ella Oppenheimer[12] at Johns Hopkins in an article entitled "Congenital aneurysmal dilatation of the aorta associated with arachnodactyly" and by Etter and Glover[13] in an article entitled "Arachnodactyly complicated by dislocated lens and death from rupture of dissecting aneurysm of the aorta."

Of course, earlier reports of what certainly must have been Marfan syndrome can be found, such as the very instructive case reported by William George MacCallum[14] at Johns Hopkins in 1909 with a superb drawing of the pathological specimen which is unmistakably the chronic aortic dissection of Marfan syndrome (see Fig. 3-3, 1972).[4]

Another clear early description of the Marfan syndrome was that given by Bronson and Sutherland[15] in 1918 in the case of a 6-year-old child with aneurysm of the ascending aorta that ruptured into the pericardium. "The unusual shape of his head and ears and the looseness of his joints attracted attention early in infancy". Inguinal hernia was repaired surgically at the age of 2 years, and left diaphragmatic hernia was discovered by x-ray examination. "He was always undernourished but was sensitive and mentally advanced for his age, with a quaint way of expressing himself and a sense of humor of his own. The forehead was high and full, the palate highly arched. The ears were large and without the normal folds of the pinnae. The joints were lax, the limbs were flaillike, and the elbows showed definite subluxation. There was lordosis and pigeon breast with an increased prominence of the right side of the chest, which showed better expansion than the left side."

My studies of the Marfan syndrome began in 1950 during a previous incarnation as a cardiologist.[16] The 28-year-old cardiac patient who brought Marfan syndrome forcefully to my attention worked as a riveter at Glenn L. Martin Aircraft Company near Baltimore. In April 1951 while at his riveting job, he noted the rather sudden onset of severe steady pain in his right chest radiating down the right arm. The pain disappeared in a few hours and he was essentially asymptomatic thereafter, but was aware of profuse sweating, particularly of the hands and feet. Examination revealed profound manifestations of aortic regurgitation. There was no history of rheumatic fever or syphilis and no evidence of bacterial endocarditis. He gave a story that in his work as a riveter the instrument he held in front of his chest had, on several occasions, slipped, striking his chest forcefully. The possibility of traumatic rupture of an aortic cusp was considered so likely by his physicians that with their assistance the patient succeeded in making a $4,000 settlement with his employer! The history of his having been turned down nine years previously for insurance because of aortic regurgitation was not elicited until after his death.

When the patient was admitted 6 months later, it was noted by the resident that he had iridodonesis bilaterally and that the edge of ectopic lenses could be seen with the ophthalmoscope. The diagnosis of Marfan syndrome was made and superannuated dissecting aneurysm of aorta was considered likely. He died 16 months later. The last two years of the patient's life were characterized by severe attacks of sweating, substernal pain, and orthopnea. At no time were there signs of right-sided heart failure.

Clinical examination on first presentation showed a height of 5'7"; at autopsy he was measured to be 5'6". Fingertip-to-fingertip span was 5'11"; pubic symphysis to heel was 34 inches. The patient came from pyknic Polish stock. Although he looked very different from his brothers and sisters, who were 5'3" tall and heavily muscled, with short powerful limbs and stubby fingers, when he was seen alone at the time of first hospitalization his habitus was not immediately impressive.

The diffuse dilation of the ascending aorta found at autopsy, rather than dissecting aneurysm, was of the type reported by Baer et al in 1943.[12] (See Fig. 8C, 1956[4] for the Leon Schlossberg illustration of the specimen from this patient.) The aorta was found to narrow sharply beyond the mouth of the left subclavian in a typical, although only partial (about 40%), stenosis of the isthmus ("partial coarctation").

X-ray studies of the chest on initial presentation of this patient did not immediately suggest dilatation of the aortic root within the cardiovascular shadow, although in later stages an increase in size of the ascending aorta became evident. This initial patient impressed me with the facts that:

1. the first part of the aorta to undergo dilatation in Marfan syndrome is the root and proximal part of the ascending aorta within the cardiovascular silhouette, and
2. progressive dilatation goes on over a period of many years, often with development of aortic regurgitation, before changes on the standard chest x-ray are evident.

Tutelage in the Principle of Pleiotropism

My first major publication was an early description of a genetic disorder, the syndrome of polyps and spots. At the end of my internship on the Osler Medical Service of the Johns Hopkins Hospital in June of 1947, a teenage patient, Harold Parker, came under my care who had melanin spots of the lips and buccal mucosa in association with intestinal intussusception due to jejunal polyps. In the next year or so, four other patients came to my attention, three of whom were members of the same family, indicating autosomal dominant inheritance. Harold Jeghers, first in Boston and then after 1948 the first full-time Professor and Chairman of Medicine at Georgetown University in Washington, D.C., also had five patients. Jeghers and I, together with a colleague of Jeghers at Boston City Hospital, wrote up the 10 cases for publication in the *New England Journal of Medicine* in two successive issues in December 1949.[17] We pointed to Jonathan Hutchinson's description in 1896[18] of identical twins with the characteristic pigmentary anomaly-supporting the genetic nature of the disorder-and were able to trace the family of the twins named Howard and determine that one twin died of intussusception at age 20 and the other of breast cancer at age 52.[19] We also made reference to the Dutch family reported by Peutz in 1921.[20] In a paper on the polyps-and-spots syndrome, emanating from the Mayo Clinic in 1954, the designation Peutz-Jeghers syndrome was introduced.[21]

Professor Bentley Glass at Johns Hopkins University coached me on the genetic interpretation of the polyps-and-spots syndrome, making it quite clear that the association was almost certainly not the result of chromosomal linkage of separate genes but rather pleiotropism, i.e., multiple phenotypic expression of a single mutant gene. Because of that tutoring in the principle of pleiotropism, if was easy for me to appreciate the likelihood that the ocular, skeletal, and cardiovascular features of the Marfan syndrome were manifestations of a defect in one element of connective tissue wherever it occurred in the body. I collected all the patients I could find in the records of the relevant departments of the Johns Hopkins Hospital, particularly the Wilmer Eye Institute, Pediatric and Adult Cardiology, Orthopaedics, and Pathology. Follow-up of patients and tracing of families were then undertaken.

I first published on the Marfan syndrome in *Circulation* in March 1955 in an article entitled "The cardiovascular aspects of Marfan's syndrome: a heritable disorder of connective tissue".[16] (It was not until later that year that I adopted the consistent use of the non-possessive form of eponyms. The 1955 article represented the first use of the designation **heritable disorders of connective tissue. Heritable** was chosen because the disorder in individual cases, while capable of being inherited, was often, of course, the result of new mutation.) The 1955 report pointed out the natural history of the aortic involvement as learned from my first patient and confirmed by the larger study: six patients studied because of ectopia lentis were found to have aortic regurgitation with no symptoms and no clear radiologic changes of aortic aneurysm.

I was prompted to search for other disorders that might also be legitimately considered heritable disorders of connective tissue and to investigate them in the same manner as the

Marfan syndrome. In this way, I settled on Ehlers-Danlos syndrome, osteogenesis imperfecta, pseudoxanthoma elasticum (PXE), and Hurler syndrome (the prototype of the mucopolysaccharidoses). These four, along with the Marfan syndrome, were covered in the five major chapters in my monograph *Heritable Disorders of Connective Tissue* (HDCT) first published in 1956.[4] Subsequent editions were published in 1960, 1966, and 1972[4] with a multi-author 5th edition in 1993, edited by my former fellow, Peter Beighton.[5]

Clinical Delineation of Marfan Syndrome and Three Main Principles of Clinical Genetics

Clinical delineation of the Marfan syndrome involved consideration of three main principles of clinical genetics: pleiotropism (already mentioned), genetic heterogeneity, and variability.

Pleiotropism is important to clinical genetics because one feature of a syndrome may be a clue to the presence of serious internal features with prognostic implications. For example, ectopia lentis and the skeletal features of Marfan syndrome are clues to aortic aneurysm. The principle of pleiotropism was useful in working out the range of manifestations in the Marfan syndrome because it provided plausibility when specific findings were considered for inclusion in the syndrome. These included hernia, varicose veins, mitral valve prolapse and dural ectasia. Pulmonary emphysema or cysts with "spontaneous" pneumothorax became recognized features. The characteristic anterior chest deformity, pectus excavatum or carinatum, was recognized as the consequence of excessive linear growth of the ribs comparable to that occurring in the "round bones" of the limbs. In the third edition of HDCT,[4] striae of the skin were pointed to as a frequent feature of Marfan patients, and publications (in English) in 1963 and 1964 and (in German) in 1936 were referenced.

Genetic heterogeneity originates from the fact that mutation in different genes can cause the same or a very similar phenotype. These are called genocopies; a phenocopy is a phenotypic mimic with a non-genetic cause. Homocystinuria and congenital contractual arachnodactyly are major genocopies of the Marfan syndrome.

Homocystinuria

The late Professor Charles Dent, well-known student of inborn errors of metabolism at University College Hospital in London, visited Johns Hopkins in 1964 to deliver the Thayer Lectures. It was with great eagerness that he told me of the Marfan "look-alike" that he had been studying in conjunction with Nina Carson of Belfast, North Ireland. In a urine chromatography survey of "mentally backward" individuals in Ireland, most of them institutionalized, she and her colleagues[22] discovered homocystinuria. In 1963,[23] they pointed out that the patients had Marfanoid skeletal features and dislocated lenses. They also became aware of the proneness to disastrous thrombotic episodes, venous and arterial, often accompanying surgery.

The information Dent gave me brought to mind the family I had studied two years earlier because of the question of Marfan syndrome. Two of three sibs were deceased by the time of Dent's visit; the urinary cyanide nitroprusside test (as used for diagnosis of cystinuria) was found to be positive in the surviving sib. Thus, before Dr. Dent returned to London, it was possible to tell him that we too had a family with homocystinuria.

Neal Schimke, one of my post-doctoral fellows, and I then undertook as extensive a survey as possible of all the cases of "Marfan syndrome" or suspected Marfan syndrome in our collection as well as cases of non-traumatic ectopia lentis in the records of the Wilmer Eye Institute at Johns Hopkins and other large eye clinics around the US. We contacted patients and had urine samples mailed to the Moore Clinic at Johns Hopkins Hospital for testing. (Such contacting would not be permitted today, or might be achieved only with difficulty.) In this way we collected a total of 38 cases in 20 families which were reported to the Association of American Physicians in the spring of 1965[24] and published in full in the JAMA later that year.[25] At a point when we had approximately 35 homocystinuric families, we estimated that we had surveyed about 700 families in all, indicating that about 5% of non-traumatic ectopia lentis occurs on the basis of homocystinuria.

Because of the means of ascertainment, all of Carson's patients were mentally retarded. In our series, some were of normal intelligence or only mildly retarded. One patient had a Ph.D. degree. We recognized that the character of the ectopia lentis differed from that in Marfan syndrome: the lens dislocation in the Marfan syndrome (if it is going to occur at all and it is perhaps found in only 50% of cases) is likely to be present, at least in some degree, at birth (although it may increase over time), and the dislocation tends to be upward. In homocystinuria, the dislocation usually is not present at birth but comes on only later and is presaged by progressive myopia. Furthermore, the dislocation is downward rather than upward.

In homocystinuria, the ectopia lentis is "acquired"; it develops only after some years, sometimes not until age 20 or later. If Marfan syndrome can be considered a primary defect of connective tissue, homocystinuria might be considered a secondary defect because it results from damage to connective tissue fibers by the accumulated metabolites in that inborn error of metabolism. (This line of thinking justified inclusion of a chapter on alkaptonuria, another "secondary" disorder of connective tissue, in editions of HDCT beginning in 1966.)[4]

The Schimke reports[24,25] documented the Marfanoid skeletal features and indicated the occurrence of osteoporosis with increased fractures in homocystinuria. They also corroborated the fact that major cardiovascular complications occur with homocystinuria as with Marfan syndrome, but the complications are **thrombotic** in nature, not **aortic**.

Congenital Contractual Arachnodactyly

Congenital contractual arachnodactyly (CCA) was described and so named by Beals and Hecht in 1971,[7] in two families showing a pattern of inheritance consistent with autosomal dominant transmission. Epstein et al[26] had described a similar family in 1968. Beals and Hecht[7] suggested that Marfan's original patient, Gabrielle, had CCA, not the Marfan syndrome as we now define it. Phenotypic overlap between CCA and "true" Marfan syndrome is not unexpected since they are caused by mutations in paralogous genes, *FBN1* and *FBN2*.

Variation is a third principle of clinical genetics. It is a feature of all genetic disease, indeed all disease, and a feature with which the clinician must cope. Whatever the etiologic agent —a mutant gene or a pathogenic bacterium— variation in the clinical picture occurs. Clinical medicine would be, relatively speaking, child's play were it not for variation in the clinical picture produced in different persons by one and the same etiologic agent. Learning medicine is, to a large extent, learning to cope with variability.

Variability can be both interfamilial and intrafamilial. In the third edition of HDCT (1966),[4] two brothers with the strikingly disparate manifestations of Marfan syndrome were presented (Fig. 3-21); the older brother, age 17, had no detectable ocular abnormality, even refraction error, and no intrinsic cardiovascular anomaly then detectable by non-invasive methods, although skeletal features, especially unusual height, pectus excavatum, and scoliosis, as well as cutaneous striae were striking. The younger brother, age 13, had ectopia lentis, more marked skeletal abnormalities, and hernia, as well as severe mitral regurgitation. The brothers inherited the self-same gene from their father who died of complications of Marfan syndrome at the age of 36 years. The differences in manifestations were presumably the result of differences in genetic background; brothers share half their genes in common, and differences in the other genes may account for the differences between brothers, i.e., intrafamilial variability.

Interfamilial variability can be due to different specific mutations in the same gene. When the fibrillin-1 gene was identified as the site of mutations causing Marfan syndrome and the full repertoire of *FBN1* mutations was correlated with phenotype, this allelic source of variability was obvious. These analyses revealed that formes frustes of the Marfan syndrome are indeed that, including "monosymptomatic" cases such as "isolated" ectopia lentis, and the MASS syndrome.

The Evolution of Clinical Management of the Marfan Syndrome

Medical genetics was institutionalized at the Johns Hopkins Hospital July 1, 1957 when Dr. A. McGehee Harvey asked me to take over direction of a multifaceted chronic disease clinic that had been developed by Dr. J. Earle Moore. The deal I made with Dr. Harvey was that I be permitted to develop, based in the clinic we renamed the Moore Clinic, a division of Medical Genetics which would function like the other divisions of the Department of Medicine (Department of Internal Medicine in the terminology of many institutions), i.e., that it would do patient care, research and teaching in relation to a particular category of disease, in this case, genetic disease. Success of the clinic was assured by excellent funding including NIH funding, and by a tradition and mechanism for long-term follow-up and for epidemiologic research through linkages with the Johns Hopkins School of Public Health. The large cadre of patients with heritable disorders of connective tissue got the clinic off to a running start from the point of view of its clinical function. During the first 10 years or more of its existence, the Moore Clinic had funding for post-doctoral fellows without restriction as to citizenship. A very large number of trainees during that period went on, in the U.S. and abroad, to distinguished careers in medical genetics, including Malcolm Ferguson-Smith, David Weatherall, Alan Emery, David Rimoin, Judy Hall, Peter Harper, and many others. Whatever their main research, and all had a personal research arbeit, most of the fellows were involved in the clinical management of the patients with heritable disorders of connective tissue and several conducted specific projects in this area.

The second edition of HDCT (1960, p. 52-53)[4] contained graphs of upper segment/lower segment ratio (US/LS) in white and African American children. These normative data were collected in the Baltimore schools by research fellows Malcolm Ferguson-Smith, James T. Leeming and Carmen F. Merryman. Also in the second edition,[4] it was stated that the data were "to be published"; they never were. An interesting finding was that the mean US/LS values were lower at all ages and for both sexes for African American children than for European American children. This was shown to be due to longer legs and shorter trunk in the case of African-American children. Comparison of the normal standards with measurements in Marfan patients indicated that most Marfan values for US/LS fell in an abnormally low range. Previous measurements of US/LS had been provided by Engelbach in 1932.[27] Our measurements reported in 1960 gave curves lower than the curve of Engelbach for both ethnic groups.

Two of my postdoctoral fellows in the Moore Clinic, J. Lamont Murdoch and Brian A. Walker, made three specific contributions:

1. a life table analysis of the Marfan syndrome published in the *New England Journal of Medicine* in 1972;[28]
2. a study of parental age effect on the occurrence of new mutations for the Marfan syndrome published in the *Annals of Human Genetics* in the same year;[29]
3. a report in 1970 of the wrist sign as a useful physical finding in the Marfan syndrome.[30] (The Steinberg sign,[31] introduced by radiologist I. Steinberg in 1966, involved the extension of the clasped thumb beyond the border of the palm of the hand.)

As discussed elsewhere in this monograph, special considerations in the management of ectopia lentis and of the spinal problems of Marfan syndrome have evolved over the years. There have also been important developments in the diagnosis and surgical management of the major life-threatening feature of the Marfan syndrome, the aortic involvement.

In the late 1950s, Henry T. Bahnson in the Department of Surgery at Johns Hopkins operated on two cases of idiopathic cystic medial necrosis of the aorta leading to diffuse dilatation of the ascending aorta and profound aortic regurgitation. He plicated the dilated aorta and surrounded the ascending aorta as far proximally as possible with a splinting sleeve of synthetic fabric. A happy, although surprising, result of the procedure was a pronounced decrease in the aortic regurgitation as evidenced by diminution in the diastolic murmur and manifestations of left ventricular strain and by rise in diastolic pressure to almost normal levels. His cases were historic from another point of view: it was the first time the diagnosis of Erdheim cystic medial necrosis had been established histologically in vitam. These two patients with idiopathic Erdheim cystic medial necrosis had a syndrome with bicuspid aortic valve as another feature, as reported by me with others including Bahnson.[32] Bahnson had the impression that the ascending aorta is different in the cases of idiopathic cystic medial necrosis than in Marfan syndrome. Specifically, the sinuses of valsalva seemed to be involved less strikingly in the former condition. His plication operation encountered difficulties in the case of a patient with Marfan syndrome operated on in 1959; progressive dilatation of the aorta beginning at the ring was subsequently complicated by dissection of the media during the operation. In later cases in 1960 Bahnson replaced the ascending aorta with an autograft in one case and with a synthetic fabric graft in a second.[33]

Similar surgical replacement of the aorta above the junction between the tubular ascending aorta and the sinuses of Valsalva was performed by others in the Marfan syndrome in the 1950s and 1960s with unsatisfactory results because a defective part of the aorta remained. Yehuda Kesten, a journalist in Tel Aviv, born in 1926, published his personal experience with Marfan syndrome in Diary of a Heart Patient in 1968.[34] Michael DeBakey replaced the ascending aorta in May 1964, and because of continuing marked aortic regurgitation, later replaced the aortic valve with a Starr-Edwards prosthetic valve. The patient died a few years later.

Mr. Hugh Bentall[35] introduced his composite graft operation in 1969 and this rather quickly became the standard method of surgical repair of the aorta. Among the many surgeons who adopted the procedure was Vincent Gott at Johns Hopkins who had one of the most extensive and successful experiences.[36] The complete story of the development of surgical procedures for Marfan syndrome is presented elsewhere in this monograph.

The use of ultrasonography (echocardiography) for detecting dilatation of the root of the aorta started about 1970. The method was powerfully useful for detecting the dilation of the aortic root which the 1955[16] paper pointed to as the characteristic first change in the aorta in Marfan syndrome. In the 1950s and 1960s, aortograms could, of course, be performed by retrograde insertion of an arterial catheter but the use of the non-invasive method of ultrasonography was a great boon to both diagnosis and follow-up surveillance.

Medical treatment to stay the progression of aortic dilation and to prevent dissection of the aorta began in the late 1960s with the use of reserpine. Reserpine had been used with success in a naturally occurring dissecting aneurysm that had high frequency in turkeys, and was used in the treatment of human dissecting aneurysm of non-Marfan etiology. It was concluded that the beneficial effect was not solely, or even mainly, on the basis of lowering of blood pressure. Reduction in the abruptness of ventricular ejection, i.e., prolongation of dP/dT, seemed to be an important factor. The beta blocker isoproterenol (Isuprel) was substituted for reserpine. Later other beta blockers were substituted for isoproterenol.

In 1977, Reed E. Pyeritz came to Johns Hopkins from Harvard as a resident on the Osler Medical Service of which I was chief from 1973 to 1985. After a year of residency, he became a post-doctoral fellow in the Moore Clinic and thereafter rose steadily through the ranks of the faculty becoming a full professor in 1990. He identified immediately with the large number of patients with Marfan syndrome, surveyed the entire collection, and wrote a classic review published in the *New England Journal of Medicine* in 1979.[37] He delineated previously underemphasized clinical features such as dural ectasia, which through his work became recognized as an important diagnostic criterion of Marfan syndrome. He brought close attention to the question of whether "beta blockers" had a role in Marfan syndrome, and his voice was a leading one in determining the present, almost universal use of beta blockers in an effort to prevent progression of aortic dilatation. Pyeritz was also important in delineating formes frustes of the Marfan syndrome, such as the so-called MASS syndrome.

The Basic Defect

In 1956 in the first edition of HDCT,[4] I wrote, "What the suspensory ligament of the lens has in common with the tunica media of the aorta is obscure. If this common factor were known, the basic defect of the Marfan syndrome might be understood". The quest for the basic defect in Marfan syndrome went on for another 35 years before the precise mutational basis was discovered. Early reports suggested abnormalities in various connective tissue elements so that I would revise a famous pictorial "pedigree of causes", (ref. 4 see HDCT, 1960) indicating that the defect was in elastin or collagen or some other element of connective tissue (always with an accompanying question mark) until exclusion of these causes led to it being left simply as "defect of connective tissue".

In the early 1980s, David Hollister, then in Los Angeles, developed a monoclonal antibody that identified a novel human basement membrane component in the dermo-epidermal junction of skin. In 1986, Sakai et al[38] showed that the antigen is a microfibrillar protein they termed fibrillin and that it had a ubiquitous distribution, including the suspensory ligament of the lens and the aortic media. Perhaps because of my comments of three decades earlier, tissues from the Marfan syndrome were studied. Hollister and his colleagues[39] in 1990 described a striking abnormality of microfibrillin in skin and those produced by cultured dermal fibroblasts from Marfan patients. A single patient who made a particularly strong case for the etiologic role of fibrillin was an Italian girl with unilateral Marfan syndrome affecting only the left side of the body, with longer limbs on that side and subluxated lens in her left eye only. In this patient described by Burgio et al[40] Godfrey et al[41] found abnormalities of microfibrils in the skin and cultured dermal fibroblast on immunofluorescence studies only on the left side of the body. Skin and fibroblast from the right side of the body were similar to those of normal age-matched controls.

In the same year, 1990, Kainulainen et al[42] in Finland used linkage analysis to locate the Marfan syndrome phenotype to the long arm of chromosome 15, a mapping that was confirmed and refined by Dietz et al[43] and Tsipouras et al[44] in 1991 by using an intragenic DNA polymorphism. Also in 1991, Lee et al[45] and Maslen et al[46] isolated partial clones of the *FBN1* gene, and by in situ hybridization Lee et al[45] showed that this gene maps to the same region as the Marfan syndrome on 15q.

The denouement of the search for the defect in the Marfan syndrome[47] was reported in the July 25, 1991 issue of *Nature* (which had on its cover a picture of President Abraham Lincoln towering above his generals at a civil war camp). Using clones of the fibrillin gene isolated by Maslen et al as reported in the same issue of *Nature*, Dietz et al[48] identified de novo mutations in the fibrillin gene in patients with sporadic Marfan syndrome. Though a majority of mutations in *FBN1* subsequently discovered were private mutations occurring in single cases or families, the first-to-be-identified mutation occurred in two unrelated patients of my large collection, one Caucasian and one African American. Both were of a severe clinical type.

Marfan Support Groups

My patients have obviously played a central role in my personal odyssey through Marfanland in the last 50+ years. They have taught me much. Organization of the Marfan genetics support group formalized relationships not only among patients and their families but also between Marfan patients as a group and their physicians, cardiologists, surgeons, and others.

According to legend, the National Marfan Foundation (NMF) had its simple beginning in a Sunday afternoon meeting in my living room in 1979. The attendees included patients Elaine Stein and Mary Ann Rooney, social worker Joan Weiss, and colleague Reed Pyeritz. The organization was formally incorporated in Baltimore in 1980-81 and held its first "national" meeting at the Enoch Pratt Library in Baltimore in 1981. The second meeting was at the Johns Hopkins Hospital in 1984. That year, the NMF headquarters moved to Long Island, N.Y., first to the home of Priscilla Ciccariello, then to offices in Port Washington that it occupied for the next 16 years.

The third "patients conference" of the NMF was held in New York City in 1987; the 19th NMF conference, combining a professional and a patient gathering, was in Chicago in 2003. These conferences give an opportunity for professionals to confer with each other, for patients to consult experts concerning their individual problems, and for patients to learn from each other.

The NMF became a member organization of the National Organization of Rare Diseases (NORD) in 1989, and the following year, a member of the Alliance for Genetic Support Groups (which was directed for a decade or more by Joan Weiss who for 20 years previously had been genetics social worker in the Moore Clinic).

The International Federation of Marfan Syndrome Organizations (IFMSO) was initiated in San Francisco in 1991. Since then there have been five more meetings including a memorable meeting in Davos, Switzerland, in 1996.

Public Awareness of Marfan Syndrome

The tragic death of prominent persons from Marfan syndrome has in several instances brought the disorder to public attention. In 1986, Flo Hyman, a member of the championship 1984 Olympic US volleyball team, died on the court in Japan. Reed Pyeritz, who examined other affected members of Flo's family, appeared on the "Good Morning America" TV show, following which the NMF office received more than 3,000 letters from people seeking information about Marfan syndrome. An article about Flo Hyman and Marfan syndrome by Demak[49] in *Sports Illustrated* stirred public awareness.

Jonathan Larson (1960-1996) died of aortic rupture from Marfan syndrome within 24 hours before the opening of his award winning Broadway musical *Rent*. His symptoms had been severe enough that he was taken from his busy schedule to the emergency department of a leading Manhattan medical center, where the underlying diagnosis of Marfan syndrome, let alone the aortic complication, was not recognized and a trivial explanation such as gastroenteritis was proposed. One consequence of this tragedy was a major drive mounted by the National Marfan Foundation to educate emergency room physicians concerning Marfan syndrome, specifically in the recognition of dissection in the aorta and impending rupture, and the importance of special imaging and consideration of emergency aortic surgery.

But it is, of course, the possibility of Marfan syndrome in U.S. President Abraham Lincoln that has stimulated the most attention to the disorder. There are many who have supported the diagnosis but a few who have considered it very unlikely. On February 12, 1991, Dr. Marc Micozzi, then director of the National Museum of Health and Medicine in Washington, held a public conference entitled "The patient is Abraham Lincoln" to examine the question and plan for possible ways to establish or refute the diagnosis. This museum, on the grounds of the Walter Reed Army Hospital, is part of the AFIP (Armed Forces Institute of Pathology). It houses Lincoln material from which DNA might be recovered: scraps of bone from the dressing of the head, blood stains on the shirt cuffs of Dr. Curtis who probed and dressed the wound, and hair.

After the conference of February 12, Micozzi organized two committees, one to consider whether DNA testing **should** be done and one to evaluate whether DNA testing of the available material for Marfan syndrome **could** be done technically. The meeting of each committee, as well as that of February 12, was funded in large part by the Foundation of the AMA, of which Dr. James Todd, a relative of Mrs. Lincoln, was executive director. I chaired both committees.

The first committee included Lincoln scholars, legal scholars, historians, and ethicists. The second committee included clinical geneticists and molecular geneticists as well as connective tissue experts.

The first committee, meeting in late March, 1991, concluded that there existed no moral or legal impediment to conduct DNA analyses, especially in light of the fact that there are no living descendants and that disinterment is not necessary.

The second committee, meeting in early May, 1991, concluded that it should be technically possible to test for Marfan syndrome using available DNA but not at that time. Less than three months later, the description of the first mutation in the fibrillin gene by Dietz et al[48] made testing Lincoln DNA a realistic undertaking. Difficulties that have come to light as more cases are studied from the molecular point of view include the fact that the mutations in this very large gene are usually "private" ones occurring in only a single family. The best approach would be to sequence the coding parts and neighboring intronic sections of the entire gene. If a change is found that is not a normal variation, that would prove Marfan syndrome. If none is found, that would not completely exclude the possibility because the change might involve controller or intronic elements not covered in the sequencing.

An appeal to the Board of Regents of the AFIP for permission to test the Museum's Lincoln material was rejected in 1992 and several times thereafter. That material is the responsibility of the Department of Defense. In early 2000, I wrote to the then Secretary of Defense, William Cohen, a fellow native of Maine, arguing the case for DNA studies. Despite my eloquent request, a subordinate responded in the negative, arguing that testing would destroy precious and irreplaceable material and was, furthermore, an invasion of historic privacy.

It was of interest to note the public reactions when the proposals for testing Lincoln's DNA were discussed in the media in the Spring of 1991. Much of the adverse reaction was probably based on the mistaken impression that disinterment would be required. Even without that necessity, bioethicist Arthur Caplan in his syndicated column spoke out against it, characterizing it as voyeurism, an effort at exposé. This bought an angry letter from one reader in California, "I have the Marfan syndrome...The fact that Lincoln may have had Marfan syndrome shows those of us with genetic syndromes that we too can contribute something of value to society. It's time that all people, especially medical ethicists, realize that having the Marfan syndrome is not shameful—it's just darned inconvenient."[50]

I[50] suggested that Lincoln's Marfan syndrome might do for genetic disorders what Roosevelt's poliomyelitis did for that disease. The vaccine for polio certainly would not have been developed by 1955 were it not for the President's polio and the founding of the National Foundation for Infantile Paralysis in 1937. Indeed, much of the beginnings of molecular biology were financed by the polio foundation as part of its study of viruses in search of a vaccine.

Epilogue

It will be evident that this is a highly personal and Johns Hopkins-oriented account of the development of knowledge about Marfan syndrome in the last 50 years. Many not mentioned here have obviously contributed heavily to the understanding of Marfan syndrome and its clinical management. I apologize to them and hope that my egocentric survey may provide useful perspective.

References

1. Williams E. Rare cases, with practical remarks. Trans Am Ophthalmol Soc 1873-1879; 2:291.
2. Marfan AB. Un cas de déformation congénitale des quatre membres plus prononcée aux extrémités charactérisée par l'allongement des os avec un certain degré d'amincissement. Bull Mém Soc Méd Hôp Paris 1896; 13:220-226.
3. Marfan AB. Sur la ponction du péricarde et en particulier sur la ponction par voie épigastrique sous-xiphoidienne. Arch Mal Coeur 1936; 29:153.
4. McKusick, VA Heritable Disorders of Connective Tissue. St. Louis: CV Mosby; 1956 (1st ed.); 1960 (2nd ed.); 1966 (3rd ed.); 1972 (4th ed.).
5. Beighton P, editor. McKusick's Heritable Disorders of Connective Tissue 5th ed. St. Louis: CV Mosby; 1993.
6. Méry H, Babonneix L. Un cas de déformation congénitale des quatre membres: hyperchondroplasie. Bull Mem Soc Méd Hôp Paris 1902; 19:671.
7. Beals RK, Hecht F. Congenital contractural arachnodactyly: a heritable disorder of connective tissue. J Bone Joint Surg 1971; 53A:987-993.
8. Achard C. Arachnodactylie. Bull Mém Soc Méd Hôp Paris 1902; 19:834.
9. Salle V. Ueber einen Fall von angeborener abnormen Grösse der Extremitäten mit einen an Akronemegalia erinnerden Symptomenkomplex. Jahrb Kinderheilk 1912; 75:540-550.
10. Boerger F. Ueber zwei Fälle von Arachnodaktylie. Monatsschr Kinderheilk 1914; 13:335.
11. Weve H. Ueber Arachnodaktylie (Dystrophia mesodermalis congenita, typus Marfanis). Arch F Ophthal 1931; 104:1-46.
12. Baer RW, Taussig HB, Oppenheimer EH. Congenital aneurysmal dilatation of the aorta associated with arachnodactyly. Bull Johns Hopkins Hosp 1943; 72:309-331.

13. Etter LE, Glover LP. Arachnodactyly complicated by dislocated lens and death from rupture of dissecting aneurysm of the aorta. JAMA 1943; 123:88-89.
14. MacCallum WG. Dissecting aneurysm. Bull Johns Hopkins Hosp 1909; 20:9.
15. Bronson E, Sutherland GA. Ruptured aortic aneurysms in childhood. Br J Child Dis 1918; 15:241.
16. McKusick VA. The cardiovascular aspects of Marfan's syndrome. Circulation 1955; 11:321-342.
17. Jeghers H, McKusick VA, Katz KH. Generalized intestinal polyposis and melanin spots of the oral mucosa, lips and digits. N Engl J Med 1949; 241:993-1105, 1031-1036.
18. Hutchinson J. Pigmentation of lips and mouth. Arch Surg 1896; 7:290.
19. McKusick VA. The clinical observations of Jonathan Hutchinson. Am J Syph 1952; 36:101-126.
20. Peutz JLA. Very remarkable case of familial polyposis of mucous membrane of intestinal tract and nasopharynx accompanied by peculiar pigmentations of skin and mucous membrane [Dutch]. Nederl Maandschr Geneesk 1921; 10:134-146.
21. Bruwer A, Bargen JA, Kierland RR. Surface pigmentation and generalized intestinal polyposis (Peutz-Jeghers syndrome). Proc Staff Meet Mayo Clin 1954; 29:168-171.
22. Carson NAJ, Neill DW. Metabolic abnormalities detected in a survey of mentally backward individuals in Northern Ireland. Arch Dis Child 1962; 37:505-513.
23. Carson NAJ, Cusworth DC, Dent CE et al. Homocystinuria: a new inborn error of metabolism associated with mental deficiency. Arch Dis Child 1963; 38:425-436.
24. Schimke RN, McKusick VA, Pollack AD. Homocystinuria simulating the Marfan syndrome. Trans Assoc Am Physicians 1965; 78:60.
25. Schimke RN, McKusick VA, Huang T et al. Homocystinuria: a study of 38 cases in 20 families. JAMA 1965; 193:711-719.
26. Epstein CJ, Graham CB, Hodgkin WE et al. Hereditary dysplasia of bone with kyphoscoliosis, contractures, and abnormally shaped ears. J Pediatr 1968; 73:379-386.
27. Engelbach E. Endocrine Medicine. Springfield, Ill: Charles C Thomas; 1932.
28. Murdoch JF, Walker BA, Halpern BL et al. Life expectancy and causes of death in the Marfan syndrome. New Engl J Med 1972; 286:804-808.
29. Murdoch JL, Walker BA, McKusick VA. Parental age effects on the occurrence of new mutations for the Marfan syndrome. Ann Hum Genet 1972;35:331-336.
30. Walker VA, Murdoch JL. The wrist sign: A useful physical finding in the Marfan syndrome. Arch Intern Med 1970;126:276-277.
31. Steinberg I. A simple screening test for the Marfan syndrome. Am J Roentgenol Radium Ther Nucl Med 1966;97:118-124.
32. McKusick VA, Logue RB, Bahnson HT. Association of aortic valvular disease and cystic medial necrosis of the ascending aorta. Circulation 1957; 46:188-194.
33. Bahnson HT, Spencer FC. Excision of aneurysm of ascending aorta with prosthetic replacement during cardiopulmonary bypass. Ann Surg 1960; 151:879-890.
34. Kesten Y. Dairy of a Heart Patient. New York: McGraw Hill, 1968.
35. Bentall HH, De Bono A. A technique for complete replacement of the ascending aorta. Thorax 1968; 23:338-339.
36. Gott VL, Cameron DE, Alejo DE et al. Aortic root replacement in 271 Marfan patients: a 24-year experience. Ann Thorac Surg 2002; 438-443.
37. Pyeritz RE, McKusick VA. The Marfan syndrome: diagnosis and management. N Engl J Med 1979; 300:772-777.
38. Saki LY, Keene DR, Engvall E. Fibrillin, a new 350-KD glycoprotein, is a component of extracellular microfibrils. J Cell Biol 1986; 103:2499-2509.
39. Hollister DW, Godfrey M, Sakai LV et al. Immunohistologic abnormalities of the microfibrillar-fiber system in the Marfan syndrome. N Engl J Med 1990; 323:152-159.
40. Burgio RG, Martini A, Cetta G et al. Asymmetric Marfan syndrome. Am J Med Genet 1988; 30:905-909.
41. Godfrey M, Olson S, Burgio RG et al. Unilateral microfibrillar abnormalities in a case of asymmetric Marfan syndrome. Am J Hum Genet 1990; 46:661-671.
42. Kainulainen K, Pulkkinen L, Savolainen A et al. Location on chromosome 15 of the gene defect causing Marfan syndrome. N Engl J Med 1990; 323:935-939.
43. Dietz HC, Pyeritz RE, Hall BD et al. The Marfan syndrome locus: confirmation of assignment to chromosome 15 and identification of tightly linked markers at 15q15-q21.3. Genomics 1991; 9:355-361.
44. Tsipouras P, Sarfarazi M, Devi A et al. Marfan syndrome is closely linked to a marker on chromosome 15q1.5-q2.1. Proc Natl Acad Sci USA 1991; 88:4486-4488.
45. Lee B, Godfrey M, Vitale E et al. Linkage of Marfan syndrome and a phenotypically related disorder to two different fibrillin genes. Nature 1991; 352:330-334.

46. Maslen CL, Corson GM, Maddox BK et al. Partial sequence of a candidate gene for the Marfan syndrome Nature 1991; 352:334-337.
47. McKusick VA. The defect in Marfan syndrome. Nature 1991; 352:279-281.
48. Dietz HC, Cutting GR, Pyeritz RE et al. Marfan syndrome caused by a recurrent de novo missense mutation in the fibrillin gene. Nature 1991; 352:337-339.
49. Demak R. Marfan syndrome: A silent killer. Sports Illustrated 1986; 64:30-35.
50. McKusick VA. Abraham Lincoln and Marfan syndrome. Nature 1991; 352:338.

CHAPTER 1

Diagnosis and Treatment of Marfan Syndrome—A Summary

Anne H. Child, Luitgard Neumann and Peter N. Robinson

Overview

Marfan syndrome (MFS) is a dominantly inherited disease of connective tissue with diverse manifestations (Table 1), involving primarily the skeletal, ocular, and cardiovascular systems. Affected individuals may have a characteristic habitus including tall stature and dolichostenomelia, arachnodactyly, joint laxity, scoliosis, and pectus deformities (Figs. 1 and 2). Patients may have murmurs of aortic or mitral regurgitation. Prompt diagnosis may prevent premature death from aortic dissection and/or rupture, as well as unnecessary visual loss and skeletal deformity.[1] Regular examination by appropriate specialists should then be coordinated by a single physician (Table 3) and based on a timetable of care (Table 4).

The first step in management is to make a proper diagnosis, which may present difficulties due to uncertainty about the definition of the Marfan syndrome. Along with increasing understanding of the pathogenesis of the syndrome, the criteria for diagnosis have been critically refined. The diagnostic criteria for the Marfan syndrome which are widely used today are codified in the revised diagnostic criteria for the Marfan syndrome,[2] which are occasionally referred to as the "Ghent Nosology", after the city in Belgium where they were developed (Table 2). These are based on whether or not a first-degree relative is affected by Marfan syndrome, the three main systems which may be involved (cardiovascular, ocular and skeletal), and on weighting of particular clinical features. For example, dilatation of the aorta is a relatively specific sign. In contrast, the high prevalence of mitral valve prolapse in the general population makes it a relatively unspecific sign.

Similarly, lens dislocation is a much more specific sign than myopia, and many people with long limbs, joint laxity, or slight scoliosis do not have Marfan syndrome. Other evidence may also need to be considered: Striae distensae, papyraceous scars, inguinal hernia, joint hyperextensibility, arthralgia, myalgia and ligamentous injury are relatively frequent findings,[3] and spontaneous pneumothorax is not a rare occurrence.[4] Recent studies have demonstrated dural ectasia in the spinal canal of 60% of Marfan patients. This is a useful diagnostic sign, detectable by magnetic resonance imaging.[5]

Diagnostic Criteria for the Marfan Syndrome[2]

If a first-degree relative is affected by the Marfan syndrome, the person under consideration must have:
- Involvement of at least two systems (skeletal, cardiovascular, ocular);
- At least one major manifestation e.g., ascending aortic aneurysm; ectopia lentis

Marfan Syndrome: A Primer for Clinicians and Scientists, edited by Peter N. Robinson and Maurice Godfrey. ©2004 Eurekah.com and Kluwer Academic / Plenum Publishers.

Table 1. Salient features of the Marfan syndrome

Skeletal:	Tall, thin physique with long limbs, spinal curvature, flattening of chest (with pectus deformity), joint hypermobility.
Cardiovascular:	Dilation of ascending (and sometimes descending) aorta, incompetence of aortic (and sometimes mitral) valve, dissection and/or rupture of aorta.
Ocular:	Dislocation of lens, myopia and refraction problems (often unstable), detachment of retina.
Dental:	High arched narrow palate, crowding of teeth.
Genetic:	Autosomal dominant
Variability:	In severity and pattern of features affected

If no first degree relative is unequivocally affected by the Marfan syndrome the proband must have:
* Major manifestations in two systems
* Involvement of at least one other (skeletal, cardiovascular, ocular)

Dural Ectasia

Lumbosacral dural ectasia is one of the major criteria, and the demonstration of dural ectasia by magnetic resonance or computer-tomographic imaging can help establish the diagnosis

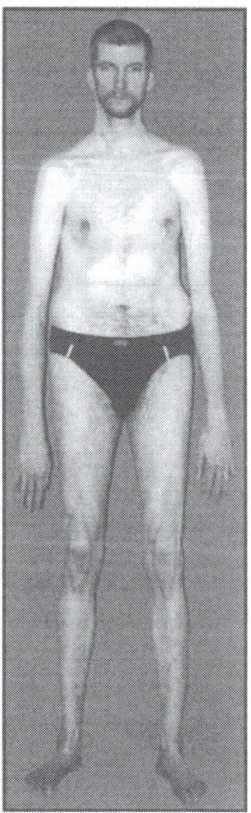

Figure 1. Typical Marfan habitus with dolichostenomelia, arachnodactyly, increased leg length (reduced upper-to-lower segment ratio), sparse subcutaneous tissue.

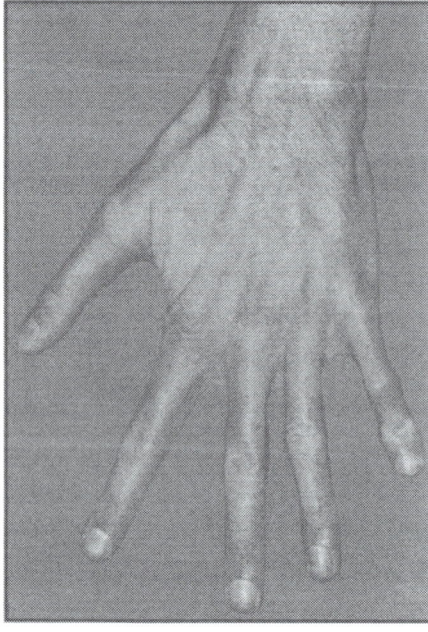

Figure 2. Arachnodactyly in a person with typical features of MFS.

Table 2. Diagnostic criteria for Marfan syndrome

Skeletal System
A *major criterion* is defined by the presence of at least four of the following:

Pectus carinatum

Pectus excavatum severe enough to require surgery

Reduced upper to lower segment ratio [< 0.85] or arm-span-to height ratio [> 1.05] (both in absence of severe scoliosis)

Positive wrist *and* thumb signs

Reduced extension of the elbows (< 170°)

Medial displacement of the medial malleolus associated with pes planus

Protrusio acetabuli of any degree (ascertained on radiographs, CT or MRI)

Involvement of the skeletal system is defined by the presence of two of the preceding features *or* the presence of one of the preceding features and two of the following: pectus excavatum not requiring surgery; joint hypermobility; high-arched palate; facial features – at least two of dolichocephaly, malar hypoplasia, enophthalmos, retrognathia, down-slanting palpebral fissures.

Ocular System
A *major criterion* is defined by ectopia lentis of any degree.

Involvement of the ocular system is defined by the presence of at least two of the following: flat cornea; increased axial length of the globe (> 23.5mm); hypoplastic ciliary muscle causing decreased miosis.

Cardiovascular
A *major criterion* is defined by dissection of the ascending aorta with or without aortic regurgitation and involving at least the sinuses of Valsalva.

Involvement of the cardiovascular system requires the presence of at least one of the following: mitral valve prolapse with or without mitral regurgitation; dilatation of the main pulmonary artery, in the absence of valvular or peripheral pulmonic stenosis, under the age of 40 years; calcification of the mitral anulus under the age of 40 years; dilatation or dissection of the descending thoracic or abdominal aorta under the age of 50 years.

Pulmonary System
Involvement is defined by either spontaneous pneumothorax or radiographic evidence of apical blebs.

Skin and Integument
Involvement is defined by either striae distensae or a recurrent or incisional hernia.

Dura
A *major criterion* is the presence of lumbosacral dural ectasia.

Family/Genetic History
A *major criterion* is defined by one of the following: a first-degree relative who independently meets these criteria; the presence of a mutation in the *FBN1* gene that is likely to be pathogenic; or the presence of a haplotype around the *FBN1* locus inherited by descent and unequivocally associated with diagnosed Marfan syndrome in the family.

of Marfan syndrome. Dural ectasia is one of the most common clinical manifestations of Marfan syndrome and may be used as a major criterion in the establishment of the diagnosis of Marfan syndrome.[6] The reader is referred to the chapter by Sponseller and Shindle in this volume for further information.

Table 3. Overview of management

Regular examination by one coordinating physician

Periodical multidisciplinary evaluation:
  Ophthalmological – between birth and 6 months; annually thereafter and as needed for adults
  Orthopedic – scoliosis, pectus deformity, flat feet, to age 20
  Cardiovascular – newborn; every one to two years to age 11; annually thereafter
  Genetic counseling – at first diagnosis, and again with partner before marriage

Periodic ECG and echocardiogram, focusing on dysrhythmias, mitral valve prolapse and aortic root dilation

Chest radiograph, focusing on apical blebs – tall pubertal males at greatest risk of pneumothorax

CT or MRI scan of entire aorta – baseline assessment in all adults and periodically in patients with aneurysm or dissection – annually after aortic surgery

Endocarditis prophylaxis during dental care involving bleeding

Beta-adrenergic blockade

Restriction of strenuous activities, especially contact sports and isometric exercises, e.g. weightlifting

Consider prophylactic surgical replacement of ascending aorta at 5 cm dilatation (or earlier if a family history of dissection/rupture at diameter < 5 cm is present)

Life-long regular monitoring of all unreplaced aortic segments and heart

Clinical Management

Table 3, Table 4 and the following paragraphs offer an overview of the most important aspects of the clinical management of individuals with Marfan syndrome. Specific guidelines for children with Marfan syndrome have been published by the American Academy of Pediatrics[7]. More detailed accounts are available in the subsequent chapters. Optimal care of persons with Marfan syndrome requires a generalist with broad experience in Marfan syndrome who can refer to specialists as appropriate and ensure that indicated examinations are performed.[1] Genetic counseling for discussion of inheritance and other issues should be arranged.

Cardiological evaluation including baseline echocardiographic measurements for aortic root widening (which almost always precedes dissection) and mitral valve prolapse should be undertaken at regular intervals throughout adult life. Echocardiography of all possibly affected family members should be performed at the time of diagnosis of an index case. The reader is referred to the chapters on cardiology and cardiovascular surgery in this volume for further information.

Examination by slit lamp detects lens dislocation, may permit therapy to prevent retinal detachment and allow correction of myopia resulting from the increased length of the eyeball. The chapter on "Ophthalmological Aspects" of Marfan syndrome in this volume presents further information.

Referral to an orthopedic surgeon for scoliosis, pectus deformities or flat feet will allow for bracing, corrective orthoses and shoes or surgery at a critical time in development. The chapter "Orthopaedic Problems in Marfan Syndrome" in this volume presents further information.

Table 4. Timetable of care

	Newborn	Pre-School	Pubertal Years	Adulthood
Cardiac	Echocardiography of newborn and parents if first affected child	Repeat echocardiography Consider Beta-blockers	Repeat echocardiography every 1-2 years Restriction of strenuous exercise and contact sports	Regular echocardiography Pre, intra- and postpartum echos for females. Avoid pregnancy if aortic root > 4.2 cm
Ocular	Examine for dislocated lenses, strabismus	Glasses if required	Advice about glaucoma if dislocated lenses	See regularly if ocular problems. If none, see only when necessary
Skeletal	Orthopedic examination	Orthopedic examination	Stop scoliosis check at age 18	Joint pain, early osteoarthritis require NSAIDs
Lung			Age 18- chest X ray for blebs, fibrosis	
Genetic counseling and psychosocial issues	Explain mode of inheritance, prenatal diagnosis Help family handle guilt, grief, anger	Help coping with teasing, anger	Explain inheritance and variability, offer counseling session with partner Career choices	Consider *FBN1* mutation screening Further counseling as required

Affected individuals should consider carrying some form of emergency identification stating that they have Marfan syndrome and summarizing other important aspects including cardiovascular problems and the possibility of aortic dissection. Physicians should inform their patients of the forms of identification available in their country.

Inheritance

Marfan syndrome is a dominantly inherited connective tissue disease. Almost all the identified *FBN1* mutations have been family specific, and the rate of new mutations is high.

The Marfan syndrome follows an autosomal dominant mode of inheritance. Population-based studies have estimated the minimal prevalence of MFS in the population to be between 4-6 per 100,000,[8] 7.03 per 100,000,[9] and 17.2 per 100,000.[10] However, given the lack of a simple method of clinical diagnosis for MFS, it is possible that the true prevalence is higher, and estimates have ranged as high as 2-3 per 10,000 persons.[1] About a quarter of all cases are due to de novo (sporadic) mutations.[9]

The MFS is characterized by a high degree of clinical variability, both within and between families.[11,12] Although the penetrance of MFS appears to be extremely high, it should be noted that not all *FBN1* mutation carriers meet the Ghent diagnostic criteria for MFS. For instance, a proportion of premature-truncation codon (PTC) mutation carriers do not meet the Ghent diagnostic criteria for MFS although they are still at high risk for aortic dissection.[13] It does seem to be extremely rare (if at all) that carriers of an *FBN1* mutation causing disease in another individual show no clinical signs of MFS.

Rare cases of germ-line mosaicism for an *FBN1* mutation have been reported. In two affected sisters identical de novo *FBN1* mutations were identified. These were not found in the leukocyte DNA of the unaffected parents.[14] In two affected brothers identical *FBN1* mutations were found, somatic and germ-line mosaicism was demonstrated in the father who only had mild aortic dilation.[15] In another MFS patient with an identified *FBN1* mutation, somatic mosaicism was verified in his unaffected mother.[16]

Evolving Nature of Marfan Syndrome Phenotype

Most children with MFS show evolution of signs over time, with the mean age of initial diagnosis of many manifestations of MFS being around puberty.[17] This age-dependent evolution of the signs and symptoms of MFS adds to the diagnostic difficulties resulting from the extreme intra- and interfamilial variability of this disorder. This is especially so if a sporadic case of MFS is suspected. Common presenting features may involve the musculoskeletal system, but in other cases, children may first come to medical attention owing to ectopia lentis. In one study, the average age of initial diagnosis of aortic dilatation was 11 years.[17] The diagnosis of MFS cannot therefore be ruled out in children presenting with some manifestations of MFS but not (yet) fulfilling the criteria of the Ghent nosology. Individual management and careful follow-up are required if there is doubt as to the diagnosis.

Newborns

Twenty-five percent of babies with MFS will be the first one in the family. The diagnosis is made on the basis of the family history, the clinical appearance, and mutation analysis if considered helpful. Babies may be mildly to severely affected. Those that are severely affected may have dislocated lenses, deep set eyes, large simple ears, high palate, and be long babies with long fingers and feet, and loose joints. Muscle tone may be reduced (floppy child). For this reason, they may have trouble feeding. An eye examination after a few days, an echocardiogram, and mutation analysis can all be instituted. Physical milestones may be somewhat delayed for severely affected children (sitting, crawling, standing and walking a bit late).

School-Related Problems

Children generally do well in a normal school, but if weakness or teasing is a problem, a school with small class size may need to be selected. If eyesight is poor, large print books, using a computer as teaching aid, and obtaining extra one-on-one classroom help may be indicated. Extra time can be requested for examinations, if the child is slow at writing because of joint laxity. Teasing and bullying can be a problem, and should be discussed with the teachers.

Sporting activities should be encouraged. However, choice of sports to specialize in should be informed. Activities involving prolonged exertion at peak capacity are discouraged, such as cross-country running, which requires a great deal of stamina. Sprinting, football (soccer), cycling especially on the flat, swimming, golf, are all good choices. Boxing and sports such as karate are contraindicated because of the danger of dislocated lenses or detached retina. Basketball and netball (volleyball), when played as noncontact sports, are recommended. Joint supports may have to be worn during games. Trampolining is not recommended, as it is stressful for joints.

Special Problems

Easy fatigability is a feature of MFS at all ages, and leads to lack of stamina. The child needs to keep fit with regular exercise, which is within his or her capability. Joint pains and joint laxity may respond to rest, medication and support. Good posture should be encouraged, together with supportive seating and a desk which is large enough for the child.

Height should be monitored annually and a growth curve prepared. When the child reaches 150cm height, a decision should be made whether to refer for excessive stature, and consideration of hormone therapy.

Marfan Syndrome and Pregnancy

Pregnant women with MFS have an increased risk of aortic dissection and other cardiovascular complications, and the risk appears to be higher if the diameter of the aortic root is greater than 4 cm at the start of pregnancy or increases rapidly.[18,19] Close monitoring at a

Table 5. Differential diagnosis of Marfan syndrome

Disease	OMIM Code	Marfan-Like Features	Features Not Generally Found in Marfan Syndrome
Disorders with Overlap Primarily in the Skeletal System			
Congenital contractural arachnodactyly CCA	121050 (AD)	arachnodactyly	contractures
Shprintzen-Goldberg craniosynostosis syndrome	163212 (AD)	arachnodactyly dolichostenomelia	craniosynostosis, exophthalmos, mental retardation
Disorders with Overlap Primarily in the Ocular System			
Homocystinuria	3253 (AR)	ectopia of the lenses in about 60%	mental retardation
Weill-Marchesani syndrome	277600 (AR) (AD)	ectopia of the lenses	brachydactyly, short stature
Ectopia lentis, isolated	129600 (AD)	ectopia of the lenses	—
Ectopia lentis et pupillae	225200 (AR)	ectopia of the lenses	ectopia of the pupils
Stickler syndrome		high myopia	mild spondyloepiphyseal dysplasia
type I	108300 (AD)		
type II	604841 (AD)		
type III	184840 (AD)		
Disorders with Overlap Primarily in the Cardiovascular System			
Familial thoracic aortic aneurysms and dissections	607086 (AD) 607086 (AD) 132900 (AD)	See Chapter 9 by Hasham and Milewicz in this volume	
MASS phenotype (mitral valve, aorta, skeleton, and skin)	604308 (AD)	long limbs, mitral valve prolapse, striae atrophicae, mild dilatation of aortic root	—
Mitral valve prolapse syndrome	157700 (AD)	dolichostenomelia, high arched palate, mild pectus excavatum	—

Genetic disturbances showing some degree of phenotypic overlap with the MFS. The most significant similarities and distinguishing features are shown. Readers are referred to the website Online Mendelian Inheritance in Man (OMIM) for references and more details (http://www.ncbi.nlm.nih.gov/entrez/query.fcgi?db=OMIM). AD= autosomal dominant; AR= autosomal recessive

specialized center is recommended, and echocardiography should be performed every 6-8 weeks during pregnancy and for six months after delivery.

The Differential Diagnosis of Marfan Syndrome

There are many conditions with overlapping clinical manifestations to be considered in the differential diagnosis of MFS. Affected individuals should be thoroughly examined according to the criteria of the Ghent nosology in order to establish the diagnosis of MFS. One of the main difficulties in the differential diagnosis of MFS involves the high prevalence of many of the manifestations of MFS in the general population, especially those manifestations listed under the minor criteria of the diagnostic criteria (see Table 3). There are numerous genetic conditions that show some degree of phenotypic overlap with the MFS (see Table 5), but in some cases the question may be whether an individual with a more or less "marfanoid" appearance actually has any disorder at all. It is important to rule out homocystinuria in each family

Table 6. Type-1 fibrillinopathies

Syndrome	Clinical Features	References
Marfan syndrome	See text	22-24
Neonatal MFS	Severe manifestations (see text)	25
Shprintzen-Goldberg syndrome	Craniosynostosis, mental retardation, marfanoid habitus (see text)	26, 27
Familial arachnodactyly	Arachnodactyly, dolichostenomelia, no cardiovascular manifestations	28
Ectopia lentis	Bilateral ectopia lentis, in some cases scoliosis, no or late-onset cardiovascular manifestations, no aortic dissection	29, 30
Familial thoracic aortic aneurysms and dissections	See chapter by Hasham and Milewicz in this book	31
MASS phenotype	Mitral valve prolapse, aortic root dilatation without dissection, skeletal and skin abnormalities	32
New variant of MFS	Skeletal features of MFS with joint contractures and knee joint effusions. Ectopia lentis, no cardiovascular manifestations	33, 34
Isolated skeletal features	Tall stature, scoliosis, pectus excavatum, arachnodactyly	35
Weill Marchesani syndrome (AD)	Short stature, brachydactyly, joint stiffness, and microspherophakia, ectopia lentis, severe myopia, and glaucoma	24
Progressive kyphoscoliosis	Kyphoscoliosis, some skeletal but no ocular or cardiac manifestations	36

in light of the effective therapy available for this disorder. Suspected cases should have a plasma amino acid analysis in the absence of pyridoxine supplementation to rule out the diagnosis of homocystinuria. This disorder is characterized by a deficiency of the enzyme cysthathionine synthetase, with connective tissue defects, marfanoid habitus, slim skeletal build, arachnodactyly, pectus excavatum or carinatum, kyphoscoliosis, hypermobile joints and subluxation of the lens (usually downward). About 60% of patients have mental retardation, and patients are prone to vascular thromboses with risk of recurrent strokes.[20] In addition to the conditions listed in Table 5, there are several other conditions that often are characterized by a habitus reminiscent of that seen in persons with MFS, but other clinical aspects generally allow the delineation of these conditions from MFS without difficulty (for instance, multiple endocrine neoplasia type IIb).

Other Type-1 Fibrillinopathies

In addition to mutations found in persons with classical MFS as defined by the Ghent nosology, *FBN1* mutations have also been identified in individuals with a series of related disorders termed type-1 fibrillinopathies.[21] Some type-1 fibrillinopathies present with a subset of the findings characteristic of classical MFS, but others, such as Weill-Marchesani syndrome, are characterized by anomalies not typically found in MFS. In general, type-1 fibrillinopathies other than MFS are rarer than classical MFS, and in some cases have only been reported in one individual or family. The reader is referred to the publications cited in Table 6 for further information.

Neonatal Marfan Syndrome

Neonatal MFS represents the most severe end of the clinical spectrum of MFS. The clinical findings typical of this condition include long body length, dolichocephaly, a distinctive face with deep set eyes, crumpled ears, severe arachnodactyly, and contractures of large joints. There may be muscular hypotonia and a lack of subcutaneous adipose tissue.[25] Many of the phenotypic abnormalities of the musculoskeletal system are similar to those of congenital contractural arachnodactyly, but ocular and cardiovascular abnormalities are generally not found in the latter disorder.

The ocular symptoms of the neonatal MFS include lens subluxation or dislocation, megalocornea, iridodenesis, and cataract. The cardiovascular features are severe with early, progressive dilatation of aortic root, pronounced atrioventricular valve dysfunction, and congestive heart failure,[37-39] and death invariably occurs in the first years of life, most often within the first 12 months.[25] It should be noted that the timepoint of diagnosis in the neonatal period alone does not justify the diagnosis of neonatal MFS. The neonatal MFS is characterized not only by the severity of the manifestations but also by a series of features rare in classical MFS such as crumpled ears, contractures, and loose skin.[25] To date all reported cases of neonatal MFS have been sporadic. The mutations that cause nMFS occur in a relatively small short region of the gene, between exons 24-32 of the fibrillin-1 (*FBN1*) gene.[25, 40]

Shprintzen-Goldberg Craniosynostosis Syndrome

This craniofacial dysmorphism is characterized by craniosynostosis and a marfanoid habitus. Mitral valve prolapse and aortic root dilatation may also occur.[27] One mutation in *FBN1* has been reported in an individual with manifestations of Shprintzen-Goldberg syndrome and MFS,[26] but a further *FBN1* sequence change found in an individual with Shprintzen-Goldberg syndrome originally reported to be a mutation was later shown to be a polymorphism.[41] To date, no further *FBN1* mutations have been identified in Shprintzen-Goldberg syndrome[42-45] and it remains unclear if variants in *FBN1* contribute to the phenotype of SGS directly, indirectly, or at all. In general, SGS can be differentiated from MFS by the presence of exopthalmus, craniosynostosis, and mental retardation.

Conclusions

Since the discovery that mutations in the gene for fibrillin-1 cause MFS in 1991[46] great progress has been made in understanding the molecular pathogenesis of MFS. Advances in medical care including cardiovascular sugery[47] and prophylactic treatment with beta blockers[48] have contributed towards significantly increasing the life expectancy of affected individuals since the early 1970s. Many affected individuals are diagnosed with MFS only after significant complications have ensued. On the other hand, it is not rare for individuals with other named or unnamed disorders of connective tissue to receive a mistaken diagnosis of MFS. Correct and timely diagnosis of MFS is essential in order to provide optimal medical care and prevent serious complications. This chapter has presented an overview of the most important issues involving making a diagnosis of MFS and has summarized some of the other conditions with which clinicians may be confronted in the differential diagnosis of MFS. The following chapters in this section will cover the clinical care of affected individuals from the various perspectives of the ophthalmologist, orthopedic surgeon, cardiologist and cardiovascular surgeon.

References

1. Pyeritz RE. The Marfan syndrome. Annu Rev Med 2000; 51:481-510.
2. De Paepe A, Devereux RB, Dietz HC et al. Revised diagnostic criteria for the Marfan syndrome. Am J Med Genet 1996; 62(4):417-426.
3. Grahame R, Pyeritz RE. The Marfan syndrome: Joint and skin manifestations are prevalent and correlated. Br J Rheumatol 1995; 34(2):126-131.

4. Hall JR, Pyeritz RE, Dudgeon DL et al. Pneumothorax in the Marfan syndrome: Prevalence and therapy. Ann Thorac Surg 1984; 37(6):500-504.
5. Ahn NU, Sponseller PD, Ahn UM et al. Dural ectasia is associated with back pain in Marfan syndrome. Spine 2000; 25(12):1562-1568.
6. Rose PS, Levy HP, Ahn NU et al. A comparison of the Berlin and Ghent nosologies and the influence of dural ectasia in the diagnosis of Marfan syndrome. Genet Med 2000; 2(5):278-282.
7. Health supervision for children with Marfan syndrome. American Academy of Pediatrics Committee on Genetics. Pediatrics 1996; 98(5):978-982.
8. Pyeritz RE, McKusick VA. The Marfan syndrome: diagnosis and management. N Engl J Med 1979; 300(14):772-777.
9. Gray JR, Bridges AB, Faed MJ et al. Ascertainment and severity of Marfan syndrome in a Scottish population. J Med Genet 1994; 31(1):51-54.
10. Sun QB, Zhang KZ, Cheng TO et al. Marfan syndrome in China: a collective review of 564 cases among 98 families. Am Heart J 1990; 120(4):934-948.
11. Dietz HC, Pyeritz RE, Puffenberger EG et al. Marfan phenotype variability in a family segregating a missense mutation in the epidermal growth factor-like motif of the fibrillin gene. J Clin Invest 1992; 89(5):1674-1680.
12. Pereira L, Levran O, Ramirez F et al. A molecular approach to the stratification of cardiovascular risk in families with Marfan's syndrome. N Engl J Med 1994; 331(3):148-153.
13. Schrijver I, Liu W, Odom R et al. Premature termination mutations in FBN1: Distinct effects on differential allelic expression and on protein and clinical phenotypes. Am J Hum Genet 2002; 71(2):223-237.
14. Rantamaki T, Kaitila I, Syvanen AC et al. Recurrence of Marfan syndrome as a result of parental germ-line mosaicism for an FBN1 mutation. Am J Hum Genet 1999; 64(4):993-1001.
15. Collod-Beroud G, Lackmy-Port-Lys M, Jondeau G et al. Demonstration of the recurrence of Marfan-like skeletal and cardiovascular manifestations due to germline mosaicism for an FBN1 mutation. Am J Hum Genet 1999; 65(3):917-921.
16. Montgomery RA, Geraghty MT, Bull E et al. Multiple molecular mechanisms underlying subdiagnostic variants of Marfan syndrome. Am J Hum Genet 1998; 63(6):1703-1711.
17. Lipscomb KJ, Clayton-Smith J, Harris R. Evolving phenotype of Marfan's syndrome. Arch Dis Child 1997; 76(1):41-46.
18. Lind J, Wallenburg HC. The Marfan syndrome and pregnancy: a retrospective study in a Dutch population. Eur J Obstet Gynecol Reprod Biol 2001; 98(1):28-35.
19. Dean JC. Management of Marfan syndrome. Heart 2002; 88(1):97-103.
20. Carson NAJ, Cusworth DC, Dent CE et al. Homocystinuria: A new inborn error of metabolism associated with mental deficiency. Arch Dis Child 1963; 38:425-436.
21. Hayward C, Brock DJ. Fibrillin-1 mutations in Marfan syndrome and other type-1 fibrillinopathies. Hum Mutat 1997; 10(6):415-423.
22. Robinson PN, Booms P, Katzke S et al. Mutations of FBN1 and genotype-phenotype correlations in Marfan syndrome and related fibrillinopathies. Hum Mutat 2002; 20(3):153-161.
23. Collod-Beroud G, Le Bourdelles S, Ades L et al. Update of the UMD-FBN1 mutation database and creation of an FBN1 polymorphism database. Hum Mutat 2003; 22(3):199-208.
24. Faivre L, Gorlin RJ, Wirtz MK et al. In frame fibrillin-1 gene deletion in autosomal dominant Weill-Marchesani syndrome. J Med Genet 2003; 40(1):34-36.
25. Booms P, Cisler J, Mathews KR et al. Novel exon skipping mutation in the fibrillin-1 gene: Two 'hot spots' for the neonatal Marfan syndrome. Clin Genet 1999; 55(2):110-117.
26. Sood S, Eldadah ZA, Krause WL et al. Mutation in fibrillin-1 and the Marfanoid-craniosynostosis (Shprintzen-Goldberg) syndrome. Nat Genet 1996; 12(2):209-211.
27. Greally MT, Carey JC, Milewicz DM et al. Shprintzen-Goldberg syndrome: a clinical analysis. Am J Med Genet 1998; 76(3):202-212.
28. Hayward C, Porteous ME, Brock DJ. A novel mutation in the fibrillin gene (FBN1) in familial arachnodactyly. Mol Cell Probes 1994; 8(4):325-327.
29. Lönnqvist L, Child A, Kainulainen K et al. A novel mutation of the fibrillin gene causing ectopia lentis. Genomics 1994; 19(3):573-576.
30. Comeglio P, Evans AL, Brice G et al. Identification of FBN1 gene mutations in patients with ectopia lentis and marfanoid habitus. Br J Ophthalmol 2002; 86(12):1359-1362.
31. Hasham SN, Guo DC, Milewicz DM. Genetic basis of thoracic aortic aneurysms and dissections. Curr Opin Cardiol 2002; 17(6):677-683.
32. Dietz HC, McIntosh I, Sakai LY et al. Four novel FBN1 mutations: Significance for mutant transcript level and EGF-like domain calcium binding in the pathogenesis of Marfan syndrome. Genomics 1993; 17(2):468-475.

33. Stahl-Hallengren C, Ukkonen T, Kainulainen K et al. An extra cysteine in one of the non-calcium-binding epidermal growth factor-like motifs of the FBN1 polypeptide is connected to a novel variant of Marfan syndrome. J Clin Invest 1994; 94(2):709-713.
34. Black C, Withers AP, Gray JR et al. Correlation of a recurrent FBN1 mutation (R122C) with an atypical familial Marfan syndrome phenotype. Hum Mutat 1998; Suppl 1:S198-200.
35. Milewicz DM, Grossfield J, Cao SN et al. A mutation in FBN1 disrupts profibrillin processing and results in isolated skeletal features of the Marfan syndrome. J Clin Invest 1995; 95(5):2373-2378.
36. Ades LC, Sreetharan D, Onikul E et al. Segregation of a novel FBN1 gene mutation, G1796E, with kyphoscoliosis and radiographic evidence of vertebral dysplasia in three generations. Am J Med Genet 2002; 109(4):261-270.
37. Geva T, Sanders SP, Diogenes MS et al. Two-dimensional and Doppler echocardiographic and pathologic characteristics of the infantile Marfan syndrome. Am J Cardiol 1990; 65(18):1230-1237.
38. Buntinx IM, Willems PJ, Spitaels SE et al. Neonatal Marfan syndrome with congenital arachnodactyly, flexion contractures, and severe cardiac valve insufficiency. J Med Genet 1991; 28(4):267-273.
39. Bresters D, Nikkels PG, Meijboom EJ et al. Clinical, pathological and molecular genetic findings in a case of neonatal Marfan syndrome. Acta Paediatr 1999; 88(1):98-101.
40. Kainulainen K, Karttunen L, Puhakka L et al. Mutations in the fibrillin gene responsible for dominant ectopia lentis and neonatal Marfan syndrome. Nat Genet 1994; 6(1):64-69.
41. Watanabe Y, Yano S, Koga Y et al. P1148A in fibrillin-1 is not a mutation leading to Shprintzen-Goldberg syndrome. Hum Mutat 1997; 10(4):326-327.
42. Katzke S, Booms P, Tiecke F et al. TGGE screening of the entire FBN1 coding sequence in 126 individuals with marfan syndrome and related fibrillinopathies. Hum Mutat 2002; 20(3):197-208.
43. Biggin A, Holman K, Brett M et al. Detection of thirty novel FBN1 mutations in patients with Marfan syndrome or a related fibrillinopathy. Hum Mutat 2004; 23(1):99.
44. Stoll C. Shprintzen-Goldberg marfanoid syndrome: A case followed up for 24 years. Clin Dysmorphol 2002; 11(1):1-7.
45. Loeys B, Nuytinck L, Delvaux I et al. Genotype and phenotype analysis of 171 patients referred for molecular study of the fibrillin-1 gene FBN1 because of suspected Marfan syndrome. Arch Intern Med 2001; 161(20):2447-2454.
46. Dietz HC, Cutting GR, Pyeritz RE et al. Marfan syndrome caused by a recurrent de novo missense mutation in the fibrillin gene. Nature 1991; 352(6333):337-339.
47. Gott VL, Cameron DE, Alejo DE et al. Aortic root replacement in 271 Marfan patients: A 24-year experience. Ann Thorac Surg 2002; 73(2):438-443.
48. Shores J, Berger KR, Murphy EA et al. Progression of aortic dilatation and the benefit of long-term beta-adrenergic blockade in Marfan's syndrome. N Engl J Med 1994; 330(19):1335-1341.

CHAPTER 2

Orthopaedic Problems in Marfan Syndrome

Paul Sponseller and Michael Shindle

Overview

The clinical findings which are most apparent in Marfan Syndrome (MFS), and which in fact often lead to the diagnosis, involve the skeleton (arachnodactyly, scoliosis, dolichostenomelia, sternal deformities, and joint laxity). Taken individually, these findings are not very specific, since they can be seen in the general population. The physician examining the skeletal system must be aware of the manifestations and able to help recognize affected individuals and make the diagnosis, but also avoid overdiagnosis of those who do not have it. This chapter will address general skeletal issues first, and then focus on specific regions and problems.

Since Marfan syndrome takes into account a wide clinical spectrum of disease, it is best viewed as a continuum. One end represents patients who are only mildly affected while the other end represents patients with a severe neonatal form who may develop cardiovascular complications during the first year of life. For many years, observers felt the neonatal form represented a different syndrome due to the extreme phenotype;[1] however, in 1994 it was concluded that the neonatal Marfan syndrome was also the result of mutations in the fibrillin gene on chromosome 15 and was part of the Marfan spectrum.[2] In addition, it has now been shown that mutations that cause the classic manifestation of Marfan syndrome occur throughout the *FBN1* gene[3,4] while the mutations resulting in the severe neonatal form of Marfan syndrome cluster in exons 24-32 of the gene.[2,5-7] Recently, Pepe et al[8] have provided some further insight into the Marfan continuum by postulating that the phenotypic severity is directly proportional to the amount of mRNA expression of the abnormal *FBN1* allele.

Diagnosis

In 1996, the diagnostic criteria for Marfan syndrome were revised from the "Berlin" criteria into a more stringent "Ghent" formulation.[9,10] The revisions took place to account for molecular data, and allowed for skeletal manifestations to be included as major diagnostic criteria. According to the Ghent criteria, if a first-degree relative has Marfan syndrome, the diagnosis can be made with the presence of one major criterion in an organ system in addition to involvement of a second organ system. If there are no affected relatives, than the diagnosis is made if major criteria are present in two different organ systems in addition to involvement of a third organ system (see Chapter 1 of this volume, "Diagnosis and Treatment of Marfan Syndrome – A Summary"). Rose et al[11] evaluated 73 patients under the Ghent criteria and concluded that 20% of patients who had been diagnosed with MFS using the Berlin criteria were appropriately excluded from the diagnosis of Marfan syndrome using the Ghent criteria. In addition, it was noted that dural ectasia is valuable in the evaluation of Marfan syndrome and established the diagnosis under the Ghent criteria in 25% of the patients. Conditions that are most often considered in the differential diagnosis include: homocystinuria, familial or isolated mitral valve prolapse syndrome, familial or isolated annuloaortic ectasia (Erdheim disease), congeni-

Marfan Syndrome: A Primer for Clinicians and Scientists, edited by Peter N. Robinson and Maurice Godfrey. ©2004 Eurekah.com and Kluwer Academic / Plenum Publishers.

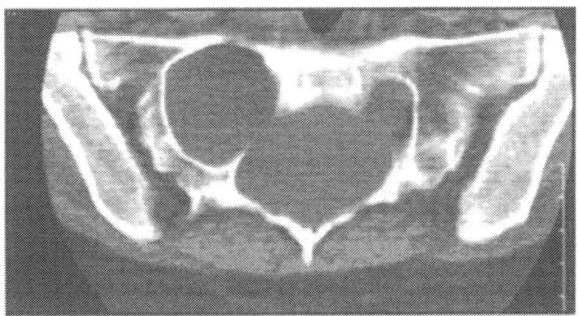

Figure 1. Osseous scalloping in Marfan syndrome.

tal contractural arachnodactyly, Stickler syndrome, Ehlers-Danlos syndrome, MASS phenotype, familial ectopia lentis, and Shprintzen-Goldberg syndrome.[10]

The most worrisome clinical manifestation of Marfan syndrome is dilatation of the aorta.
Early diagnosis is of the utmost importance in order to initiate prophylactic β-adrenergic blockade
therapy, which has been shown to be effective in slowing the rate of aortic dilatation and
reducing the development of aortic complications in patients with Marfan syndrome.[12] Thus,
an increased awareness of the skeletal malformations that occur in childhood combined with
molecular techniques can lead to earlier diagnosis and proper cardiovascular monitoring which
is enabling Marfan patients to have near normal life expectancies.

Spinal Abnormalities

Vertebral Morphology and Dural Ectasia

Tallroth et al have noted a high prevalence of biconcave vertebrae, transitional vertebrae,
and lengthened transverse processes.[13] Sponseller et al[14] found a reduction in both the lumbar
pedicle width and laminar thickness. One-fourth of Marfan individuals have pedicles of L5
which are too narrow to accept even the smallest pedicle screw (5 mm). Many patients have
significant scalloping of the sacrum inside the spinal canal (Fig. 1). Therefore, careful preoperative planning and routine computed tomography scans are recommended for any surgeon
operating on the Marfan spine.[14] Ahn et al[15] concluded that the lumbar interpediculate distance
in Marfan patients is widened compared to the normal population at all levels of the lumbar
spine. In addition, cut-off values were determined for the interpediculate distances in the lumbar spine and can be used as a screening tool for MFS with a specificity of 95% (Table 1).[16,17]

Table 1. Screening tool for marfan syndrome using interpediculate distance (IPD)
values. IPD greater than the "cut off value" is highly indicative of Marfan
syndrome.

Level	"Cut-Off Value" (mm)	Specificity (%)	Sensitivity (%)
L1	29.4	95	51
L2	30.1	95	53
L3	31.4	95	61
L4	32.9	95	75
L5	37.1	95	70

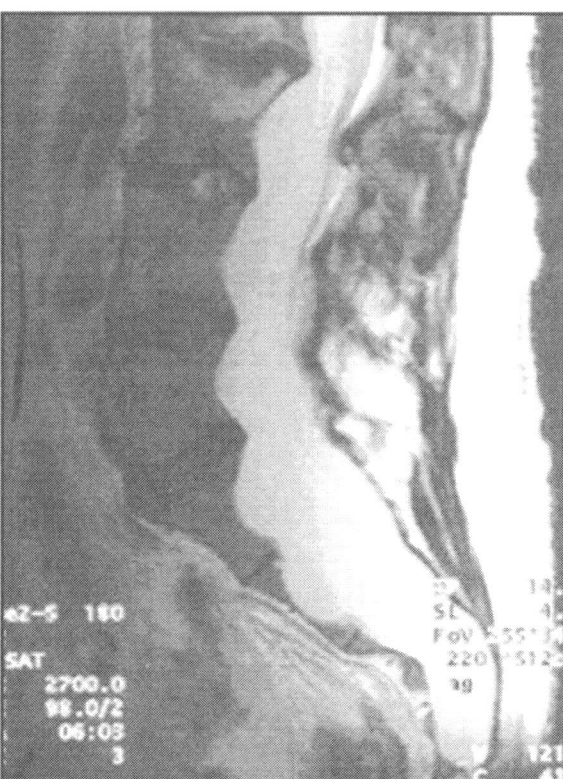

Figure 2. Dural ectasia in Marfan syndrome.

Dural ectasia (DE), a ballooning or significant widening of the dural sac or root sleeves, is rare in the normal population but has a high prevalence ranging from 56% to 92% in the Marfan population[18-20] (Fig. 2). The consequences of DE include bony erosion, anterior meningocele and/or posterior meningocele. In Marfan patients, the enlarged sac is primarily below the level of L5[18,19,21-24] which has led to the hypothesis that the weakened dural tissues cause ectasia because the fluid pressure is greatest in the most caudal portion of the spine.[22] In addition, dural ectasia may be associated with back pain and headaches[16,23-27] or neurologic deficits.[19,22,28] However, a high prevalence of dural ectasia (41%) exists, even in Marfan patients without back pain.[21] Thus, back pain and dural ectasia are related, but the presence of dural ectasia does not guarantee that the patient will present with symptoms.

Defining dural ectasia is a matter of some variability. Fattori used morphologic criteria consisting of bulging of the dural sac to obliterate the epidural fat and contact at least one posterior vertebral body, or the presence of radicular cysts.[29] Villeirs[30] devised a definition of dural ectasia using computed tomography of the spine. Analyzing controls and Marfan individuals, he devised a dural sac index according to the following formula: 4(transverse diameter of dura at L1+L2+L3+L4+L5+S1)/ Transverse vertebral body diameter L1+L2+L3+L4. Values greater than 3.75 were rarely seen in the general population, but were seen in 70% of Marfan persons. Oosterhof et al[31] found that abnormal dural sac ratio values (the dural sac diameter corrected for vertebral size) at L3 or S1 can be used to identify Marfan syndrome with 95% sensitivity and 98% specificity. The dural sac ratio (DSR) is calculated by dividing the dural sac diameter by the vertebral body diameter at the midcorpus level of L1 through S1. However, this method requires advanced imaging which led Ahn et al to develop simpler criteria to determine whether or not dural ectasia is present on CT and MRI in adult patients.[21] A sagittal width of the dural sac at S1 or below greater than the width of the dural sac above L4, or the presence of an anterior meningocele is required to fulfill a major criterion.[21] In addition, both minor criteria are required which include a nerve root sleeve at L5 > 6.5 mm in diameter or scalloping at S1 > 3.5 mm. Although CT can accurately detect dural ectasia, MRI was superior with a sensitivity exceeding 85% and a specificity of 95%.

MRI and CT are the gold standards for diagnosing dural ectasia; however, there is also a role for conventional radiographs because they can detect dural ectasia with a very high specificity (91.7%) but a low sensitivity (57.1%).[15] One of the following criteria needs to be met in order to be consistent with dural ectasia: interpediculate distance at L4 ≥ 38.0 mm; sagittal diameter at S1 ≥ 18.0 mm; or scalloping value at L5 ≥ 5.5 mm. Therefore, conventional radiographs may be useful in the initial evaluation of the Marfan patient with back pain. However, satisfying the criteria does not necessarily mean the patient has dural ectasia, but the physician should perform more advanced radiologic studies.

The symptoms seen with dural ectasia vary. Although some patients are asymptomatic, most have one or more of the following: pain in the very low back or legs or pelvis, hypoesthesia in the legs or perineum, and postural headaches. There is currently no treatment for this symptom which has a high degree of success.

Scoliosis and Kyphosis

Scoliosis in the Marfan population has been reported from 40 to 70 percent.[32-34] In a large cross-sectional study, there was a prevalence of scoliosis of 62% which was equal between the sexes, and 96% of the patients has the thoracic portion of the curve convex to the right.[32] In this and some other ways, the curve pattern was similar to that seen in idiopathic scoliosis. Differences were seen in the sagittal plane, however (see below). Despite the frequency of spinal deformity in MFS, however, many of the curves were minor and only 10-20% require treatment of any kind. Scoliosis progresses (worsens) at a faster rate in Marfan syndrome than in the general population. Virtually all curves larger than 30 degrees in immature patients will reach at least 40 degrees at maturity. Curves larger than 40 degrees progress during adulthood, at a slightly higher rate than idiopathic scoliosis. Curves greater than 50 degrees progressed at a mean rate of 3 degrees per year in adulthood. Marfan patients with scoliosis were found to have more back pain than those who did not have scoliosis.

Brace treatment is commonly used to control curves of idiopathic scoliosis in growing children. However, Sponseller et al reported a 17% success rate for bracing scoliosis curves in Marfan syndrome, which is lower than that reported for idiopathic scoliosis. Thus, beginning bracing is mainly recommended for curves in the range of 15-25°. Patients with curves 25-45° may be offered the option of using the brace, but physicians should be aware that there is only a 1 in 5 chance of successfully controlling the curve. Bracing is not recommended for curves > 40° as they can only be controlled with surgical correction.[35]

Infantile scoliosis in the Marfan population is defined as a curve of more than 10° by age 3. It is seen in approximately 2% of persons with MFS.[36] Sponseller et al[36] note that bracing has a limited role in this population especially in smaller curves. It is mainly useful in promoting upright posture in patients with a coexistent kyphosis. In addition, surgery is not usually recommended on patients younger than 4 years of age because instrumentation is bulky and cardiac status has not often been sorted out yet. For patients with no significant kyphosis, instrumentation without fusion should be considered. The use of a growing rod employing iliac fixation distally, and strong proximal anchorage has been more successful recently (Fig. 3).

Sponseller et al also found that the mean kyphosis of the Marfan population was greater than that in the general population. In addition, 41 percent of the Marfan patients had a kyphosis that was more than 50 degrees. Five subtypes of sagittal alignment were found. There was a tendency for longer kyphoses extending through the thoracolumbar junction.[32]

Due to the prevalence of scoliosis and kyphosis, and to the relative ineffectiveness of bracing in the Marfan population, spine surgery is sometimes required (13% to 21%).[32,33] There is also a higher incidence of complications following spine surgery in the Marfan population.

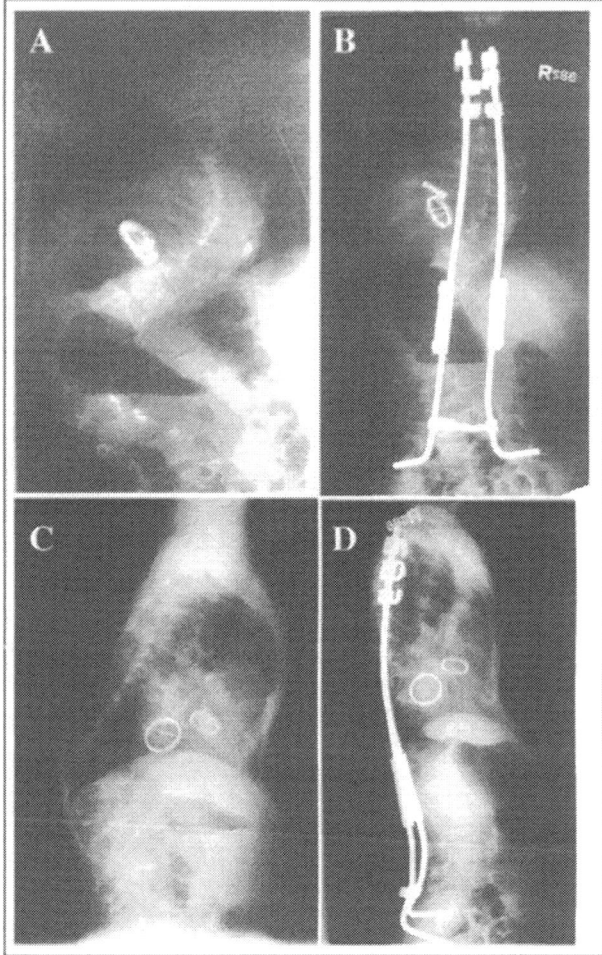

Figure 3. Infantile scoliosis in Marfan syndrome A) preop AP; B) postop AP; C) preop lateral; D) postop lateral.

Intra-operative complications include increased bleeding and cerebrospinal fluid leak. Postoperative complications include failure of fixation, adding on of curvature, and pseudarthrosis.[33,37] Therefore, if correcting spinal deformity it is important not to fuse too short a segment of a spine (Fig. 4). It is also important to take into account the any unusual kyphosis which may be present.

Spondylolisthesis

The prevalence of spondylolisthesis in the general population is approximately 3 percent, and, of this population, the mean slip is approximately 15 per cent.[38] Sponseller et al found that 6 percent of the Marfan population had spondylolisthesis of the fifth lumbar on the first sacral vertebra with a mean slip of 30 percent.[32] Sponseller et al have hypothesized that the frequency of spondylolisthesis in Marfan patients may not be markedly higher than the general population. However, if a slip is present, the altered properties of the ligament and shear resis-

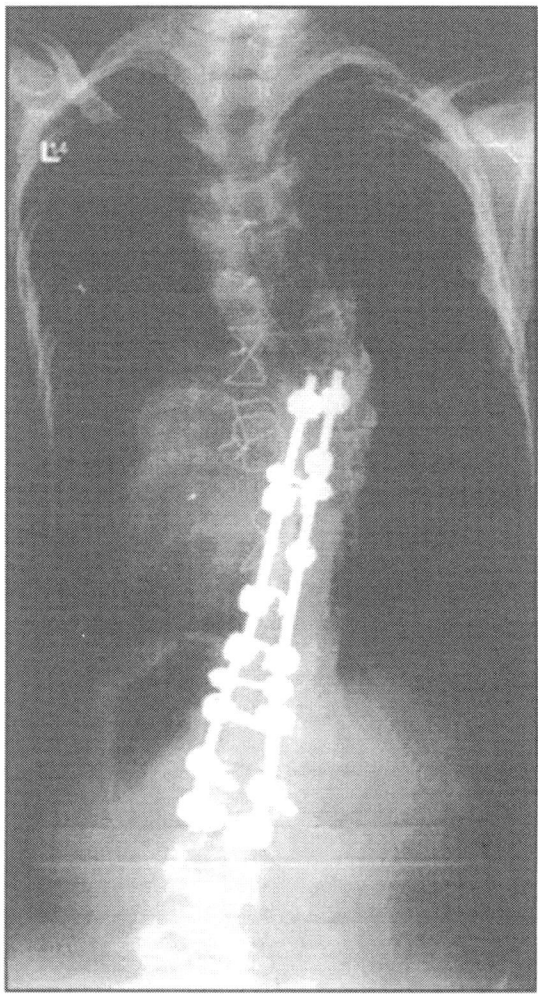

Figure 4. Adding-on deformity after correction of scoliosis in a twelve-year old.

tance of the disc allow greater forward slip to occur. It is at this level where the dural ectasia, sacral scalloping, and potential cerebrospinal fluid leaks have the greatest clinical importance.

Cervical Spine Problems

Hobbs et al[39] conducted a radiographic analysis of the cervical spine in patients with Marfan syndrome, which revealed an increased prevalence of focal kyphosis, and a slightly increased atlantoaxial translation with flexion and extension. Marfan patients may occasionally suffer unusual cervical spine injuries. The cervical kyphosis may predispose to vertebral fracture. Herzka et al[40] report three cases that suggest that Marfan children may be at risk for atlanto-axial rotatory subluxation when muscle tone is attenuated by general anesthesia or muscle relaxants. Special attention to intubation and positioning, both intraoperatively and postoperatively, may be warranted and rotatory subluxation should be included in the differential diagnosis for Marfan patients with neck pain after injury or surgery.

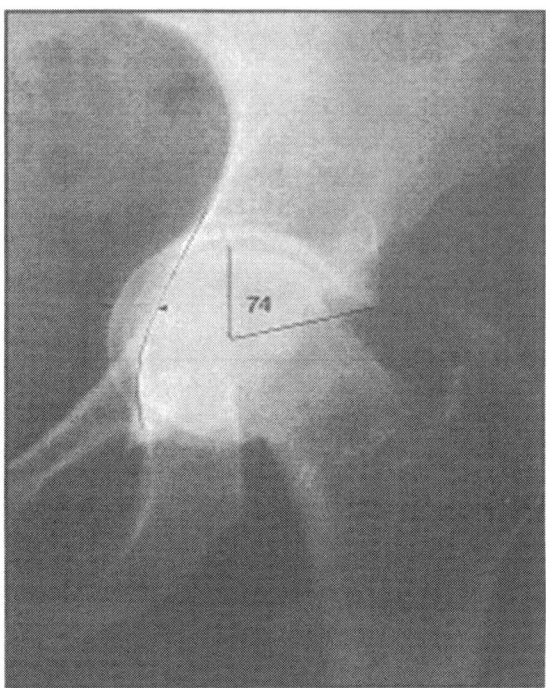

Figure 5. Protrusio acetabulae in an adult with Marfan syndrome. Indicated here are the measurment of the center-edge angle (74 degrees) and the acetabular protrusion medial to the ilioischial line.

Hip Deformity

Since first reported in 1978, many authors have confirmed the association of protrusio acetabuli and Marfan syndrome.[41-44] Wenger[44] radiographically defined protrusio acetabuli in the Marfan population by the presence of both a center edge angle of 40° or more and crossing of the acetabular teardrop by the femoral head[44] (Fig. 5). Armbuster et al[45] developed five different radiologic parameters to measure the protrusio including: the acetabular line crossing the ilioischial line on a straight anteroposterior pelvis, a center-edge (CE) angle greater than 40°, the acetabular line crossing Kohler's line, the acetabular line crossing the iliopectineal line, and an increased distance from the femoral head to the ilioischial line. The prevalence of protrusio acetabuli has been reported from 31-100%.[42,46] In 1996, Steel[42] described a technique for closure of the triradiate physis and observed that it was effective in reversing or stopping the protrusio. Steel recommended that the procedure should be reserved for children between the ages of 8 and 10 who have had documented progression of acetabular deepening. Do et al[46] found that the presence of protrusio is not related to either the bone mineral density (BMD), or to hip symptoms. Furthermore, Marfan patients who did not have surgical intervention did not have radiographic evidence of degenerative changes and two patients in their 50's were still active. This raises the issue of whether or not surgical intervention is necessary at an early age. Sponseller reviewed 350 consecutive unselected hips with Marfan syndrome and found that protrusio was seen in only one-third of adults, and degenerative joint disease was rare. Only young patients with acetabular protrusion of the medial wall of the acetabulum more than 5mm medial to the ilioischial line should be considered possible candidates for this procedure.

Sponseller et al noted a 2% occurrence of developmental dysplasia of the hip among 235 consecutive patients with Marfan syndrome.[47] Due to a narrow stable zone and knee laxity, Pavlik harness treatment was not effective. Therefore, closed reduction and spica cast treatment is recommended for the dislocated hip in patients with Marfan syndrome.[47]

Marfan patients require total hip arthroplasty very occasionally. Study of the only series revealed that the mean age at surgery was 44 years, the loosening rate was 40% at ten years after surgery, dislocation occurred in 20%, but patients had good clinical results after these complications were corrected.[48]

Joint Manifestations

The appendicular joints of Marfan patients are commonly viewed as being lax; however, normal mobility and even outright contractures are frequent.[49] Longitudinal laxity of the hand is responsible for the Steinberg sign, in which the thumb, when apposed across the palm, extends well beyond the ulnar border of the hand.[50] Although common, joint laxity has little diagnostic specificity and little disability usually results from laxity of the hands, wrists, elbows, shoulders, hips, and knees.[51] It has been postulated that the joint laxity occurs due to a marked reduction of elastin fiber content and abnormalities in the existing fibers which impairs the normal properties of the articular capsule.[52]

Laxity of the ankle and foot produce the most common problems, and has been postulated to be caused by stretching of the plantar fascia and other ligaments of the midfoot.[49] In a study of 63 individuals with Marfan syndrome, Lindsay et al[53] found that 60% of the Marfan patients had laxity of the foot by objective testing (Fig. 6). However, the ligamentous laxity did not correlate significantly in this population with either pes planus or hallux valgus. In addition, despite the predominance of objective ligamentous laxity, Lindsey found that a significant percentage (74.8%) of Marfan individuals maintain an arch that is within the range of the normal population.[53] Patients with Marfan syndrome should be offered orthotics only if symptomatic, and they should not be rigid because of the very thin subcutaneous layer in the foot. Surgery is not indicated unless pain or weakness develop.

Growth and Maturation

Young patients with Marfan syndrome often desire to know their eventual height at maturity. It is not possible to predict this using standard growth charts, since they are usually above the 95th percentile for the general population, and the pattern of their growth may not follow these lines. Erkula[54] studied this issue and found that Marfan children have their peak height velocity ("growth spurt") about two years early. Menarche occurs earlier than usual in most Marfan girls. A series of growth charts for height and weight in Marfan syndrome were developed.[54]

Bone Density

It has been postulated that the abnormal fibrillin in the Marfan population plays a role in affecting the mineralization of bone.[55] However, there is still some controversy over BMD values in the Marfan population, and it still has not been determined whether or not there is a significant relationship between the BMD values and fracture risk. For example, BMD studies in children with Marfan syndrome have reported reductions at the femoral neck.[55] In addition, women with Marfan syndrome were shown to have reduced BMD values in the lumbar spine and at the total hip.[55] However, other studies have reported normal BMD values at the distal forearm, hip and lumbar spine in the Marfan population.[56,57] Carter et al[58] have demonstrated that there is a reduced axial BMD in men and women with Marfan syndrome and postulated that this may be due to mutations of the fibrillin gene or due to environmental issues such as reduced exercise which leads to sub-optimal peak bone mass.[58]

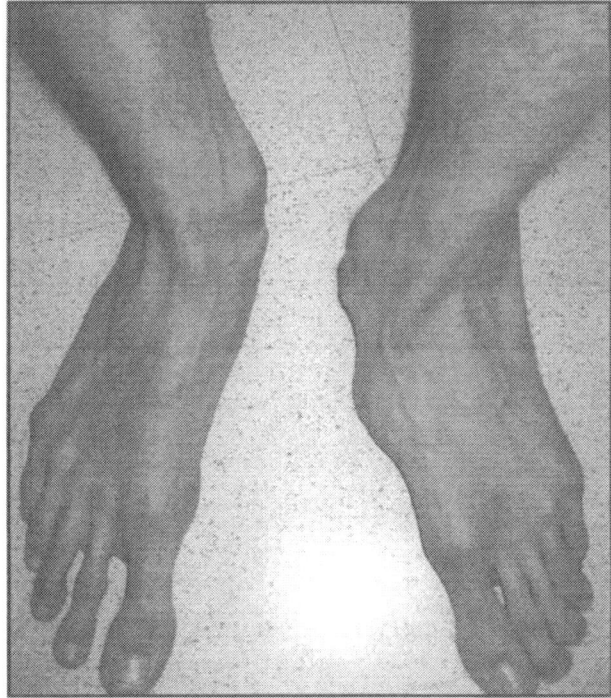

Figure 6. Pes planovalgus in a nineteen year old with Marfan syndrome.

Screening and Athletics

It is recommended that Marfan patients avoid situations of considerable emotional or physical stress because many aortic dissections occur during these events.[49,59,60] Pyeritz[49] recommends that the physician counsel on an individual basis. For example, a child with only a slightly dilated aortic root does not need outright restriction but should be counseled away from competitive athletics. On the other hand, an older patient should be advised against any sort of strenuous exertion especially activities with sudden stops such as basketball, or isometric exercises.

Although Marfan patients should avoid contact, competitive, or isometric sports, other activities such as low intensity isokinetic sports should be encouraged.

References

1. Morse RP, Rockenmacher S, Pyeritz RE et al. Diagnosis and management of infantile marfan syndrome. Pediatrics 1990; 86(6):888-895.
2. Kainulainen K, Karttunen L, Puhakka L et al. Mutations in the fibrillin gene responsible for dominant ectopia lentis and neonatal Marfan syndrome. Nat Genet Jan 1994; 6(1):64-69.
3. Dietz HC, Cutting GR, Pyeritz RE et al. Marfan syndrome caused by a recurrent de novo missense mutation in the fibrillin gene. Nature 1991; 352(6333):337-339.
4. Vincent GM. Role of DNA testing for diagnosis, management, and genetic screening in long QT syndrome, hypertrophic cardiomyopathy, and Marfan syndrome. Heart 2001; 86(1):12-14.
5. Putnam EA, Cho M, Zinn AB et al. Delineation of the Marfan phenotype associated with mutations in exons 23-32 of the FBN1 gene. Am J Med Genet 1996; 62(3):233-242.
6. Milewicz DM, Duvic M. Severe neonatal Marfan syndrome resulting from a de novo 3-bp insertion into the fibrillin gene on chromosome 15. Am J Hum Genet 1994; 54(3):447-453.
7. Wang M, Price C, Han J et al. Recurrent mis-splicing of fibrillin exon 32 in two patients with neonatal Marfan syndrome. Hum Mol Genet 1995; 4(4):607-613.

8. Pepe G, Giusti B, Attanasio M et al. A major involvement of the cardiovascular system in patients affected by Marfan syndrome: Novel mutations in fibrillin 1 gene. J Mol Cell Cardiol 1997; 29(7):1877-1884.

9. De Paepe A, Devereux RB, Dietz HC et al. Revised diagnostic criteria for the Marfan syndrome. Am J Med Genet 1996; 62(4):417-426.

10. Beighton P, de Paepe A, Danks D et al. International nosology of heritable disorders of connective tissue, berlin, 1986. Am J Med Genet 1988; 29:581-594.

11. Rose PS, Levy HP, Ahn NU et al. A comparison of the Berlin and Ghent nosologies and the influence of dural ectasia in the diagnosis of Marfan syndrome. Genet Med 2000; 2(5):278-282.

12. Shores J, Berger KR, Murphy EA et al. Progression of aortic dilatation and the benefit of long-term beta-adrenergic blockade in Marfan's syndrome. N Engl J Med 1994; 330(19):1335-1341.

13. Tallroth K, Malmivaara A, Laitinen ML et al. Lumbar spine in Marfan syndrome. Skeletal Radiol 1995; 24(5):337-340.

14. Sponseller PD, Ahn NU, Ahn UM et al. Osseous anatomy of the lumbosacral spine in Marfan syndrome. Spine 2000; 25(21):2797-2802.

15. Ahn NU, Nallamshetty L, Ahn UM et al. Dural ectasia and conventional radiography in the Marfan lumbosacral spine. Skeletal Radiol 2001; 30(6):338-345.

16. Arroyo JF, Garcia JF. Case report 582: Lumbosacral meningocele and aortic aneurysm in Marfan syndrome. Skeletal Radiol 1989; 18(8):614-618.

17. Nelson JD. The Marfan syndrome, with special reference to congenital enlargement of the spinal canal. Br J Radiol 1958; 31:561-564.

18. Fishman EK, Zinreich SJ, Kumar AJ et al. Sacral abnormalities in Marfan syndrome. J Comput Assist Tomogr 1983; 7(5):851-856.

19. Pyeritz RE, Fishman EK, Bernhardt BA et al. Dural ectasia is a common feature of the Marfan syndrome. Am J Hum Genet Nov 1988; 43(5):726-732.

20. Pyertiz RE, Fishman EK, Bernhardt BA et al. Dural ectasia: A common pleitotropic feature of the Marfan syndrome. Am J Hum Genet 1986; 39S:A75.

21. Ahn NU, Sponseller PD, Ahn UM et al. Dural ectasia in the Marfan syndrome: MR and CT findings and criteria. Genet Med 2000; 2(3):173-179.

22. Stern WE. Dural ectasia and the Marfan syndrome. J Neurosurg 1988; 69(2):221-227.

23. Strand RD, Eisenberg HM. Anterior sacral meningocele in association with Marfan's syndrome. Radiology 1971; 99(3):653-654.

24. Robinson L, Dominguez R, Cabrera J et al. Multiple meningeal cysts in Marfan syndrome. AJNR Am J Neuroradiol 1989; 10(6):1275-1276.

25. Ahn NU, Sponseller PD, Ahn UM et al. Dural ectasia is associated with back pain in Marfan syndrome. Spine 2000; 25(12):1562-1568.

26. Weir B. Leptomeningeal cysts in congenital ectopia lentis. Case report. J Neurosurg 1973; 38(5):650-654.

27. Duncan RW, Esses S. Marfan syndrome with back pain secondary to pedicular attenuation. A case report. Spine 1995; 20(10):1197-1198.

28. Smith MD. Large sacral dural defect in Marfan syndrome. A case report. J Bone Joint Surg Am 1993; 75(7):1067-1070.

29. Fattori R, Nienaber CA, Descovich B et al. Importance of dural ectasia in phenotypic assessment of Marfan's syndrome. Lancet 1999; 354(9182):910-913.

30. Villeirs GM, Van Tongerloo AJ, Verstraete KL et al. Widening of the spinal canal and dural ectasia in Marfan's syndrome: Assessment by CT. Neuroradiology 1999; 41(11):850-854.

31. Oosterhof T, Groenink M, Hulsmans FJ et al. Quantitative assessment of dural ectasia as a marker for Marfan syndrome. Radiology 2001; 220(2):514-518.

32. Sponseller PD, Hobbs W, Riley LH et al. The thoracolumbar spine in Marfan syndrome. J Bone Joint Surg Am 1995; 77(6):867-876.

33. Robins PR, Moe JH, Winter RB. Scoliosis in Marfan's syndrome. Its characteristics and results of treatment in thirty-five patients. J Bone Joint Surg Am 1975; 57(3):358-368.

34. Winter RB, Banta JV, Engler G. Screening for scoliosis. JAMA 1995; 273(3):185-186.

35. Sponseller PD, Bhimani M, Solacoff D et al. Results of brace treatment of scoliosis in Marfan syndrome. Spine 2000; 25(18):2350-2354.

36. Sponseller PD, Sethi N, Cameron DE et al. Infantile scoliosis in Marfan syndrome. Spine 1997; 22(5):509-516.

37. Birch JG, Herring JA. Spinal deformity in Marfan syndrome. J Ped Orthop 1987; 7:546-552.

38. Fredrickson BE, Baker D, McHolick WJ et al. The natural history of spondylolysis and spondylolisthesis. J Bone Joint Surg Am 1984; 66:699-707.

39. Hobbs WR, Sponseller PD, Weiss AP et al. The cervical spine in Marfan syndrome. Spine 1997; 22(9):983-989.
40. Herzka A, Sponseller PD, Pyeritz RE. Atlantoaxial rotatory subluxation in patients with Marfan syndrome. A report of three cases. Spine 2000; 25(4):524-526.
41. Steel HH. Protrusio acetubali in Marfan syndrome: A surgical approach to arresting the problem by closure of the triradiate epiphysis. Orthop Trans 1978; 2:47.
42. Steel HH. Protrusio acetabuli: Its occurrence in the completely expressed Marfan syndrome and its musculoskeletal component and a procedure to arrest the course of protrusion in the growing pelvis. J Pediatr Orthop 1996; 16(6):704-718.
43. Clement J, Lebarbier P, Cahuzac J et al. La protrusion acetabulaire dans la maladie de Marfan. Chir Pediatr 1984; 26:153-161.
44. Wenger DR, Ditkoff TJ, Herring JA et al. Protrusio acetabuli in Marfan's syndrome. Clin Orthop Mar-Apr 1980; (147):134-138.
45. Armbuster TG, Guerra Jr J, Resnick D et al. The adult hip: An anatomic study. Part I: The bony landmarks. Radiology 1978; 128(1):1-10.
46. Do T, Giampietro PF, Burke SW et al. The incidence of protrusio acetabuli in Marfan's syndrome and its relationship to bone mineral density. J Pediatr Orthop 2000; 20(6):718-721.
47. Sponseller PD, Tomek IM, Pyertiz RE. Developmental dysplasia of the hip in Marfan syndrome. J Pediatr Orthop B 1997; 6(4):255-259.
48. Foran JRH, Sponseller PD, Hungerford DS. Results of total hip arthroplasty in Marfan syndrome. 2003; In press.
49. Pyeritz RE. The Marfan syndrome. In: Royce PM, Steinmann B, eds. Connective Tissue and its Heritable Disorders. New York: Wiley-Liss & Sons, 1993:437-468.
50. Steinberg I. A simple screening test for the Marfan syndrome. Am J Roentgenol Radium Ther Nucl Med May 1966; 97(1):118-124.
51. Pennes DR, Braunstein EM, Shirazi KK. Carpal ligamentous laxity with bilateral perilunate dislocation in Marfan syndrome. Skeletal Radiol 1985; 13(1):62-64.
52. Gigante A, Chillemi C, Greco F. Changes of elastic fibers in musculoskeletal tissues of Marfan syndrome: A possible mechanism of joint laxity and skeletal overgrowth. J Pediatr Orthop 1999; 19(3):283-288.
53. Lindsey JM, Michelson JD, MacWilliams BA et al. The foot in Marfan syndrome: Clinical findings and weight-distribution patterns. J Pediatr Orthop 1998; 18(6):755-759.
54. Erkula G, Jones KB, Sponseller PD et al. Growth and maturation in Marfan syndrome. Am J Med Genet 2002; 109(2):100-115.
55. Kohlmeier L, Gasner C, Bachrach LK et al. The bone mineral status of patients with Marfan syndrome. J Bone Miner Res 1995; 10(10):1550-1555.
56. Tobias JH, Dalzell N, Child AH. Assessment of bone mineral density in women with Marfan syndrome. Br J Rheumatol 1995; 34(6):516-519.
57. Gray JR, Bridges AB, Mole PA et al. Osteoporosis and the Marfan syndrome. Postgrad Med J 1993; 69(811):373-375.
58. Carter N, Duncan E, Wordsworth P. Bone mineral density in adults with Marfan syndrome. Rheumatology (Oxford) 2000; 39(3):307-309.
59. Bain MA, Zumwalt RE, van der Bel-Kahn J. Marfan syndrome presenting as aortic rupture in a young athlete: Sudden unexpected death? Am J Forensic Med Pathol 1987; 8(4):334-337.
60. Maron BJ, Epstein SE, Roberts WC. Causes of sudden death in competitive athletes. J Am Coll Cardiol 1986; 7(1):204-214.

CHAPTER 3

Ophthalmological Aspects

Thomas Rosenberg

Introduction

The eye is involved in a majority of patients affected by Marfan syndrome. Morphological changes essentially affect the microfibrillar elements of the corneoscleral envelope, the iris, and the ciliary zonules, giving rise to clinically and diagnostically significant characteristics.

The ophthalmologist, as an indispensable member of the interdisciplinary team involved in assessment and follow-up of patients, is primarily engaged in the optical and surgical optimization of the visual functions, including the prevention and treatment of sight-threatening complications.

The Role of the Ophthalmologist

The principal aim of the ophthalmological profession in relation to Marfan syndrome (MFS) is to preserve and optimize visual functions, by means of refractive or surgical procedures or both. Furthermore, in cases with a permanent visual disability, rehabilitation measures should be initiated.

Visual complaints due to myopia are frequently the initial symptom in MFS. Therefore, the ophthalmologist is often the first medical specialist to be involved. It puts an obligation and a responsibility on the practitioners of this specialty to unravel the underlying cause. Yet, it is not as easy as it sounds, if one takes into account the fact that myopia is one of the most common causes for an ophthalmological examination, whereas the appearance of a patient with MFS is a relatively rare event in ordinary ophthalmological practice. Nevertheless, myopia is always a reason for a full eye examination, including cycloplegia, during which an even minor displacement of the crystalline lenses can be identified.

Often, the stature of the patient or the facial characteristics will suggest the presence of Marfan syndrome and lead to further diagnostic examinations.

Therefore, the role of the ophthalmologist is not confined to ophthalmological problems per se, but it is extended to join an interdisciplinary team caring for assessment, follow-up, and rehabilitation.

Key Symptoms

The international diagnostic criteria[1] include one major criterion for the ocular system, ectopia lentis (Fig. 1), as well as three minor criteria: Flat cornea, increased axial length, and hypoplastic iris or miotic pupils, which are difficult to dilate. In the presence of two of the three minor criteria the ocular system is considered as 'involved'.

Ocular Pathophysiology

A deficiency in the structure and function of fibrillin-1-containing microfibrils is generally considered to represent the fundamental defect in the development of ocular tissues in Marfan

Marfan Syndrome: A Primer for Clinicians and Scientists, edited by Peter N. Robinson and Maurice Godfrey. ©2004 Eurekah.com.

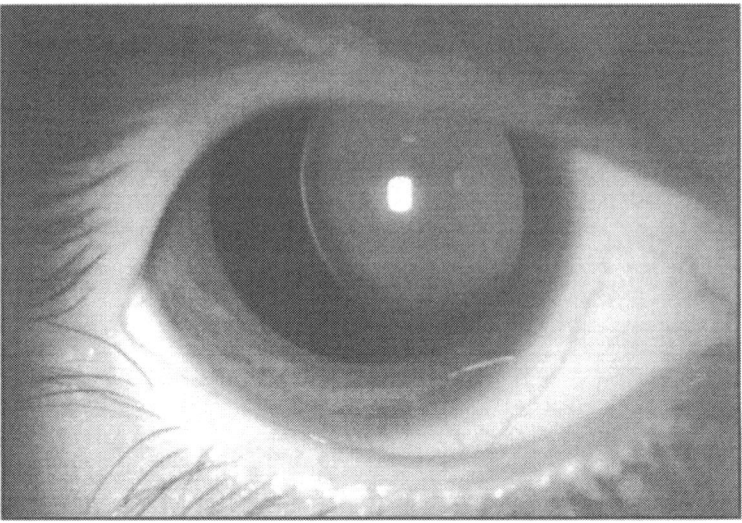

Figure 1. Marfan syndrome. Right eye with an upward and nasal dislocation of the lens.

syndrome. Fibrillin-rich microfibrils are present in the connective tissues of the eye including (among others) the corneal and scleral stroma, the zonular fibers, the iris and ciliary body stroma, Bruch's membrane, the choroid, and the vitreous.[2]

Histological examinations of the ciliary zonules, the most thoroughly investigated ocular structure involved, from patients with ectopia lentis and MFS have demonstrated morphological abnormalities as thinning, stretching, and diameter-irregularity.[3] In addition, ultrastructural examination has shown disrupted zonules with disorganized and comprehensively fragmented microfibrils in patients with ectopia lentis.[4] These observations suggest that structural weakness of the ciliary zonules is a main cause of lens dislocation.

Increased susceptibility to the proteolytic activity from serine proteases and matrix metalloproteinases has also been implicated as possible contributory pathogenetic factors (see ref. 5 and references herein). Wheatley and coworkers stressed the possible pathogenetic role played by defective zonular anchoring to the lens capsule and the ciliary processes.[2]

Marked variations in the clinical expression of ocular manifestations within as well as between families are poorly understood and indicate a rather complex molecular basis for lens dislocation. Genotype-phenotype correlations have not been able to account for this differential pattern of expression. Nevertheless, in a recent article it was suggested that MFS patients with nonsense mutation have a lesser chance of developing EL than patients with cysteine-missense mutations.[6,7]

In this connection it is noteworthy that about one third of patients fulfilling the diagnostic Ghent criteria do not present any sign of ocular involvement. In those with ocular involvement, on the other hand, a spectrum of manifestations involving nearly all structures of the globe may be present.

Since mutation-specific changes in the function of fibrillin-1 do not seem to account for the spectrum of ophthalmological features, lens displacement of even low grade may be the primary pathogenetic event, triggering a chain of secondary structural and functional changes including all refractive components, reduced visual acuity, amblyopia, and vision-threatening complications like retinal detachment, glaucoma, and cataract.

Clinical Signs

Facial Characteristics

The inspection of the area around the eye may elucidate some facial characteristics consisting of moderate enophthalmos, antimongoloid lid axis, malar hypoplasia, and a beaked nose. Due to weakened mimetic muscles the facial expression has been characterized as myopathic.[8]

Refraction

Refractive errors are common in MFS. Maumenee, in her monograph,[8] analyzed the refractive power in 160 patients and found a much wider distribution of the spherical equivalent among the Marfan patients than in the normal population. Among her patients 16% had myopia of -7.00 D or more compared to 0.3% in a Swedish population study on refractive power in conscripts. Myopia exceeding -15 D was encountered among 4.5% of the Marfan patients. The myopia in MFS is either due to an axial length elongation[9] or an elevated refractive power of the crystalline lens,[10] or both.

The refractive power is highly correlated with age. It is noteworthy that Maumenee found all patients in the age group of 0 to 3 years to be virtually emmetropic on retinoscopy.[8]

Due to a high prevalence of aphakia in her patient group a significant proportion of high hypermetropia was noted as well.

The frequent occurrence of astigmatism is due to the combined effects of corneal toricity and lens dislocation, which also includes various degrees of tilt.

Visual Acuity

In most patients with MFS and without lens dislocation the visual acuity is normal. Provided optimal correction of refractive errors, vision might even be normal in the presence of ectopia lentis. In a study, approximately one fourth of the eyes had a visual acuity of 20/30 or better and about 60% had an acuity of at least 20/70.[11] In another patient group nearly 85% of the patients had a visual acuity of 20/40 or more in the better eye.[8] On the other hand, about 5% of the patients in earlier materials had visual acuities below 20/200, mainly due to retinal detachment or glaucoma. Recent reports on visual acuity in large samples are not available.

Only sporadic information on binocular vision in MFS has been published. Speedwell, in a diagnostically mixed group of 27 children with ectopia lentis, observed poor stereoacuity in all but one child despite the presence of straight eyes.[12] Other reports give solid evidence of a high percentage of ocular misalignment in MFS, preferentially exotropia, anisometropia, and amblyopia, even in the absence of ectopia lentis.[13,14]

Binocular amblyopia in MFS is not uncommon but it is unresolved whether this is due to a deficiency of complete correction of complicated refractive errors or real perceptive amblyopia.

Cornea

The overall size of the cornea tends to be larger than normal and in some cases megalocornea with a corneal diameter of 13 mm or above is present. Corneal thickness as measured by pachymetry is normal.

In the series of Maumenee[8] keratometry disclosed a considerable corneal flattening of 2 to 2.5 D at average, with a much wider distribution than normal. Despite a significant positive association between corneal flattening and the presence of ectopia lentis, a flat cornea is included as a separate trait among the international diagnostic criteria (see above).

A high degree of corneal astigmatism is present in MFS, and about a quarter of the eyes had a toricity of more than 2 D.[8]

Specular microscopic examinations in MFS-patients with as well as without ectopia lentis, have revealed significant morphological endothelial abnormalities consisting of decreased cell density, pleomorphism, guttata formation, and "black spots". The significance of these changes is uncertain but may indicate an increased vulnerability to surgical trauma and anoxia.[15]

Anterior Chamber and Iris

The anterior chamber is deep and the angle is open in MFS, but various degrees of iris strands bridging the angle are common.[16] A retroposition of the iris attachment to the ciliary body has been described histologically.[17] and Maumenee[8] mentions a case with a dialysis-like recess of the iris.

The iris surface may appear dull due to inconspicuous ridges and crypts of the anterior leaf. The posterior leaf may also be thinned as demonstrated by peripheral radial transillumination. The pupil is often miotic, sluggish, and difficult to dilate. This observation is explained by histopathologically proven absence or hypoplasia of the dilator muscle.[18] Iris hypoplasia is counted as a minor diagnostic criterion, and alternatively a 'hypoplastic ciliary muscle causing decreased miosis'.[1] This is presumably a misrepresentation of the above-mentioned clinical observation of miotic pupils, which are difficult to dilate.

The symptom of iridodonesis or trembling of the iris is striking when present, and is often observed with the naked eye. In addition slit lamp examination or gonioscopy might reveal even minute shaking of the iris.

The Crystalline Lens

Ectopia lentis (Fig. 1) is the most prominent ophthalmic feature in MFS and present in between 50% and 80% of all affected individuals,[8,11,19] possibly depending on differences in the selection of patients. In a retrospective population-based national survey including 390 subjects with ectopia lentis, a nosologic classification was possible in only 69%. Among these, nearly 70% was due to MFS.[20] Mostly, both eyes are affected, yet at different degrees. Only occasionally a unilateral lens dislocation is reported.[21]

The displacement of the lens, also named dislocation or subluxation, is a progressive condition due to an increasing elongation of the zonular fibers. Among 80 patients with ectopia lentis and MFS, Cross and Jensen[11] reported 50% diagnosed in the age group 0-5 years followed by a steep rise over the next ten years. By the end of the third decade 95% of the dislocations had been diagnosed. Slit lamp observation of a mobile lens, phakodonesis, is sometimes possible. If the pupils are widely dilated, the zonules become visible in the direction opposite to the dislocation of the lens. They are mostly seen to be stretched but intact (Fig. 2).

The size of the lens is normal in most cases, but the shape is often irregular due to a crenellated edge. A notch at the lens margin, sometimes referred to as lens coloboma, is a common finding, which possibly has a certain diagnostic specificity. The zonules are often missing in the notch area. Microspherophakia is an uncommon finding in MFS.

An increased curvature of the lens surface due to laxity of the zonules has been postulated to be responsible for the presence of myopia in eyes with normal axial length. This hypothesis is easily testable but still needs to be confirmed by clinical studies.

In the frontal plane the typical direction for lens dislocation in MFS is upward and temporally. Only 15% of the dislocations occurred in a downward direction.[8] One may wonder about the preferential upward displacement opposite to the action of the gravitational force. A retroposition of the lens may be observed as a gap between the pupillary border and the lens surface. Tilting of the lens is indicated by different size of the gap in different directions. In a few percent of the patients, the lens becomes free floating. In these cases the lens occasionally migrates into the anterior chamber. Most often it inclines into the vitreous cavity. This state may be initially uneventful, but the risk for later complications is greatly enhanced.

Retina

The condition of the retina is of major concern in the management of patients with MFS. Pathological changes consist in peripheral lattice degeneration, "white without pressure", and hole formation which all are highly correlated with the overall enlargement of the globe secondary to lens dislocation. In contrast, myopic degenerations in the posterior pole of the eye such as Fuchs spot and posterior staphyloma are rarely present. The retina, in other words, is

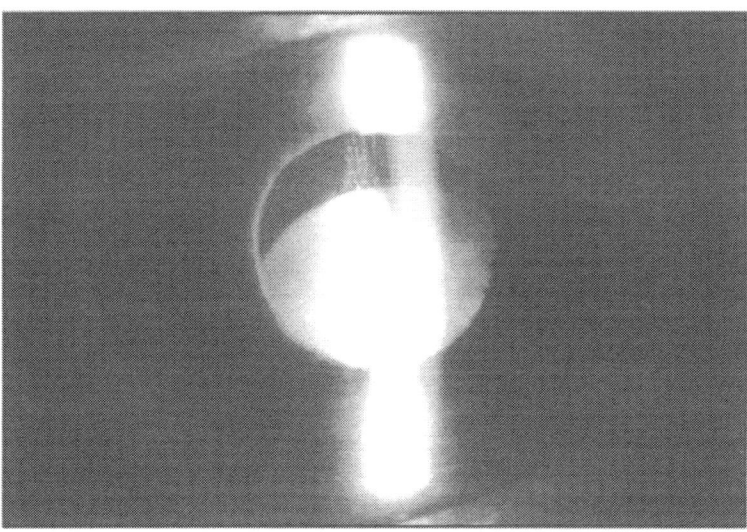

Figure 2. Marfan syndrome. Slit lamp photo showing a dislocated and cataracteous lens. Notice the stretched but intact zonular fibres.

usually normal in eyes devoid of dislocated lenses and high myopia. Electroretinography in 11 patients with MFS demonstrated normal electroretinograms (ERGs) in 6 and reduced ERGs in five patients due to high myopia, retinal detachment or glaucoma.[22]

Unlike many other connective tissue disorders, vitreoretinal degeneration is rarely present in MFS.[23]

Ophthalmological Differential Diagnosis

The demonstration of ectopia lentis in a patient should always give rise to etiological and diagnostic considerations. Among non-genetic causes, trauma is by far the most common, which not always may be evident from history. A traumatic dislocation is mostly unilateral and in the absence of relevant information, additional signs of blunt force like pupillary defects and iris dialysis should be looked for.

Lens dislocation may occasionally complicate other eye disorders, among others persistent primary hyperplastic vitreous, hypermature cataracts, uveitis, and pseudoexfoliation. Furthermore, anterior ocular malformations are sometimes complicated by a secondary displacement of the lenses, e.g., megalocornea, cornea plana, Rieger anomaly, congenital glaucoma, iris coloboma, and aniridia. These causes are all well known to the ophthalmologist.

The simultaneous occurrence of ectopia lentis and pigmentary retinal dystrophy has been observed in a few instances.[24-26] It is not yet clarified whether this association is incidental or due to extremely rare, possibly autosomal recessive traits. The association of dominantly inherited blepharoptosis, high myopia, and ectopia lentis (OMIM #110150) has been reported in a single family.[27]

Among connective tissue diseases other than MFS, ectopia lentis has been sporadically reported in Kniest dysplasia (OMIM #156550),[28] and congenital contractural arachnodactyly (OMIM #121050).[29] A pedigree with a Marfan-like syndrome with lens involvement, hyaloideoretinal degeneration, anterior chamber angle, facial, dental, and skeletal anomalies was reported by Cotlier and Reinglass,[30] and a case with bilateral aniridia, dental anomalies and Marfanoid skeletal features has been proposed representing a new association.[31] A few reports on lens dislocation in Ehlers-Danlos syndrome are suggested to be of traumatic origin.[32]

Table 1. Genetic conditions typically associated with ectopia lentis

Ocular conditions	OMIM[38]
Simple ectopia lentis	#129600
Ectopia lentis et pupillae	#225200
Late subluxation of lenses with angle closure glaucoma[39,40]	#185450
Systemic conditions	
Marfan syndrome	#154700
Homocystinuria	#236200
Weill-Marchesani syndrome	#277600
Sulfite oxidase deficiency	#272300

Craniosynostosis with ectopia lentis is a separate disease entity (MIM #603595) as well as a part of Shprintzen-Goldberg syndrome (MIM #182212). A rare syndrome consisting of ectopia lentis and craniofacial dysmorphism (MIM #601522) has been reported in the Lebanese Druze population.[33,34] A report of a family with mandibulofacial dysostosis with ectopia lentis[35] may represent a rare feature of Stickler syndrome with Pierre Robin sequence. These rare instances are irrelevant in most situations, but should be kept in mind.

The relative frequency of ectopia lentis taken into consideration, the incidental simultaneous occurrence with other hereditary disorders is not surprising. The coincidence of ectopia lentis with hyperlysinemia[36] and Refsum syndrome[37] may possibly belong to this category.

For all practical purposes, the differential diagnosis is confined to the conditions listed in Table 1.

The ophthalmological examination may yield several diagnostic hints, but alone it is often insufficient for a specific diagnosis.

Apart from a thorough family history, including information on intermarriage and retrieval of medical reports on all possibly affected family members, systemic examinations by medical consultants or pediatricians including biochemical tests for homocystinuria (Fig. 3) are mandatory. Both the absence and the presence of systemic manifestations are instrumental for the final conclusion.

It should be noted that despite meticulous clinical assessments in patients with ectopia lentis, a final diagnosis is unattainable in a considerable number of cases, especially in those who have several features in common with MFS. This may reflect rigidity of the present diagnostic criteria, but also demonstrates that a number of overlapping features are common to a number of even lesser delineated Marfanoid connective tissue disorders. Nevertheless, the many undiagnosed cases represent a dilemma, not least for the patients who feel unhappy with the state of no definite diagnosis.

Complications

Retinal detachment is a potentially sight-threatening ocular complication. A study on hospitalized cases with dislocated lenses of various categories included 14 detachments from 10 patients with MFS. In 8 of these cases the detachment occurred in eyes which had not undergone surgery.[13] From the same clinic, Cross and Jensen a few years later described retinal detachments in 12 of 134 (9%) eyes with dislocated lenses, and in 9 of 47 (19%) surgically aphakic eyes from patients with MFS.[11] The mean age of detachment was in the mid twenties. In a third study from the Wilmer Eye Institute on 160 patients with MFS, a retinal detachment was found in 16 eyes from 13 patients with dislocated lenses.[8] Modern surgical lensectomy techniques have reduced the incidence of postoperative retinal detachments substantially.[41]

Despite improved visual results after retinal reattachment surgery, spontaneous retinal detachment in Marfan patients with ectopia lentis remains a serious complication, which often occurs in children and adolescents. Avoidance of blows to the head (boxing, high diving) and

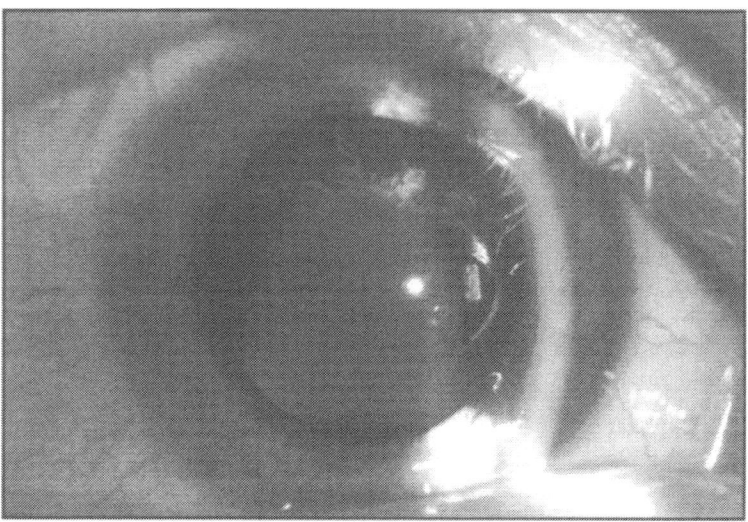

Figure 3. Homocystinuria. Slit lamp photo showing a dislocated lens. Notice the broken and crumpled zonular fibres.

globe (racquet sports) is recommended as a prophylactic measure. On the other hand there is no ophthalmological evidence that would support more rigorous restrictions on common activities.

In a classical study, glaucoma was diagnosed in 7.7% of 181 eyes with ectopia lentis in MFS and in 14.8% of 47 aphakic eyes.[11] In Maumenee's material, glaucoma occurred in only five eyes of five patients. All of these had concomitant or previous retinal detachment and dislocated lenses. The elevated intraocular pressure has mostly been attributed to pupillary block in combination with luxation into the anterior chamber or secondary to a free floating lens in the vitreous.

Today the risk for the development of glaucoma seems to be substantially reduced due to an earlier surgical intervention in patients with ectopia lentis. On the other hand, the prolonged life-span for patients with MFS may possibly increase the incidence of open angle glaucoma in the future and the supervision of intraocular tension is still a mandatory element in the ophthalmological care for patients with MFS. Large diurnal fluctuations have been described and may indicate regular measurements of IOP at different times of the day.[42] The same authors pointed out that position-dependant variations in the intraocular pressure in eyes with displaced lenses are useful as a provocation test of subluxated lenses.

Early lens opacification is often observed in patients with MFS and dislocated lenses.

Management

Optical Rehabilitation

Patients with subnormal vision due to subluxation of the lenses are often a demanding and time-consuming challenge to the refractionist. Routine objective methods as retinoscopy or refractometry often fail, and the practitioner is thrown on her or his skills in subjective refraction. This is perhaps the reason why the obvious importance of optimal spectacle correction, phakic or aphakic, in patients with subluxated lenses has been underlined repeatedly.[8,43]

Optimally, the resulting spectacle correction should correct distance as well as near vision accounting for anisometropia of an at times considerable magnitude. Sometimes, when phakic as well as aphakic vision in the same eye is possible this opportunity should be exploited.[44] If binocular vision is unattainable, it may be advantageous to correct one eye for distance and the other eye for near.[43]

In children, accommodation is often spared despite considerable dislocation of the lenses and stretched zonules. The resulting loss of accommodation should be taken into account when lensectomy is considered.

Contact lenses have been shown to be extremely helpful in Marfan syndrome by improving visual acuity due to reduction of the distortion brought about by high lens powers.[45] Provided the necessary support, contact lenses may be especially useful in children at risk for the development of amblyopia.

Due to frequent shifts in refraction, children with dislocated lenses should be checked regularly, especially throughout the amblyogenic period.

In cases of irreversible visual loss, low vision aids as high addition bifocals and high magnification telescopes should be considered along with electronic devices.

Surgery

A few decades ago, surgical lens removal was discouraged due to poor results and high complication rates.[10,13,46] This situation changed dramatically with the introduction of closed-system lensectomy using vitrectomy instruments and pars plana approach.[47-49] Recently, reports on combined vitreolensectomy and intraocular lens implantation have emerged.[41,50-53] Hubbard[41] also addressed the issue of intraoperative prophylactic retinal laser treatment in Marfan syndrome. Long-term follow-up on the latest developments of lens surgery in MFS is still missing and it seems reasonable to expect more treatment modalities to be developed in the future.

The improved post-operative results after clear lens removal in MFS have tended to broaden the indications for surgery. Nevertheless, a conservative attitude including maximal effort in non-surgical visual enhancement still seems to be good practice, especially in children and young adults. Accordingly, the main indications for lensectomy are either reduced visual acuity due to uncorrectable refractive errors, retinal detachment or glaucoma. If the visual impairment is due to cataract the need for lens extraction is obvious.

Retinal detachment surgery in Marfan syndrome does not differ from this type of surgery in general, including scleral buckling, vitrectomy, laser/cryotherapy and gas-tamponade, or silicone oil application in different combinations. Re-attachment rates are generally between 75% and 90%, although the visual results may be disappointing due to the post-operative development of proliferative vitreoretinopathy.[41,54,55]

References

1. De Paepe A, Devereux RB, Dietz HC et al. Revised diagnostic criteria for the Marfan syndrome. Am J Med Genet 1996; 62(4):417-426.
2. Wheatley HM, Traboulsi EI, Flowers BE et al. Immunohistochemical localization of fibrillin in human ocular tissues. Relevance to the Marfan syndrome. Arch Ophthalmol 1995; 113(1):103-109.
3. Mir S, Wheatley HM, Hussels IE et al. A comparative histologic study of the fibrillin microfibrillar system in the lens capsule of normal subjects and subjects with Marfan syndrome. Invest Ophthalmol Vis Sci 1998; 39(1):84-93.
4. Kielty CM, Davies SJ, Phillips JE et al. Marfan syndrome: fibrillin expression and microfibrillar abnormalities in a family with predominant ocular defects. J Med Genet 1995; 32(1):1-6.
5. Ashworth JL, Kielty CM, McLeod D. Fibrillin and the eye. Br J Ophthalmol 2000; 84(11):1312-1317.
6. Schrijver I, Liu W, Odom R et al. Premature termination mutations in FBN1: distinct effects on differential allelic expression and on protein and clinical phenotypes. Am J Hum Genet 2002; 71(2):223-237.
7. Schrijver I, Liu W, Brenn T et al. Cysteine substitutions in epidermal growth factor-like domains of fibrillin-1: distinct effects on biochemical and clinical phenotypes. Am J Hum Genet 1999; 65(4):1007-1020.
8. Maumenee IH. The eye in the Marfan syndrome. Trans Am Ophthalmol Soc 1981; 79:684-733.
9. Nelson LB, Maumenee IH. Ectopia lentis. Surv Ophthalmol 1982; 27(3):143-160.

10. Hindle NW, Crawford JS. Dislocation of the lens in Marfan's syndrome. Its effect and treatment. Can J Ophthalmol 1969; 4(2):128-135.
11. Cross HE, Jensen AD. Ocular manifestations in the Marfan syndrome and homocystinuria. Am J Ophthalmol 1973; 75(3):405-420.
12. Speedwell L, Russell-Eggitt I. Improvement in visual acuity in children with ectopia lentis. J Pediatr Ophthalmol Strabismus 1995; 32(2):94-97.
13. Jarrett WH, II. Dislocation of the lens. A study of 166 hospitalized cases. Arch Ophthalmol 1967; 78(3):289-296.
14. Izquierdo NJ, Traboulsi EI, Enger C et al. Strabismus in the Marfan syndrome. Am J Ophthalmol 1994; 117(5):632-635.
15. Setala K, Ruusuvaara P, Karjalainen K. Corneal endothelium in Marfan syndrome. A clinical and specular microscopic study. Acta Ophthalmol (Copenh) 1988; 66(3):334-340.
16. Von Noorden GK, Schultz RO. A gonioscopic study of the chamber angle in Marfan's syndrome. Arch Ophthalmol 1960; 64:929-934.
17. Wachtel JG. The ocular pathology of Marfan's syndrome, including a clinicopathological correlation and an explanation of ectopia lentis. Arch Ophthalmol 1966; 76(4):512-522.
18. Sautter H. Aplasie des dilatator pupillae beim Marfanschen symptomkomplex. Klin Monatsbl Augenheilkd 1949; 114:449-453.
19. Fuchs J. Marfan syndrome and other systemic disorders with congenital ectopia lentis. A Danish national survey. Acta Paediatr 1997; 86(9):947-952.
20. Fuchs J, Rosenberg T. Congenital ectopia lentis. A Danish national survey. Acta Ophthalmol Scand 1998; 76(1):20-26.
21. Rasooly R, BenEzra D. Unilateral lens dislocation and axial elongation in Marfan syndrome. Ophthalmic Paediatr Genet 1988; 9(2):135-136.
22. Wu LZ, Wu DZ, Ma QY et al. The electroretinogram in Marfan's syndrome. Yan Ke Xue Bao 1989; 5(1-2):39-43.
23. Maumenee IH. Vitreoretinal degeneration as a sign of generalized connective tissue diseases. Am J Ophthalmol 1979; 88(3 Pt 1):432-449.
24. Halpern BL, Sugar A. Retinitis pigmentosa associated with bilateral ectopia lentis. Ann Ophthalmol 1981; 13(7):823-824.
25. Simonelli F, De Crecchio G, Testa F et al. Retinal degeneration associated with ectopia lentis. Ophthalmic Genet 1999; 20(2):121-126.
26. Sato H, Wada Y, Abe T et al. Retinitis pigmentosa associated with ectopia lentis. Arch Ophthalmol 2002; 120(6):852-854.
27. Gillum WN, Anderson RL. Dominantly inherited blepharoptosis, high myopia, and ectopia lentis. Arch Ophthalmol 1982; 100(2):282-284.
28. Maumenee IH, Traboulsi EI. The ocular findings in Kniest dysplasia. Am J Ophthalmol 1985; 100(1):155-160.
29. Ramos Arroyo MA, Weaver DD, Beals RK. Congenital contractural arachnodactyly. Report of four additional families and review of literature. Clin Genet 1985; 27(6):570-581.
30. Cotlier E, Reinglass H. Marfan-like syndrome with lens involvement. Hyaloideoretinal degeneration with anterior chamber angle, facial, dental, and skeletal anomalies. Arch Ophthalmol 1975; 93(2):93-106.
31. Sachdev MS, Sood NN, Kumar H et al. Bilateral aniridia with Marfan's syndrome and dental anomalies-a new association. Jpn J Ophthalmol 1986; 30(4):360-366.
32. Cross HE. Ectopia lentis in systemic heritable disorders. Birth Defects Orig Artic Ser 1974; 10(10):113-119.
33. Shawaf S, Noureddin B, Khouri A et al. A family with a syndrome of ectopia lentis, spontaneous filtering blebs, and craniofacial dysmorphism. Ophthalmic Genet 1995; 16(4):163-169.
34. Haddad R, Uwaydat S, Dakroub R et al. Confirmation of the autosomal recessive syndrome of ectopia lentis and distinctive craniofacial appearance. Am J Med Genet 2001; 99(3):185-189.
35. Kirkham TH. Mandibulofacial dysostosis with ectopia lentis. Am J Ophthalmol 1970; 70(6):947-949.
36. Smith TH, Holland MG, Woody NC. Ocular manifestations of familial hyperlysinemia. Trans Am Acad Ophthalmol Otolaryngol 1971; 75(2):355-360.
37. Karyofilis A, Berneaud-Kotz G, Jacobs I. [Heredopathia atactica polyneuritiformis]. Fortschr Neurol Psychiatr Grenzgeb 1970; 38(7):321-329.
38. Online Mendelian Inheritance in Man, OMIM (TM). Johns Hopkins University, Baltimore, MD. World Wide Web URL: http://www.ncbi.nlm.nih.gov/omim
39. McCulloch C. Hereditary lens dislocation with angle closure glaucoma. Can J Ophthalmol 1979; 14(4):230-234.

40. Malbran ES, Croxatto JO, D'Alessandro C et al. Genetic spontaneous late subluxation of the lens. A study of two families. Ophthalmology 1989; 96(2):223-229.
41. Hubbard AD, Charteris DG, Cooling RJ. Vitreolensectomy in Marfan's syndrome. Eye 1998; 12 (Pt 3a):412-416.
42. Schlote T, Volker M, Thanos S et al. Glaukom bei Marfan-Syndrom: Lageabhängige Druckmessung als diagnostiches Kriterium. Klin Monatsbl Augenheilkd 1995; 207(6):386-388.
43. Nelson LB, Szmyd SM. Aphakic correction in ectopia lentis. Ann Ophthalmol 1985; 17(7):445-447.
44. Bromley WC. Ectopia lentis with Marfan's syndrome-case with a useful optical correction. J Pediatr Ophthalmol 1976; 13(4):230-231.
45. Yeung KK, Weissman BA. Contact lens correction of patients with Marfan syndrome. J Am Optom Assoc 1997; 68(6):367-372.
46. Cross HE. Differential diagnosis and treatment of dislocated lenses. Birth Defects Orig Artic Ser 1976; 12(3):335-346.
47. Reese PD, Weingeist TA. Pars plana management of ectopia lentis in children. Arch Ophthalmol 1987; 105(9):1202-1204.
48. Plager DA, Parks MM, Helveston EM et al. Surgical treatment of subluxated lenses in children. Ophthalmology 1992; 99(7):1018-1021.
49. Halpert M, BenEzra D. Surgery of the hereditary subluxated lens in children. Ophthalmology 1996; 103(4):681-686.
50. Koenig SB, Mieler WF. Management of ectopia lentis in a family with Marfan syndrome. Arch Ophthalmol 1996; 114(9):1058-1061.
51. Omulecki W, Nawrocki J, Palenga-Pydyn D et al. Pars plana vitrectomy, lensectomy, or extraction in transscleral intraocular lens fixation for the management of dislocated lenses in a family with Marfan's syndrome. Ophthalmic Surg Lasers 1998; 29(5):375-379.
52. Siganos DS, Siganos CS, Popescu CN et al. Clear lens extraction and intraocular lens implantation in Marfan's syndrome. J Cataract Refract Surg 2000; 26(5):781-784.
53. Kazemi S, Wirostko WJ, Sinha S et al. Combined pars plana lensectomy-vitrectomy with open-loop flexible anterior chamber intraocular lens (AC IOL) implantation for subluxated lenses. Trans Am Ophthalmol Soc 2000; 98:247-251.
54. Abboud EB. Retinal detachment surgery in Marfan's syndrome. Retina 1998; 18(5):405-409.
55. Sharma T, Gopal L, Shanmugam MP et al. Retinal detachment in Marfan syndrome: clinical characteristics and surgical outcome. Retina 2002; 22(4):423-428.

CHAPTER 4

Cardiovascular Aspects of the Marfan Syndrome:
A Systematic Review

Yskert von Kodolitsch and Maike Rybczynski

The Marfan syndrome (MFS) is an autosomal dominant disorder of connective tissue caused by mutations of the fibrillin-1 gene, which codes for fibrillin, a major component of the extracellular microfibrils. The mean life expectancy in untreated MFS is 32 years with aortic dissection, aortic rupture or cardiac failure accounting for at least 90 percent of all fatalities. In severely affected patients with neonatal MFS, patients are likely to survive a few months only. According to our literature database analysis aortic dilatation is present in 83 percent, aortic regurgitation in 53 percent, mitral valve prolapse in 57 percent, and mitral valve regurgitation in 31 percent of adult males with classic MFS. We put a special focus on the pathogenesis and natural course of cardiovascular disease in MFS, including complications such as arrhythmia, sudden death, and endocarditis or complications during pregnancy. With optimal management, patients may have an acceptable life quality and almost normal life expectancy.

The Ghent nosology defines dilatation or dissection of the ascending aorta as major criteria for the diagnosis of MFS (see Chapter 2, "Diagnosis and Treatment of Marfan Syndrome—A Summary"); minor criteria comprise dilatation or dissection of the descending thoracic or abdominal aorta observed below the age of 50 years, dilatation of the main pulmonary artery and calcification of the mitral annulus, both diagnosed below the age of 40 years, and mitral valve prolapse.[1] In this chapter, we review the natural history of cardiovascular manifestations of MFS and provide an overview of current diagnostic and therapeutic options.

Methods

The literature was screened for records on cardiovascular manifestations of MFS using MEDLINE (key words: Marfan, aneurysm, aorta, sudden death, mitral valve, surgery) and literature lists provided in articles on this subject. Information was entered into an electronic databank on cardiovascular disease in MFS, which currently comprises 71 articles on 4872 cases of MFS, which were diagnosed according to established criteria.[1-3] For the purposes of this analysis, MFS diagnosed within the first months of life was classified as neonatal (nMFS, formerly referred to as infantile MFS); MFS with diagnosis at age 4 months through 11 years was juvenile, with age 12 through 16 years MFS was adolescent, and with age 17 years or more MFS was adult.[4-6] All data were analyzed with SPSS software (SPSS for Windows, Release 10.0.7, Copyright© SPSS Inc 1989-1999, Chicago, Illinois); identical patients were excluded from analysis.

Marfan Syndrome: A Primer for Clinicians and Scientists, edited by Peter N. Robinson and Maurice Godfrey. ©2004 Eurekah.com and Kluwer Academic / Plenum Publishers.

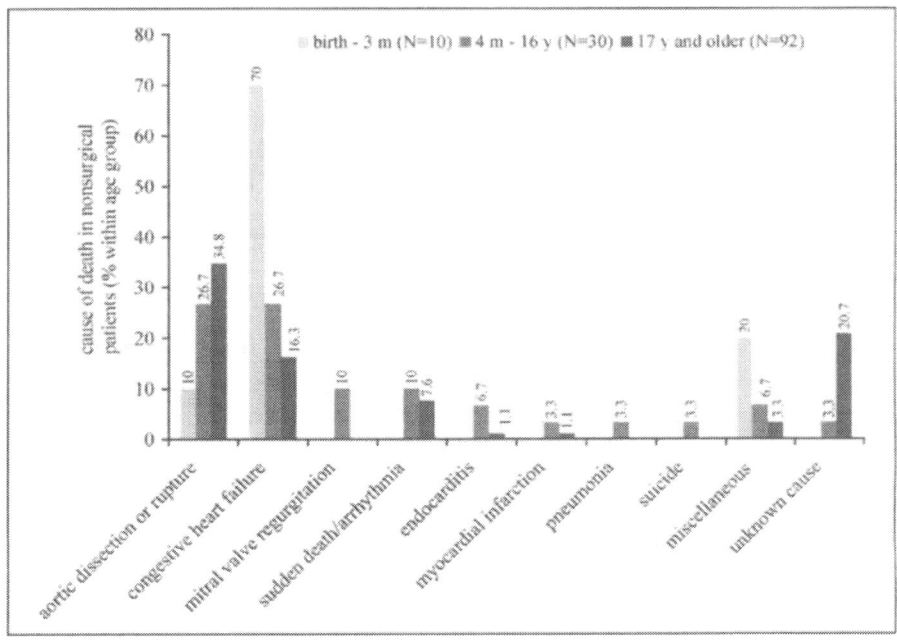

Figure 1. Causes of death in patients without cardiovascular surgery according to age at initial diagnosis of MFS. N indicates the total number of patients included in the literature databank analysis.

Prognosis and Causes of Death in Untreated MFS

The mean age at death is 32.0 ± 16.4 (SD) years in 74 deceased MFS patients; the mean age at death is 33.4 ± 16.4 years in males and 29.5 ± 17.7 years in women. However, there is a significant survival advantage for women aged 34 through 45 years.[7] In severely affected neonatal MFS, only 5 percent survive more than 12 months, whereas 50 percent of patients diagnosed at four months of age may survive the first year of life, and with diagnosis ≥ 6 months of age, most patients survive more than two years.[4] Interestingly, a study of MFS patients treated medically according to recent guidelines identifies no differences in survival of 20 children with diagnosis of MFS at age ≤ 6 years (three of these diagnosed at birth) and 30 children diagnosed at > 6 years of age.[8]

In neonatal MFS 70 percent of deaths are caused by congestive heart failure, and only 10 percent are due to aortic rupture or dissection. In juvenile and adolescent MFS heart failure and aortic complications both account for 26.7 percent of deaths, whereas aortic dissection or rupture are the leading cause of death in adults (34.8 percent). Other causes of death in juvenile MFS are mitral valve regurgitation in 10 percent, sudden death in 10 percent, endocarditis in 7.6 percent, myocardial infarction 3.3 percent, and pneumonia and suicide in 3.3 percent each. In adults, sudden death occurs in 7.6 percent of reported cases, or death is caused by endocarditis or myocardial infarction in 1.1 percent (Fig. 1).

Prevalence of Cardiovascular Manifestations

Aortic Root Disease

The classical approach to assess aortic root dimensions is to use M-mode echocardiography with measurements from the most anterior portion of the anterior aortic wall to the most anterior portion of the posterior aortic wall at end-diastole; in subjects ≥ 16 years of age dilatation of the aortic root is present with at least two of the following criteria:

1. width index of the aorta > 22 mm/m^2,
2. aortic diameter > 37 mm
3. left atrial to aortic diameter ratio < 0.7.[9]

In addition, M-mode nomograms are available to compare aortic root dimensions at the sinuses of Valsalva with body surface area.[10] More recently, two-dimensional echocardiography is used to assess aortic root dimensions at the level of the valve annulus, the aortic sinuses, the sinotubular junction and the proximal ascending aorta; such measurements are systematically larger (2 mm at the level of the aortic sinuses) than those made by M-mode echocardiography.[11] Currently, two-dimensional echocardiography is used to diagnose aortic root dilatation by means of nomograms relating aortic root size to body surface area; such nomograms are available for children < 18 years of age, for adults < 40 years of age and for adults ≥ 40 years of age;[11] the use of these nomograms is recommended by the Ghent nosology and current European guidelines.[1,12] In addition, adjusted nomograms are available for adults exceeding the 95th percentile for body height (≥ 189 cm in men; ≥ 175 cm in women)[13] and for children with suspected MFS (who are shown to present with a body surface area above the 50th percentile despite exclusion of MFS).[14]

Aortic ratios allow for comparison of individuals irrespective of age and body size.[13,15] For calculation of an aortic ratio, the observed maximum diameter of the aortic root is divided by the predicted diameter based on age and body surface area (BSA) of normal individuals. The predicted sinus diameter (cm), for instance, can be calculated using the following regression formulas:[11,15]

- in children (age < 18 years) = 1.02 + (0.98 x BSA (m^2));
- in adults (age 18-40 years) = 0.97 + (1.12 x BSA (m^2));
- in adults (age ≥ 40 years) = 1.92 + (0.74 x BSA (m^2)).

Thus, an aortic sinus ratio of 1.3 indicates a 30 percent enlargement of the aortic sinus above the mean of normal individuals of the respective age and body surface area.[15] Nomograms are less helpful in adults over 40 years of age, because obesity and aortic media degeneration account for a looser relationship between aortic size and body surface area;[13,16] as a rule of thumb, in these individuals the aortic root is normal with diameters of < 37 mm,[11] the ascending aorta is dilated with diameters ≥ 38 mm and < 50 mm, and aneurysm is present with diameters ≥ 50 mm.[12,17]

Aortic root dilatation is the most frequent cardiovascular manifestation in MFS; in neonatal MFS, the aortic root is dilated in all patients irrespective of gender. However, in MFS diagnosed after the age of three months, aortic root dilatation is more prevalent in males and is more frequent in adults than in juvenile or adolescent MFS (Fig. 2A). The mean aortic root diameters measured in MFS are displayed separately for female and male subjects (Figs. 2B and 2C); the average yearly increase of the aortic root diameter is 2.1 ± 1.6 (SD) mm in infants and 1.9 mm or 0.9 ± 0.9 mm (SD), respectively, in adults.[11,18,19]

Aortic valve regurgitation usually develops when the aortic root diameter increases; regurgitation is rare with diameters of less than 40 mm and is obligatory at diameters greater than 60 mm.[4] Receiver operating characteristic analysis (ROC) of published aortic root diameters in 152 adults with MFS reveals that 54 mm of maximum root diameter is a threshold identifying patients with increased risk of aortic valve regurgitation with a sensitivity of 91.3 percent and a specificity of 88.9 percent (Fig. 3). Aortic valve incompetence develops with age and is more prevalent in males. In neonatal MFS, 25 percent of female and 30 percent of male MFS patients exhibit aortic valve regurgitation (Fig. 3C).

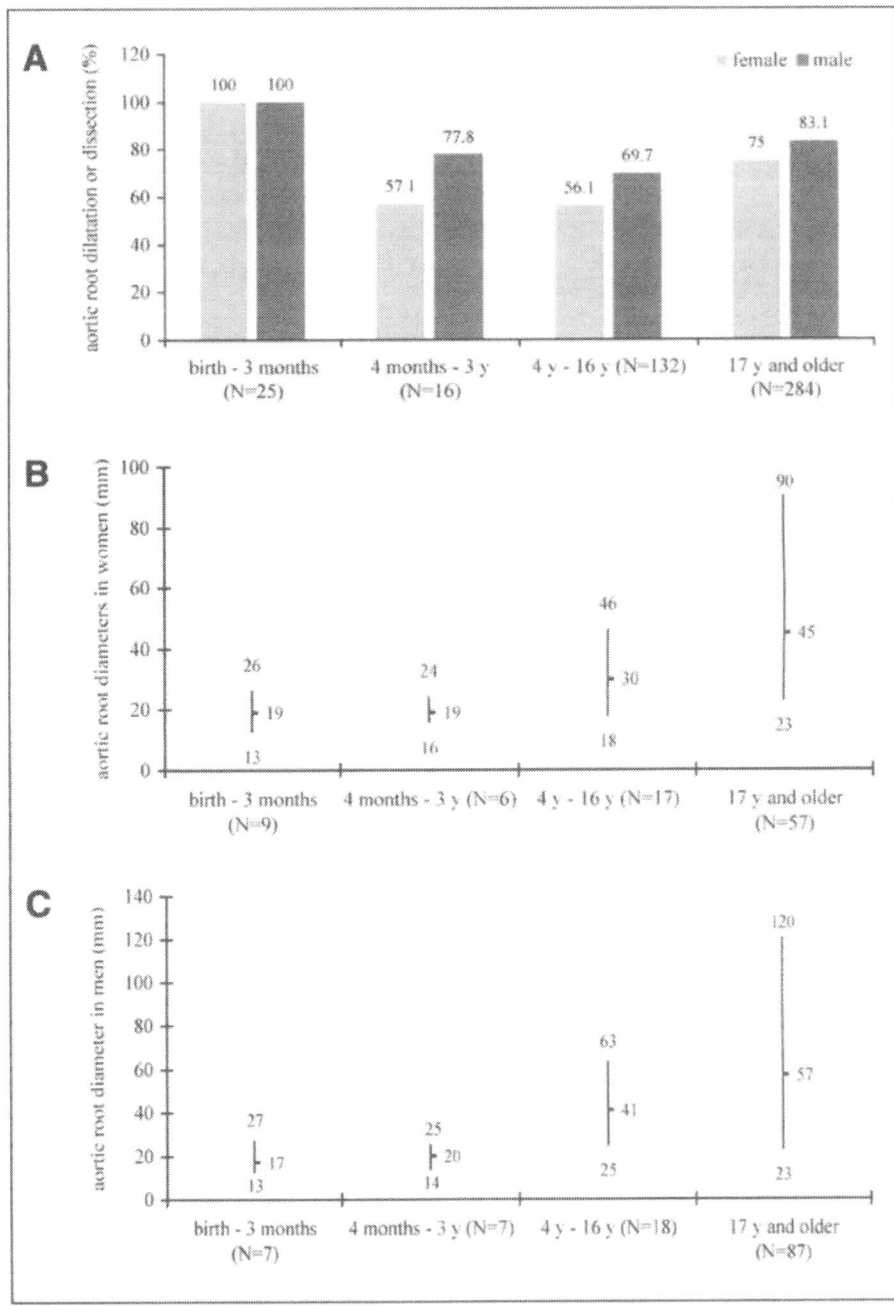

Figure 2. Literature databank analysis of the findings at the aortic root reported according to age at initial diagnosis of MFS. A) Prevalence of aortic root dilatation or dissection of the aortic root. B) Mean and range of maximum diameters of the aortic root measured in women with MFS. C) Mean and range of maximum diameters of the aortic root measured in men with MFS.

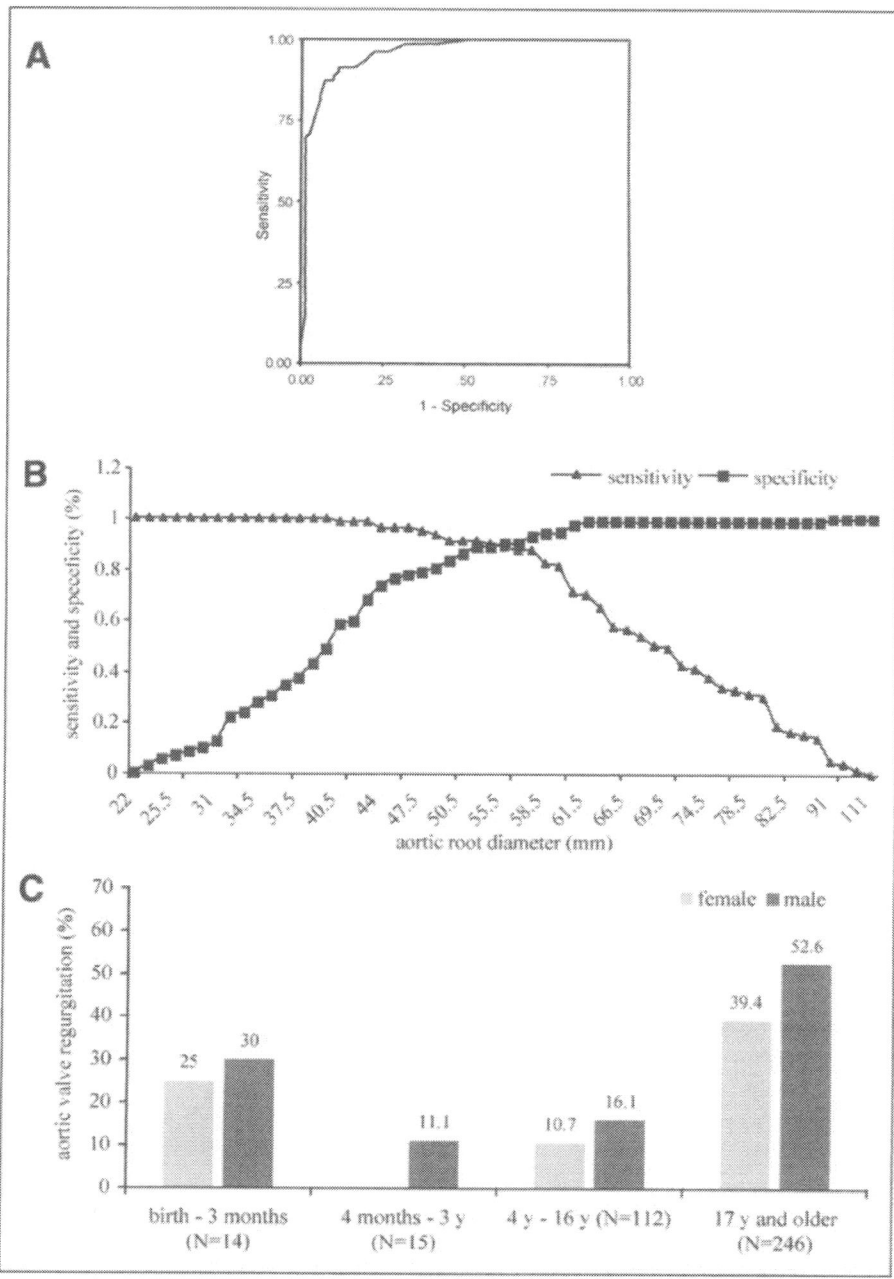

Figure 3. A) ROC analysis revealed > 54 mm of maximum aortic root diameter as a threshold for increased risk of aortic valve regurgitation. The area under the empirical ROC curve was 0.96 with an 95% confidence interval of 0.92-0.99 and *P* < .0005. B) For better identification of the cut-off diameter separating high from low probability of aortic valve regurgitation, data are displayed separately for sensitivity and specificity. C) Prevalence of aortic valve regurgitation according to age at initial diagnosis of MFS.

Mitral Valve Disease

Recent two-dimensional echocardiographic criteria for mitral valve prolapse assess the maximal superior displacement of one or both mitral leaflets during systole relative to the line connecting the annular hinge points recorded in either the parasternal or the apical long-axis view and thickness of the mitral leaflets during diastasis; classic prolapse, or "floppy mitral valve" is present with displacement > 2 mm and maximal thickness ≥ 5 mm (representing myxomatous degeneration of the valve) and nonclassic prolapse with displacement > 2 mm but maximal thickness < 5 mm.[20,21]

Female MFS patients with low body weight, low systolic blood pressure and arachnodactyly are at higher risk for mitral valve prolapse than other subjects with MFS.[22] In neonatal MFS, mitral valve prolapse is present in all females and in 91.7 percent of males. Interestingly, mitral valve prolapse is more frequent in MFS diagnosed at young age than in adults (Fig. 4A). Particularly in young MFS, mitral valve disease tends to be progressive; in neonatal MFS all females and 80 percent of males exhibit mitral valve incompetence. Almost 50 percent of juvenile MFS with mitral valve prolapse develop mitral valve regurgitation within four years, causing death in seven percent.[23] As a general rule, regurgitation of the mitral valve is more prevalent in women and in younger patients with MFS (Fig. 4B).

Tricuspid Valve and Pulmonary Valve Disease

Prolapse of the tricuspid valve is observed in all males and 87.5 percent of females with neonatal MFS. Tricuspid valve prolapse is more prevalent in MFS diagnosed at younger age than in adult MFS (Fig. 4C). The pulmonary artery root is dilated in all cases with neonatal MFS, in 7.4 percent of juvenile MFS, in 14.3 percent of adolescent MFS and in 40.5 percent of adult MFS (Fig. 4D). Pulmonary artery root dimensions relate to aortic root diameters; insufficiency of the pulmonary valve is reported in 8 of 31 cases with MFS (25.8 percent).[24] However, criteria for tricuspid valve prolapse and pulmonary artery root dilatation have not yet been established, and thus, frequencies vary across studies.

Pathogenesis of Cardiovascular Disease

Aortic Root Disease

According to the Laplace law, the integrity of the aortic wall depends on circumferential and longitudinal wall stress, which is determined by blood pressure and the aortic radius divided by the thickness of the aortic wall. Thus, aortic wall weakness, aortic dilatation and arterial hypertension are major mechanisms of dissection and rupture. Since fibrillin-1 is predominantly expressed in the proximal aorta, in MFS media degeneration and consecutive aneurysmal formation are most prevalent in the ascending segment. The aortic media may also show signs of degeneration with systemic hypertension, increased age, a bicuspid aortic valve, aortic valve stenosis or with coarctation of the aorta.[5] Histologic changes of the aortic media are particularly marked in MFS, and thus aortic complications usually develop early in life and dissection may develop already before birth in some rare circumstances. However, severe aortic complications are uncommon in children and adolescents since hemodynamic stress is usually required to act over time to cause aortic wall damage.[4]

In adult MFS, an aortic root diameter of 50 to 60 mm is associated with an increased risk of rupture or dissection; however, dissection may occur below diameters of 50 mm or even in nondilated aortas.[4,25,26] An increase of the aortic ratio of more than 5 percent per year, a family history of aortic dissection or sudden cardiac death and dilatation of the aortic sinuses with extension into the ascending aorta are independent predictors of aortic rupture or dissection.[15,19,27,28] Dissection may be initiated by laceration or tearing of the aortic intima with propagation of blood through the aortic media resulting in false lumen flow. Alternatively, dissection may primarily result from rupture of the vasa vasorum of the aortic intima with localized or extended hemorrhage of the aortic media, which may later result into rupture of

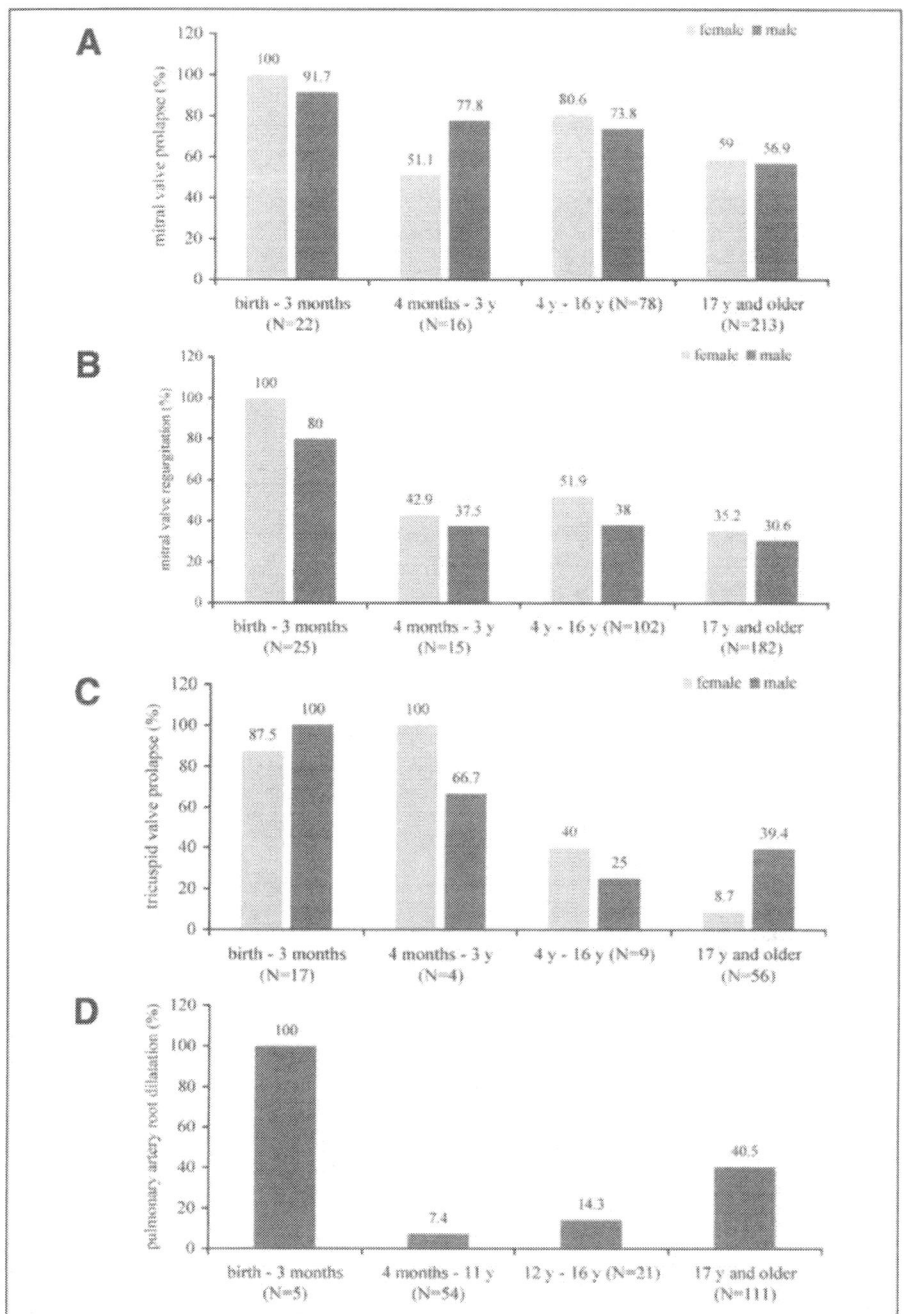

Figure 4. Literature databank analysis of the prevalence of mitral valve and tricuspid valve disease reported in MFS according to age at initial diagnosis of the syndrome. A) Prevalence of mitral valve prolapse. B) Prevalence of mitral valve regurgitation. C) Prevalence of tricuspid valve prolapse. D) Prevalence of pulmonary artery dilation.

the aortic intima and false lumen flow. Both laceration of the intima and intramural hemorrhage, may lead to direct rupture of the aorta without dissection.[17]

The major mechanism of chronic aortic valve regurgitation is dilatation of the sinutubular junction with progressive involvement of the aortic sinuses.[16] The aortic annulus is usually not dilated but there may be a varying degree of myxoid degeneration of the aortic cusps, which is usually secondary to chronic regurgitation.[4,25,29]

Mitral Valve Disease

The mitral valve is composed of three layers; the spongiosa consists of myxomatous connective tissue and is located between the outer, endothelium-covered atrialis and the central fibrosa consisting of dense collagen and occasional elastin forming the basic mechanical support of the mitral valve. In classic valve prolapse myxomatous connective tissue expands the spongiosa and separates bundles of collagen in the fibrosa and thus weakens the support of the valve; maximal disruption often occurs at sites of chordal insertion. Weakening of the leaflets leads to elongation of the leaflets and often of the chordae with doming (hooding) of the cusps into the atrium.[21] Mitral valve prolapse results from a mismatch of the size of the valve and the size of the left ventricle and in MFS, the increase or decrease of left ventricular size may cause respective appearance or disappearance of mitral valve prolapse.[30] Annular dilatation, however, is most likely not a primary manifestation of mitral valve prolapse but develops with chronic regurgitation.[21]

Mitral valve prolapse reflects a broad spectrum of abnormalities ranging from functionally normal valve variants to severe valve disease and may be either stable or progressive.[31] With "exaggeration" of the physiologic billowing of the functionally normal valve, Barlow and Pocock use the term "billowing mitral leaflets", that they still consider a "normal valve variant".[32] Conversely, patients with mitral leaflet displacement > 2 mm have an increased risk for mitral regurgitation, left atrial enlargement, leaflet thickening and infective endocarditis but, most likely, not for stroke.[20,33,34] Coaptation of the mitral valve leaflets is symmetric with the leaflet tips meeting in a common point and asymmetric with the tip of one leaflet being displaced toward the atrium relative to the other leaflet; classic mitral valve prolapse with asymmetric coaptation is particularly prone to progressive valve deterioration and possibly predisposes to chordal rupture which may lead to a flail leaflet.[21] The tip of a flail leaflet everts and becomes concave toward the left atrium rather than the left ventricle; a flail may be present with only isolated eversion of the tip but may also present with complete detachment of the chordae and unrestrained motion of the entire leaflet and severe mitral valve incompetence.[21,32] In MFS mitral valve prolapse is characterized by features of classic mitral valve prolapse; however, marked abundance of the mitral leaflets is more common, the posterior leaflet chordae more frequently arise abnormally from the posterior ventricular wall and calcification of the mitral annulus is more prevalent as compared to primary mitral valve prolapse not associated with MFS.[16,22,25] To date, the fraction of patients with classical mitral valve prolapse caused by connective tissue disorders is unknown in unselected populations[16] and neither the Ghent nosology nor classic studies on mitral valve disease in MFS consider recent diagnostic criteria of mitral valve prolapse.[1,22,23,25,30]

Concomitant Cardiovascular Disease

Review of studies with serial echocardiography or tomographic imaging identifies a congenitally bicuspid aortic valve as the most frequent associated malformation in MFS (4.9 percent). Other malformations are coarctation of the aortic isthmus (2.4 percent), atrial septal defect (1.8 percent), patent ductus arteriosus (0.9 percent), ventricular septal defect (0.7 percent), Wolff-Parkinson-White syndrome (0.5 percent), and tetralogy of Fallot (0.4 percent) (Fig. 5). The estimated prevalence in the general population is 0.9 percent for an inherited bicuspid aortic valve, 0.1 percent for aortic coarctation and 0.16 percent for the Wolff-Parkinson-White syndrome; a patent ductus arteriosus is observed in 0.06 to 0.02 percent of births.[35-39] The increased prevalence of concomitant malformations in MFS may be due to collateral effects of fibrillin-1 gene mutations or better detection due to intense diagnostic

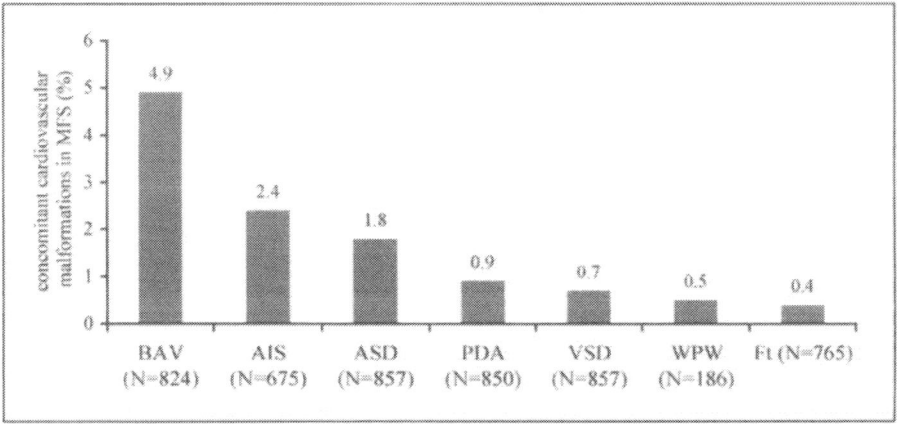

Figure 5. Literature databank analysis of concomitant cardiovascular malformations reported in patients with MFS irrespective age at initial diagnosis of the syndrome. BAV indicates bicuspid aortic valve; AIS= aortic coarctation; ASD= atrial septal defect; PDA= patent ductus arteriosus; VSD= ventricular septal defect; WPW= Wolff-Parkinson-White syndrome; Ft= tetralogy of Fallot.

work-up.[4] In neonatal MFS fibrillin-1 mutations are particularly severe and consequently, concomitant cardiovascular malformations are more frequent than in adolescent or adult MFS.[4,40]

Case records and small series have reported intracranial aneurysm in MFS;[41] however, larger series claim to exclude that MFS is associated with such aneurysm.[42,43] Aneurysms of peripheral arteries are also rare; bilateral aneurysm of the arteria iliaca communis is observed in 3 percent and aneurysm of the left subclavian artery in 1 percent of 69 MFS patients with cardiovascular surgery.[44] Aneurysm of the vena cava superior is reported in one case of 31 surgically treated MFS.[45]

Our data-base analysis reveals that 12.9 percent of 132 adult MFS exhibit coronary artery disease at angiography or open heart surgery; in medically treated patients myocardial infarction is the cause of death in 3.3 percent of adolescent MFS and in 1.1 percent of adult MFS (Fig. 1); infarction, however, may be caused by aortic dissection involving the coronary arteries. A microscopic study of extramural coronary arteries demonstrates structural changes consistent with cystic media necrosis in four of five individuals with adult MFS.[46] Interestingly, severity of coronary artery disease is associated with fibrillin-1 genotype and large-artery stiffening in individuals with coronary disease.[47]

Cardiovascular Complications

The Aortic Root and Aorta

At the time of primary aortic operation of adults with MFS, progressive dilatation of the aortic root is present in 65.6 percent and chronic or acute dissection of the ascending aorta in 14.3 and 13.7 percent, respectively. Nondissecting aneurysm of the descending aorta is present in 2.3 percent, and chronic or acute dissection of the descending aorta is found in 3.2 and 0.6 percent, respectively. Interestingly, aneurysms of the abdominal aorta not involving the descending thoracic aorta are present in 0.3 percent of adult MFS (Fig. 6). On the other hand, in 41 cases with isolated spontaneous dissection of the infrarenal abdominal aorta no patient exhibits signs of inherited connective tissue disease.[48] Thus, it remains unclear whether adult MFS patients are at increased risk for isolated thoracic or abdominal aneurysm.[16,49] According to our databank analysis, 97.2 percent of nondissecting aneurysms of the aortic root are limited to the ascending aorta, whereas 92.5 percent of aneurysm of the descending aorta extend into the abdominal aortic segment. Dissection of the ascending aorta (type A) extends into the

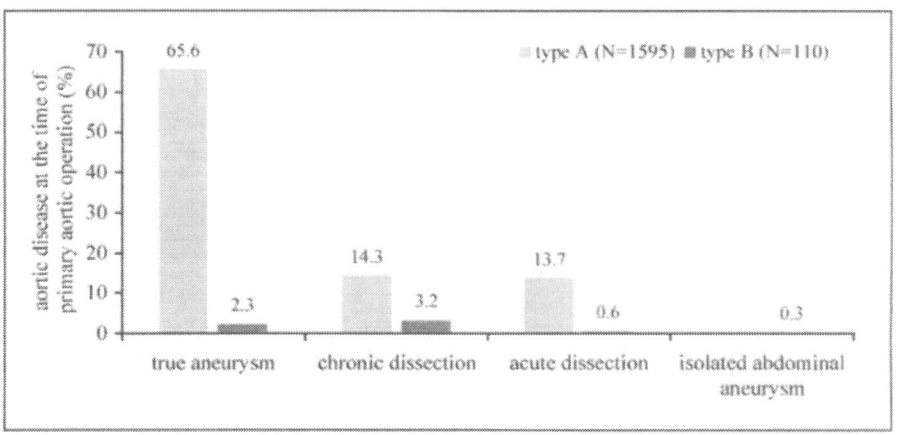

Figure 6. Literature databank analysis of the type and location of aortic disease diagnosed at the time of initial cardiovascular surgery reported in adults with MFS. Type A and type B indicate the Stanford classification with (type A) or without (type B) involvement of the ascending aorta.

aortic arch in 70.1 percent and 92.7 percent of dissections with origin in the descending aorta (type B) also affect the abdominal aorta (Fig. 7).

Chronic regurgitation of the aortic valve usually progresses when the aortic root dilates, whereas acute regurgitation usually results from prolapse of one or more aortic cusps caused by dissection of the aortic root. Prolapse of an aortic cusp without aortic dissection is rare and there is just one report of MFS with sudden death caused by acute isolated prolapse of the aortic valve[50-52] (Fig. 8).

Endocarditis

In MFS, the risk of endocarditis is increased even with macroscopically normal valves. In patients without MFS, mitral valve prolapse carries a 0.005 percent risk of endocarditis when a murmur of mitral valve regurgitance is not audible and a 0.05 percent risk when such murmur is present.[53] Autopsy of MFS without cardiovascular surgery identifies fresh or old signs of endocarditis in 0.7 percent of the aortic valves and in 5 percent of the mitral valves.[25] Moreover, endocarditis accounts for 6.7 percent of deaths in juvenile and adolescent MFS and for 1.1 percent of deaths in adult MFS (Fig. 1). Aortic valve endocarditis is diagnosed in 13 percent of adult MFS during aortic root surgery and streptococcus, especially streptococcus viridans, is the most common cause of endocarditis in MFS[45,54] (Fig. 9).

Heart Failure

Heart failure is a leading cause of death in neonatal and juvenile MFS and with growing life expectancy, there is increasing need for heart transplantation in adult MFS patients.[55] Most cases of heart failure may be caused by heart valve incompetence rather than primary myocardial tissue dysfunction.[4] However, fibrillin is expressed in heart muscle tissue, and an animal model with an abnormal fibrillin-1 gene is associated with hypertrophic cardiomyopathy.[4,56] There also is eccentric dilatation of the left ventricle in neonatal MFS and left ventricular diastolic function is altered in MFS without heart valve dysfunction or ventricular dilatation.[57-59] Moreover, abnormal insertion, elongation and spontaneous rupture of the chordae tendinae, and dilation of the annulus fibrosus in some cases with MFS suggest ventricular connective tissue abnormality.[35,59,60] In adult MFS AV-block grade I is present in 13 percent and intraventricular conduction delay in 14 percent; these electrocardiographic disturbances may relate to primary heart muscle disease;[4] inferior T-wave inversions are typical of mitral valve prolapse and are detected in 24 percent of adult MFS.[5,16]

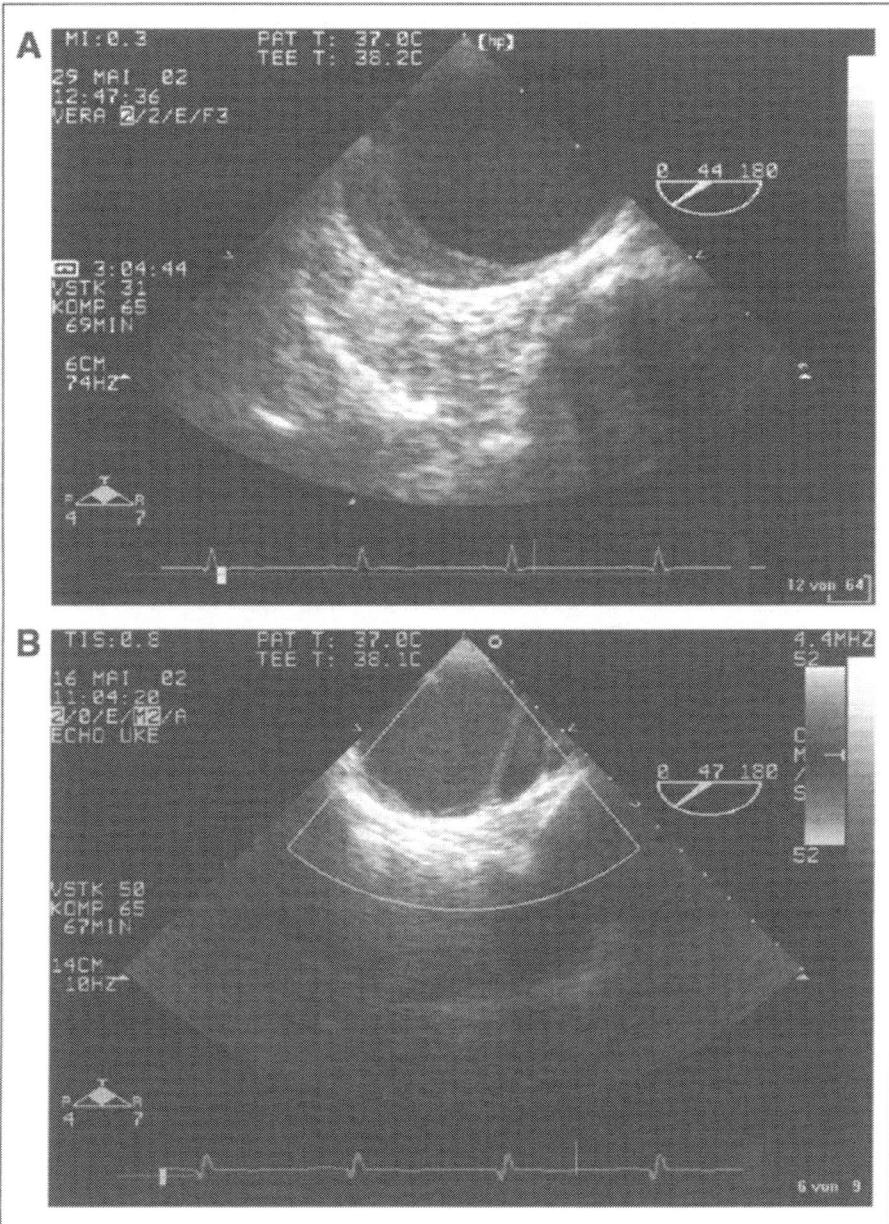

Figure 7. Transesophageal echo scans of the descending aorta in two different patients with MFS presenting with clinical signs and symptoms of acute aortic syndrome. A) The transverse scan reveals semilunar thickening of the aortic media without formation of a false lumen or entry formation representing intramural hemorrhage. B) The transverse echo scan reveals overt dissection of the descending aorta with an intimal flap separating a broad false lumen from a smaller true lumen; color Doppler signal indicates, that blood flow is faster in the true lumen.

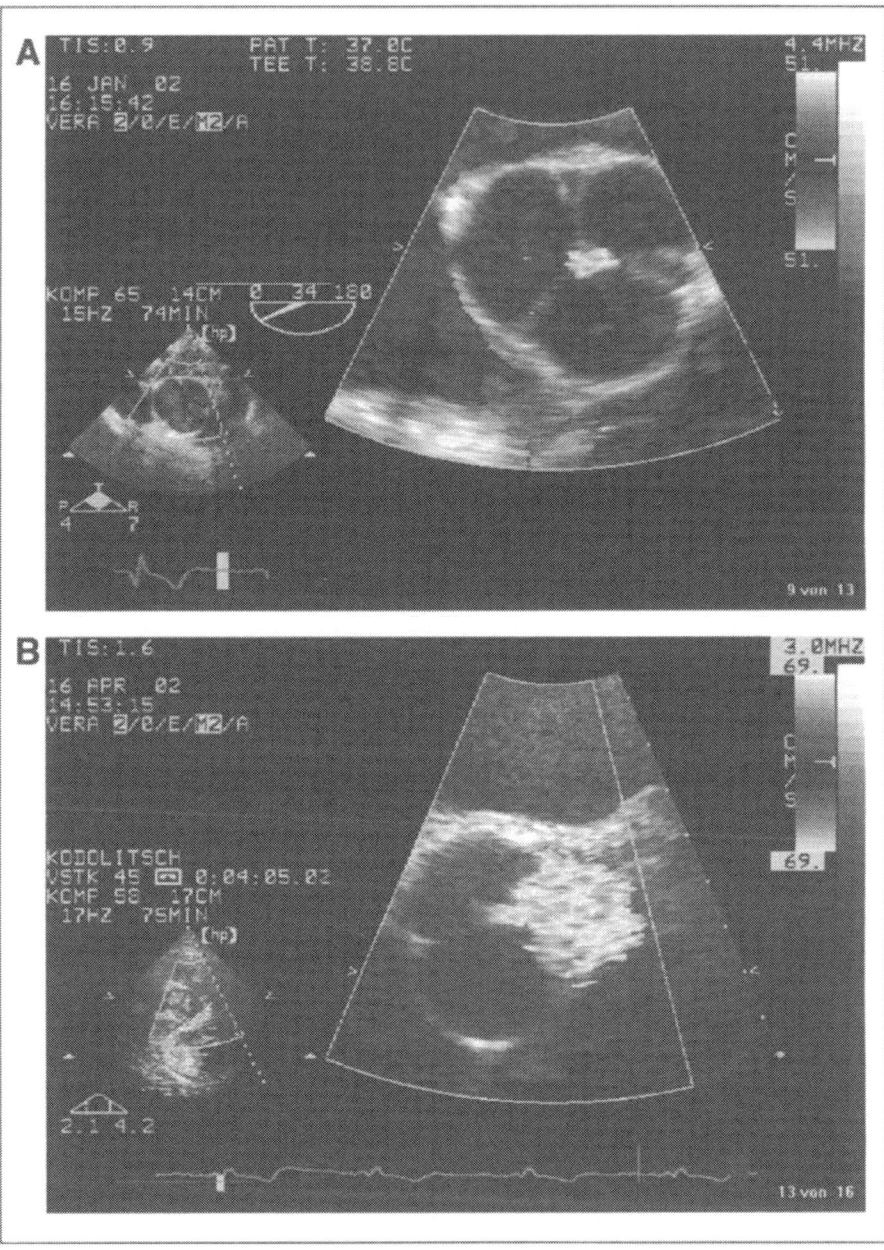

Figure 8. Transesophageal short axis view of the aortic valve in two different MFS patients with aortic valve regurgitation. A) Mild aortic valve regurgitation caused by dilatation of the aortic root resulting in incomplete coaptation of the three aortic cusps in the center of the aortic valve. B) Severe aortic valve regurgitation resulting from isolated prolapse of the noncoronary aortic cusp with regurgitation of blood through the incompetent cusp while the other two aortic cusps remain competent.

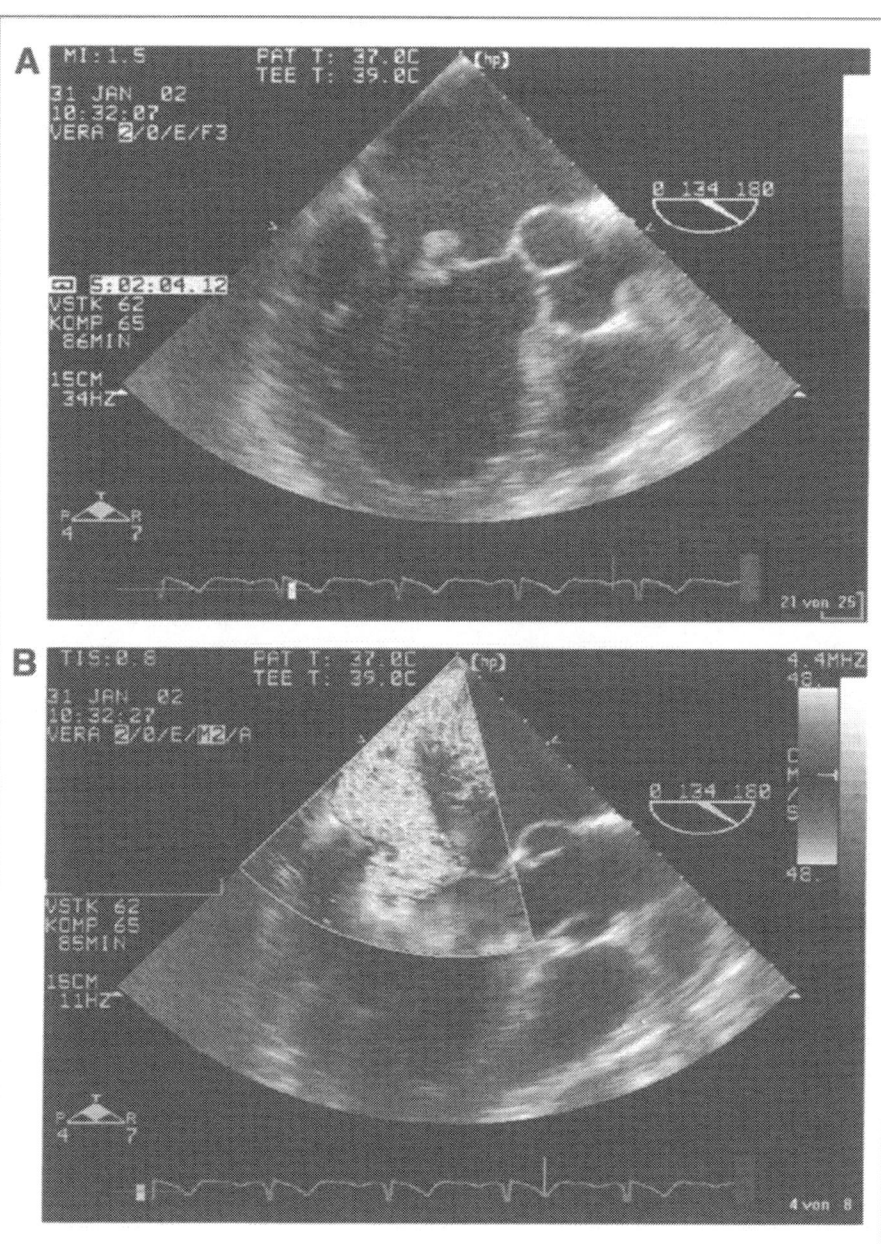

Figure 9. Transesophageal long axis views of the mitral valve in a patient with MFS presenting with septic fever and streptococcus viridans isolated in two consecutive blood cultures. Serial ultrasound examinations had previously shown an anatomically and functionally normal mitral valve. A) The 2-D-scan reveals mobile masses located mainly on the anterior mitral valve leaflet. B) Color Doppler examination demonstrates severe mitral valve incompetence. Thus, mitral valve endocarditis is suspected and the patient is taken to surgery with intraoperative confirmation of diagnosis.

Sudden Death and Cardiac Arrhythmia

There are two patients with MFS in an autopsy series of 182 cases of sudden death occurring below the age of 35 years. Similarly, two cases of MFS are identified in 158 autopsied competitive athletes with sudden death, and two MFS patients are reported in 29 athletes with sudden death below 35 years of age; with the exception of one case with mitral valve prolapse all deaths in MFS in these studies were due to aortic rupture.[61-63] Sudden death may be caused by arrhythmia due to mitral valve prolapse. For instance, 1.9 per 10,000 non-MFS patients with isolated mitral valve prolapse die suddenly and their risk increases to 94 to 188 deaths per 10,000 cases with growing degree of mitral valve incompetence.[64] Usually, lethal arrhythmias do not occur in patients with mitral valve prolapse before the valve develops incompetence.[21,65,66]

Episodes of ventricular arrhythmia are reported in children with MFS and 14 percent of adult MFS suffer giddiness or syncope. Moreover, premature atrial contractions are observed in 13 percent, premature ventricular contractions in 15 percent, paroxysmal atrial tachycardia in 3 percent, and atrial fibrillation in 11 percent of routine 12-lead-electrocardiograms obtained in adult MFS.[4,67] A recent study of 70 patients with Marfan syndrome diagnosed at birth to 52 years of age revealed that no patient died from aortic dissection, while four percent died from arrhythmias over a period of up to 24 years. Ventricular arrhythmias are identified in 21 percent of these patients and are associated with increased left ventricular size, mitral valve prolapse, and abnormalities of repolarization.[68]

Pregnancy

In women aged less than 40 years and no stigmata of MFS, 50 percent of lethal aortic dissections occur during pregnancy. Elevated serum estrogen and gestagen levels, hypervolemia and arterial hypertension, increased maternal age, multiple births, coarctation of the aorta, repaired tetralogy of Fallot and a congenitally bicuspid aortic valve may lead to aortic dissection in pregnant women without MFS.[4,69,70] A recent systematic literature review of aortic dissections occurring during pregnancy documents that 44 percent of type A dissections and 42 percent of type B dissections are associated with MFS and that 10 percent of prepartum type A dissections are related to a congenitally bicuspid aortic valve.[70]

Woman with MFS develop aortic complications in 4 to 50 percent of pregnancies.[71-77] Women without cardiac disease and aortic root diameters below 40 mm have a low risk for aortic complications. Conversely, pregnant women developing acute aortic complications exhibit preexisting cardiac disease in 66 percent of the cases with aortic dilatation in 30 percent, aortic valve incompetence in 25 percent, prolapse or insufficiency of the mitral valve in 33 percent, arterial hypertension in 17 percent and coarctation of the aorta in 8 percent.[4] One third of complications develop during the first, one third during the second and another third during the third or fourth pregnancy.[76] Maternal complications arise during the second trimenon in 8 percent, during the third trimenon in 39 percent, at the time of birth in 11 percent, and one through 8 weeks after birth in 42 percent. Cardiac complications comprise aortic dissection in 95 percent, and rupture of aortic aneurysm or heart failure in the remaining cases.[4] A literature review of type A dissections associated with pregnancy by Immer et al documents that both, maternal and fetal outcome has improved significantly over the recent two decades; interestingly, they find that the maternal mortality is lower than the fetal mortality in each time period.[70]

As a consequence, pregnancy should not be recommended to women with aortic dilatation ≥ 40 mm (or less in patients with a low body surface area[70]), with preconceptional cardiac disease or with previous cardiovascular surgery.[72] As a note of caution, however, one should be aware that aortic dissection is also observed with aortic root diameters below 40 mm and even in women without preconceptional cardiovascular abnormalities.[71,76] In patients at risk for aortic complications fetal lung maturation should be induced at 26 weeks of gestation and hospitalization should be considered between 28 and 32 weeks of gestation.[70] Women with MFS but no cardiovascular disease and normal aortic diameters may have vaginal delivery.

Since arachnoid cysts may result in considerable dilution of anesthetics women should be evaluated for dural ectasia before epidural and spinal anesthsia is administered.[16] Cesarean section under regional anesthesia should be preferred in any MFS patient with cardiovascular abnormalities including isolated aortic dilatation.[70,77] Aortic root enlargement should be repaired prepartum with diameters ≥ 40 mm.[70] When acute type A dissection occurs during pregnancy aortic surgery should be performed with the fetus in utero in cases before the 30th week of gestation, whereas immediate cesarean section should precede cardiac surgery in cases after the 30th week of gestation; cardiopulmonary bypass can already be installed during cesarean section.[70]

Diagnosis of Cardiovascular Abnormalities

Established Diagnostic Modalities

Cardiac auscultation of adult MFS is reported to detect mitral valve prolapse with a sensitivity of 69 percent and a specificity of 87 percent and moderate to severe aortic regurgitation is detected with a sensitivity of 78 percent.[5] Chest radiography with frontal and lateral projections identifies aortic root dilatation in 70 percent of patients without chest wall deformities.[5] Transthoracic echocardiography may diagnose or exclude aortic root dilatation with a sensitivity of 94 percent and a specificity of 100 percent; however, chest wall deformities may preclude MFS patients from adequate imaging in 5 to 23 percent. In addition, type A dissection is detected in only 50 percent of transthoracic ultrasound examinations.[5,23,78] In contrast, the diagnostic accuracy of transesophageal echocardiography for both dilatation and dissection of the aortic root is up to 100 percent. Since examinations can be performed at the bedside or intraoperatively transesophageal echocardiography is the method of choice for diagnosing or excluding acute aortic syndrome in unstable MFS patients. Moreover, since there is no adverse effect on the fetus, transesophageal echocardiography is optimal for serial imaging during pregnancy. However, the distal region of the ascending aorta, the proximal part of the aortic arch, and the abdominal aorta may not be visualized.[5,12]

In contrast, magnetic resonance imaging is optimal for both mapping the entire aortic vessel and for visualizing prosthetic aortic grafts; thus this modality may be employed for serial imaging of the pre- and postoperative aorta. Similarly, multislice contrast-enhanced computed tomography yields optimal results for diagnosing and excluding aortic disease. Most notably, computed tomography is most frequently employed for diagnosing acute aortic syndrome in hemodynamically stable patients. However, exposure to x-rays precludes computered tomography from serial imaging of young individuals.[12]

New Diagnostic Approaches

Invasive aortography and noninvasive transthoracic echocardiography, transesophageal echocardiography and magnetic resonance imaging are used to assess the elastic properties of the thoracic aorta in patients with MFS. Aortic distensibility (assessed as 2 x (aortic luminal area change)/(diastolic luminal area) x pulse pressure) is lower in MFS patients. Aortic stiffness index (assessed as 1n (systolic/diastolic blood pressure)/(aortic diameter change/aortic diastolic diameter), where 1n is the natural logarithm) is higher in MFS patients with aortic dilatation than in healthy, age matched patients.[79-81] However, aortic compliance is also altered in patients with systemic hypertension, coronary artery disease, atherosclerosis and aortic regurgitation.[4] Moreover, results depend on age, gender and physical activity and thus, standard values are required for healthy subjects of both genders with different age and body size.[5]

High-frequency ultrasonic tissue characterization is used to investigate aortic specimens of eleven MFS patients undergoing aortic root replacement and in eight autopsied patients without aortic disease. This method sensitively detects changes in vessel wall composition and organization manifesting by a significant decrease in integrated backscatter compared to normal aortas. To date, however, backscatter has not been assessed in vivo.[82]

General Measures of Management

Modification of life-style and physical activity is essential to improve prognosis. Patients should be informed about potential complications of the syndrome, modes of their prevention, the 50-percent likelihood of passing the causative fibrillin-1-gene mutation to any offspring, the risks of pregnancy and the need both for medical surveillance of their children and screening of their kindred for presence of MFS.[6,16,83] In adults emotional stress should be reduced[16] and physical activity should be restricted to isotonic, low-impact exercise such as swimming, biking or jogging with only moderate increase of heart rate and blood pressure. Any sports of competitive or isometric nature such as weight lifting should be avoided. Similarly, contact sports and activities with rapid acceleration and deceleration such as martial arts, football, basketball, soccer, skiing or sprinting should not be recommended to MFS patients.[5] However, although there is a report of a teenager suffering dissection during recreational weight lifting, there is little evidence that physical activity of any kind increases aortic growth rates in children.[12,84] Thus it may suffice for children to channel their interests away from competitive athletics rather than to impose outright restrictions.[16,83] Interestingly, a recent study of adherence to medication and physical activity guidelines in 174 adults with MFS revealed that 80 percent of the respondents had modified their physical activities and adhered to the prescribed medication regimens.[85]

Medical Management

Rationale for Beta Adrenergic Blocking Therapy

Beta-adrenergic blocking agents are effective for managing aortic dissection, in retarding progression of abdominal aneurysm and in preventing late progression of intramural hematoma.[17,86,87] It is unknown how beta-adrenergic blocking agents protect the aorta; however, reduction of both blood pressure and the rate of change in the central arterial pressure with respect to time (designated $\Delta P/\Delta t$, or impulse of left ventricular ejection), improvement of aortic compliance and chemical effects on the extracellular matrix of the aorta are discussed as potential mechanisms.[6,88] Moreover, elastic properties of the MFS-aorta assessed as aortic stiffness and distensibility can improve with beta-adrenergic blockade[89-91] but may also deteriorate.[79,89,91]

The use of beta-adrenergic agents in MFS is established by a prospective randomized trial, which compares the outcome of 32 MFS patients treated with high dosages of propanolol (212 ± 68 mg per day) at an average age of 15.4 years over 10.7 years to 38 control MFS patients with an average age of 14.5 years monitored over 9.3 years. The rate of aortic dilatation is significantly lower in the treatment group and clinical end points including aortic regurgitation, aortic dissection, cardiovascular surgery, congestive heart failure and death are reached in five patients in the treatment group compared to nine controls. In addition, actuarial long-term survival is better in the treatment group.[88] Retrospective studies confirm the effectiveness of beta-adrenergic blockade initiated in MFS at an average age of 10 or 15[92-94] but do not corroborate protective effects in MFS treated at an age of 21 to 28 years.[15,19,95] Moreover, beta-adrenergic blockade effectively retards progression of diameters only in patients with aortic dilatation of no more than mild or moderate degree.[6,95] Haouzi et al demonstrate that aortic stiffness indices and distensibility values normalize in eight MFS patients treated with beta-blockers but deteriorate in five MFS patients of their study group;[89] this indicates that the response of adult MFS patients to beta-adrenergic blockade may be heterogeneous, which is also observed by Rios et al.[91]

Practical Approach to Beta-Blocker Therapy

To reduce adverse effects of adrenergic beta-blockade, the initial dosage of atenolol or propanolol should be as low as 0.5 to 1.0 mg/kg/day; the dosage of atenolol should then be increased by 12.5 to 25 mg/day and propanolol by 20 to 40 mg/day until the maximum heart rate remains below 100 beats per minute after moderate exercise such as running up and down

44 steps as rapidly as possible.[88,95] This protocol is used predominantly in children with MFS; however, for adults Reed Pyeritz also recommends to gradually advance a low dose of beta-blockers until the heart rate is in the low 60s at rest and below 100 beats per minutes after a few minutes of vigorous exercise.[16]

Beta-blocker therapy is beneficial in infants and adolescents, since aortic dilatation can be retarded and thus, aortic root surgery may be performed later in life to avoid both the need for reintervention caused by aortic valve graft mismatch and the risks of indefinite anticoagulation during adolescent years.[6,96,97] However, beta-adrenergic blockade does not necessarily prevent aortic complications in children[94] and does not obviate the need for prophylactic surgery in adults.[6] Beta-blocker therapy is generally recommended after aortic surgery; however, there is no data to establish a reduced frequency of surgical reintervention.[6] Metoprolol or atenolol are the drugs of choice to protect the aortic vessel during pregnancy and should be administered up to three months postpartum;[77] however, beta-blocking agents can increase the uterine tone and contractility and decrease umbilical blood flow.[70,98] Propanolol can be associated with adverse fetal and neonatal effects and thus should only be used in selected patients.[70]

Calcium antagonists may also be effective in retarding aortic growth rates in children and adolescents with MFS.[94] Most recently, accelerated angiotensin converting enzyme-dependent angiotensin II formation and signaling via upregulated angiotensin II type 2 receptor is shown to play a role in vascular smooth muscle cell apoptosis in cystic media degeneration associated with MFS. Angiotensin converting enzyme inhibitor therapy may thus be a future strategy to prevent aortic wall degeneration.[99] A recent prospective study follows 65 children with MFS over a period of three years; this study reveals improved aortic compliance associated with a lower rate of aortic growth in individuals treated with angiotensin converting enzyme inhibitors as compared to individuals treated with beta-adrenergic blockers.[100]

Endocarditis

Since all MFS patients are at risk for endocarditis, antibiotic prophylaxis should be performed when transient bacteriemia is likely to occur. The revised guidelines of the American Heart Association recommend oral antibiotics for patients with artificial valves undergoing procedures with a high probability of transient bacteriemia.[101] However, since MFS patients with a composite graft or other heart valves prostheses are at particularly high risk for endocarditis surgeons recommend broad spectrum intravenous antibiotic prophylaxis for dental, endoscopic and surgical procedures or with infections of the upper airways or the skin.[6,26] Patients without heart surgery and macroscopically intact heart valves may be managed with oral antibiotics, but antibiotics may better be administered intravenously to patients with classic mitral valve prolapse, regurgitant heart valves or concomitant cardiovascular malformations.[6]

Indications for Surgery

Despite some promising experimental approaches to somatic gene therapy of MFS, surgical intervention currently is the only strategy, which effectively avoids or treats severe complications from aortic root disease. In adult MFS, prophylactic intervention is recommended with maximum aortic root diameters between 55 and 60 mm.[6,26] Maximum aortic diameters between 50 and 55 mm may require prophylactic intervention with an annual increase of the aortic ratio exceeding 5 percent, with dilatation of the aortic sinuses involving the ascending aorta, with severe aortic or mitral valve regurgitation, with a family history of aortic dissection, with other major surgery required in the near future, or, in women planning pregnancy.[15,19,27,102] Guidelines for surgical intervention are not established in infants. As a rule of thumb, aortic root replacement should be performed late in life to avoid reintervention for mismatch of the valve prosthesis caused by growth of the cardiovascular system.[6,103] Some authors recommend surgery with aortic root diameters between 50 and 60 mm, with marked progression of aortic dilatation, with presence of moderate aortic valve incompetence, or with an aortic root diameter of more than 50 percent of normal.[40,104,105] The European task force suggests not to routinely perform intervention with aortic root dimensions outside the upper confidence interval for the normal population[19] but recommend surgery when dimensions deviate upward from the pursued centile on follow-up echocardiograms.[12]

Dissection or intramural hemorrhage of the ascending aorta should immediately be operated upon, whereas dissection of the descending aorta can usually be managed medically.[17] Replacement of the descending aorta should be considered with aortic dilatation exceeding 55 or 60 mm, or with presence of complications such as persistent pain or organ ischemia. Alternatively, an aortic stent graft can be placed in the descending aorta to treat acute contained rupture of the descending aorta or type B dissection, which cannot be managed medically.[106] However, therapeutic results with such devices are preliminary and MFS patients with dissection may not be ideal candidates.[16] Reintervention in the ascending aorta is recommended with a diameter above 50 mm and in a distal aortic segment with diameters exceeding 60 mm.[6,107]

Following general guidelines, surgery of the mitral valve should be performed before myocardial dysfunction develops.[104,105] In patients with concomitant aortic root disease, mitral valve repair should be performed during a single intervention since, on the one hand, isolated mitral valve repair may lead to increased aortic wall stress with the risk of rupture, and, on the other hand, the postoperative course after isolated repair of the aortic root may be complicated by severe mitral valve regurgitation.[108] In children, mitral valve surgery yields unfavourable results when performed before the age of two years, and surgical intervention usually is scheduled when combined repair of the aortic root and mitral valve is indicated.[60,104,109]

Surgically Treated MFS Patients

Early Post Surgical Course

In adult MFS, initial surgical intervention is required for isolated severe aortic disease or marked aortic valve incompetence in 86.7 percent, for combined aortic disease and mitral valve regurgitation in 8.3 percent and for isolated severe mitral valve regurgitation in 5 percent (Fig. 10A). Among 1,241 cases with cardiac surgery, 72 percent of procedures were performed electively, 9.4 percent urgently (within 1 to 7 days) and 18.5 percent emergently with early death in 1.9 percent of elective procedures, in 2.6 percent of urgent procedures, and 17.4 percent with emergent surgery (Fig. 10B). Multivariable parametric hazard analysis identifies previous surgery of the ascending aorta and urgent or emergent repair as risk factors for early postoperative death.[26] The leading causes of early death are perioperative complications in 24.3 percent, low cardiac output in 15.4 percent, sudden death and arrhythmia in 14 percent, aortic dissection or rupture in 8.8 in percent, myocardial infarction in 8.8 percent, perioperative hemorrhage in 8.1 percent, and stroke or brain damage in 4.4 percent (Fig. 10C).

Late Outcome

In adult MFS, actuarial 5-year survival ranges between 71 and 90 percent, and actuarial 10-year-survival between 56 and 79 percent after initial cardiovascular surgery.[6,26] Multivariable parametric hazard analysis identifies preoperative NYHA functional class IV as the only risk factor for late postoperative death.[26] Reoperation for complications of the aortic vessel is required in 9.6 percent of survivors over a mean time interval of 6.7 years or in 27 percent of 700 patients during a 5 to 8 years interval of follow-up.[6,26] Reoperation of the aorta is performed for complications in the ascending aorta or the aortic arch in 40 percent of reoperations and for complications in the descending or abdominal aorta, or both, in 60 percent.[6] The early mortality of reintervention ranges between 4.7 und 25 percent, and emergent repair is a risk factor of early death.[110] Moreover, with improved survival in MFS, there is increasing need for heart transplantation in the late postoperative course.[55] Databank analysis of 303 adult MFS patients with late postoperative death identifies dissection or rupture of the aorta as the leading cause of death (Fig. 10D).

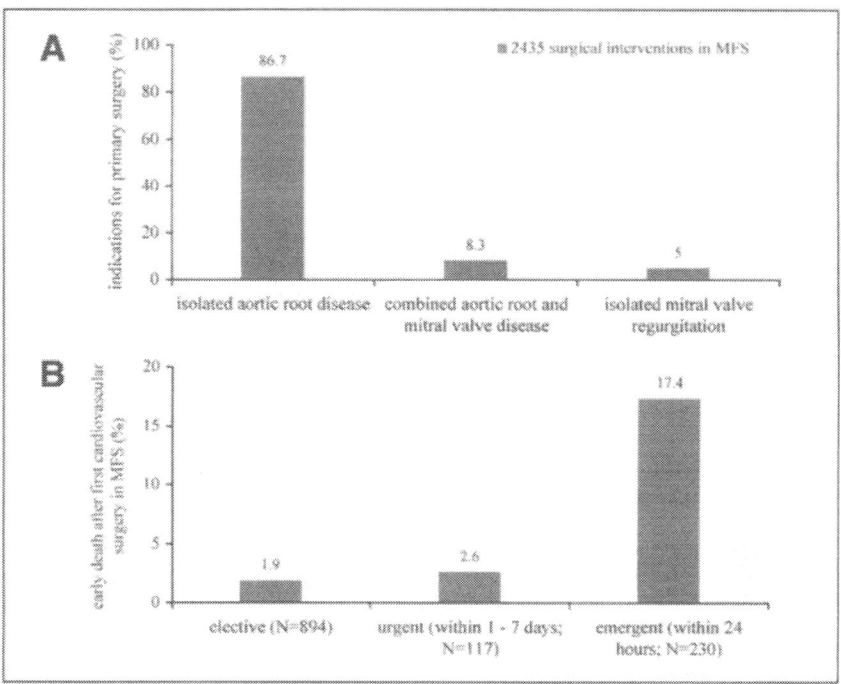

Figure 10. Literature databank analysis of the primary cardiovascular surgery reported in adult MFS. A) Indications for primary surgical intervention B) Death within 30 days after primary surgery depending on the urgency of operation. Figure continued on next page.

Strategies for Follow-Up

Adult MFS patients with an aortic root diameter of less than 40 mm and no previous cardiovascular surgery should undergo yearly imaging of the aorta. Examinations should be performed in six month intervals with an aortic diameter of more than 40 mm, increase of the aortic ratio of more than 5 percent per year, a family history of aortic dissection or sudden cardiac death or dilatation of the aortic sinuses with extension into the ascending aortic segment; intervals should be reduced to 3 to 4 months once the aortic root diameter equals 50 mm.[5,27] Prior to elective vascular surgery, magnetic resonance imaging or computer tomography is required for mapping of the entire aorta.[111,112] Preoperative coronary angiography is recommended in adults 50 years of age or older [5] (Fig. 11). Children with aortic root dilatation may have echocardiography with yearly intervals and with six months intervals once aortic regurgitation occurs.[5,40,104] Baseline magnetic resonance imaging or computer tomography should be performed within 30 days after elective aortic surgery. Serial postoperative examinations are required at six-months intervals, which may be extended to yearly intervals when the postoperative course is uneventful and aortic diameters remain stable.[5,112,113] Follow-up intervals may be shorter in patients with surgery for aortic dissection or other variants of acute aortic syndrome. Irrespective of prior cardiovascular surgery, any MFS patient with new or acute thoracic or abdominal symptoms should immediately be subjected to tomographic aortic imaging.[5,17,113]

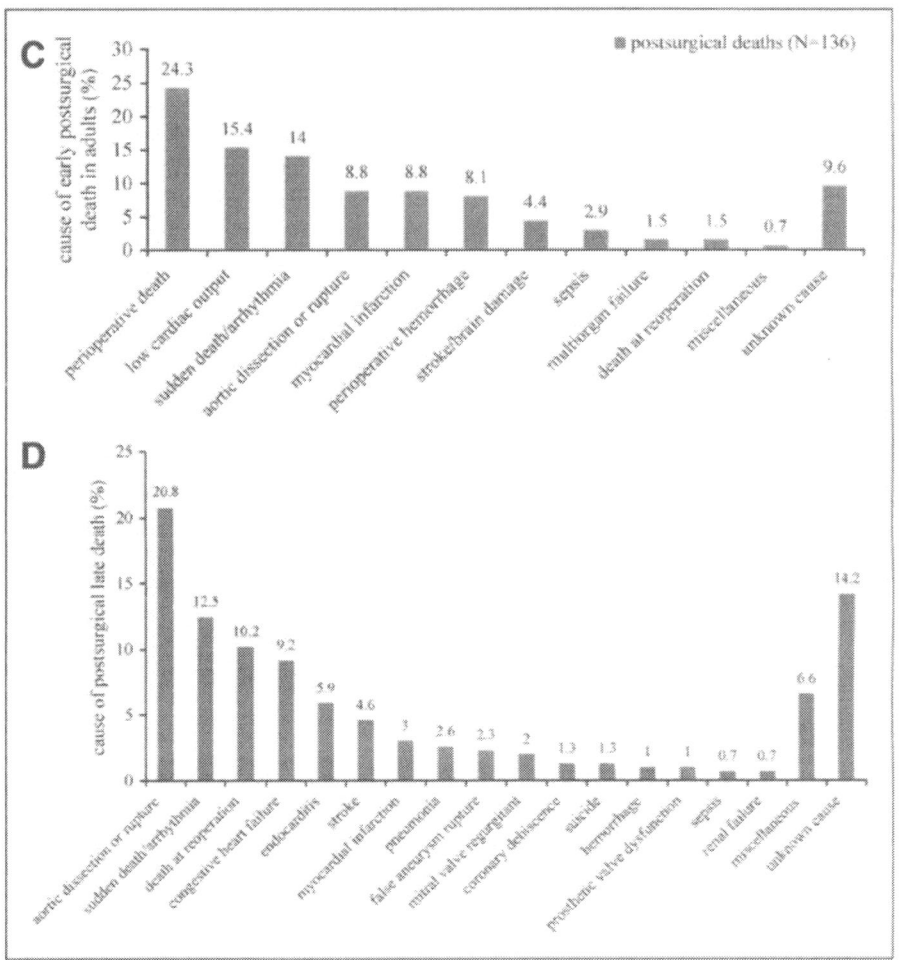

Figure 10 Continued. C) Causes of early death. D) Causes of late death reported at least 31 days after initial cardiovascular surgery.

Comment

Any review of literature from retrospective studies suffers from several potential biases. First, severe cases tend to be reported rather than milder cases even in large-scale population-based surveys. Second, the series that we review in this article use different diagnostic criteria of MFS and thus, older articles may have misclassified patients and families with familial aortic aneurysm and thoracic aneurysm associated with a bicuspid aortic valve as having MFS. Accordingly, recent studies using the Ghent criteria for MFS report presence of a bicuspid aortic valve at a rate of much less than five percent. Moreover, the diagnostic criteria for cardiovascular manifestations such as aortic root dilatation and mitral valve prolapse are not uniform across different studies. Vincent Gott and Roland Hetzer review the different techniques of cardiovascular surgery in the chapters in this book.

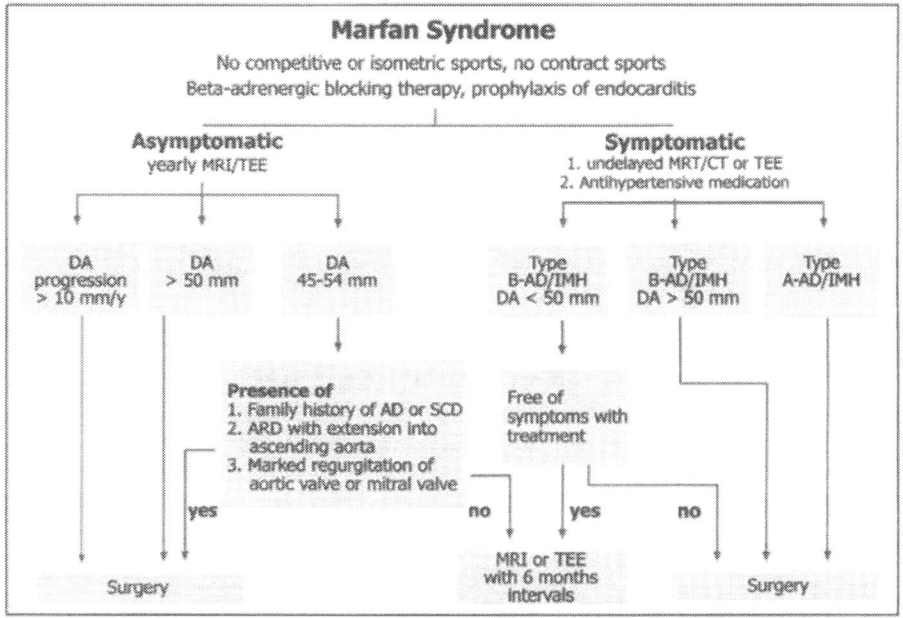

Figure 11. Algorithm for clinical management of the cardiovascular manifestations of patients with MFS. AD indicates aortic dissection; ARD= aortic root dilatation; CT= contrast enhanced computered tomography; DA= diameter of ascending aorta; MRI= magnetic resonance imaging; SCD= sudden cardiac death; TEE= transesophageal echocardiography; Type A-AD/IMH, aortic dissection or intramural hematoma involving the ascending aorta and Type B-AD/IMH is aortic dissection or intramural hematoma not involving the ascending aorta.

Acknowledgements

The authors wish to express their cordial gratefulness to Reed E. Pyeritz for his thoughtful review of the article. In particular, we owe him the idea for the final comment on this work. Moreover, we wish to thank the editors of this book, Peter Robinson and Maurice Godfrey for their comments. Finally, we wish to thank Claudia Hottendorff and Heidi Schöning for helping us in the preparation of the figures for the manuscript.

References

1. De Paepe A, Devereux RB, Dietz HC et al. Revised diagnostic criteria for the Marfan syndrome. Am J Med Genet 1996; 62:417-426.
2. McKusick VA. The cardiovascular aspects of Marfan's syndrome. Circulation 1955; 11:321-342.
3. Beighton P, De Paepe A, Danks D et al. International nosology of heritable disorders of connective tissue, Berlin, 1986. Am J Med Genet. 1988; 29:581-94.
4. von Kodolitsch Y, Raghunath M, Nienaber CA. The Marfan syndrome: prevalence and natural history of cardiovascular manifestations. Z Kardiol 1998; 87:150-160.
5. von Kodolitsch Y, Raghunath M, Dieckmann C et al. The Marfan syndrome: diagnosis of cardiovascular manifestations. Z Kardiol 1998; 87:161-172.
6. von Kodolitsch Y, Raghunath M, Karck M et al. The Marfan syndrome: therapeutic approach to cardiovascular manifestations. Z Kardiol 1998; 87:173-184.
7. Murdoch JL, Walker BA, Halpern BL et al. Life expectancy and causes of death in the Marfan syndrome. N Engl J Med 1972; 286:804-808.
8. Yetman AT, Huang P, Bornemeier RA et al. Comparison of outcome of the Marfan syndrome in patients diagnosed at age ≤ 6 years versus those diagnosed at > 6 years of age. Am J Cardiol 2003; 91:102-103.

9. Brown OR, DeMots H, Kloster FE et al. Aortic root dilatation and mitral valve prolapse in Marfan's syndrome. An echocardiographic study. Circulation 1975; 52:651-657.
10. Henry WL, Gardin JM, Ware JH. Echocardiographic measurements in normal subjects from infancy to old age. Circulation 1980; 62:1054-1061.
11. Roman MJ, Devereux RB, Kramer-Fox R et al. Two dimensional aortic root dimensions in normal children and adults. Am J Cardiol 1989; 64:507-512.
12. Erbel R, Alfonso F, Boileau C et al. Diagnosis and management of aortic dissection. Recommendations of the task force on aortic dissection, European Society of Cardiology. Eur Heart J 2001; 22:1642-1681.
13. Reed CM, Richley PA, Pullam DA et al. Aortic dimensions in tall men and women. Am J Cardiol 1993; 71:608-611.
14. Rozendaal L, Groenink M, Naeff MSJ et al. Marfan syndrome in children and adolescents: an adjusted nomogram for screening aortic root dilatation. Heart 1998; 79:69-72.
15. Legget ME, Unger TA, O'Sullivan CK et al. Aortic root complications in Marfan's syndrome: identification of a lower risk group. Heart 1996; 75:389-395.
16. Pyeritz RE. The Marfan syndrome. Annu Rev Med 2000; 51:481-510.
17. von Kodolitsch Y, Csösz K, Koschyk DH et al. Intramural hematoma of the aorta: predictors of progression to dissection and rupture. Circulation 2003; 107:1158-1163.
18. Hwa J, Richards JG, Huang H et al. The natural history of aortic dilatation in Marfan syndrome. Med J Austral 1993; 158:558-562.
19. Roman MJ, Rosen SE, Kramer-Fox R et al. Prognostic significance of the pattern of aortic root dilation in the Marfan syndrome. J Am Coll Cardiol 1993; 22:1470-1476.
20. Freed LA, Levy D, Levine RA et al. Prevalence and clinical outcome of mitral valve prolapse. N Engl J Med 1999; 341:1-7.
21. Playford D, Weyman AE. Mitral valve prolapse: Time for a fresh look. Rev Cardiovasc Med 2001; 2:73-81.
22. Pini R, Roman MJ, Kramer-Fox R et al. Mitral valve dimensions and motion in Marfan patients with and without mitral valve prolapse. Circulation 1989; 80:915-924.
23. Pyeritz RE, Wappel MA. Mitral valve dysfunction in the Marfan syndrome. Am J Med Genet 1974; 1983:797-807.
24. Nollen GJ, van Schijndel KE, Timmermans J et al. Pulmonary artery root dilatation in Marfan syndrome: quantitative assessment of an unknown criterion. Heart 2002; 87:470-471.
25. Roberts WC, Honig HS. The spectrum of cardiovascular disease in the Marfan syndrome: a clinico-morphologic study of 18 necropsy patients and comparison to 151 previously reported necropsy patients. Am Heart J 1982; 104:115-134.
26. Gott VL GP, Alejo DE et al. Replacement of the aortic root in patients with Marfan syndrome. N Engl J Med 1999; 340:1307-1313.
27. Pyeritz RE. Predictors of dissection of the ascending aorta in Marfan syndrome. Circulation 1991; 84(suppl II):II-351.
28. Silverman DI, Gray J, Roman MJ et al. Family history of severe cardiovascular disease in Marfan syndrome is associated with increased aortic diameter and decreased survival. J Am Coll Cardiol 1995; 26:1062-1067.
29. Yacoub MH, Sundt TM, Rasmi N. Management of aortic valve incompetence in patients with Marfan syndrome. In: Hetzer R, Gehle P, Ennker J, eds. Cardiovascular aspects of Marfan syndrome. 1995; Steinkopff Verlag 1995; Darmstadt: 71-81.
30. Lima SD, Lima JAC, Pyeritz RE et al. Relation of mitral valve prolapse to ventricular size in Marfan's syndrome. Am J Cardiol 1985; 55:739-743.
31. Avierinos JF, Gersh BJ, Melton LJ et al. Natural history of asymptomatic mitral valve prolapse in the community. Circulation 2002; 106:1355-1361.
32. Barlow JB, Pocock WA. Mitral valve billowing and prolapse: perspective at 25 years. Herz 1988; 13:227-234.
33. Levine RA, Stathogiannis E, Newell JB et al. Reconsideration of echocardiographic standards for mitral valve prolapse: lack of association between leaflet displacement isolated to the apical view and independent echocardiographic evidence of abnormality. J Am Coll Cardiol 1988; 11:1010-1019.
34. Marks AR, Choong CY, Sanfilippo AJ et al. Identification of high-risk and low-risk subgroups of patients with mitral-valve prolapse. N Engl J Med 1989; 320:1031-1036.
35. Roberts WC. The congenitally bicuspid aortic valve: a study of 85 autopsy cases. Am J Cardiol 1970; 26:72-83.
36. Campbell M, Baylis JH. The course and prognosis of coarctation of the aorta. Br Heart J 1956; 18:475-495.
37. Chung KY, Walsh TJ, Massie E. Wolff-Parkinson-White syndrome. Am Heart J 1965; 69:116-119.

38. Sherf L, Neufeld HN. The pre-excitation syndrome: facts and theories. New York, Yorke Med Books 1978:30:32.
39. Mitchell SC, Korones SB, Berendes HW. Congenital heart disease in 56,109 births. Incidence and natural history. Circulation 1971; 43:323-332.
40. El Habbal MH. Cardiovascular manifestations of Marfan's syndrome in the young. Am Heart J 1992; 123:752-757.
41. Schievink WI, Parisi JE, Piepgras DG et al. Intracranial aneurysms in Marfan's syndrome: an autopsy study. Neurosurgery 1997; 41:866-871.
42. Conway JE, Hutchins GM, Tamargo RJ. Marfan syndrome is not associated with intracranial aneurysms. Stroke 1999; 30:1632-1636.
43. van den Berg JSP, Limburg M, Hennekam RCM. Is Marfan syndrome associated with symptomatic intracranial aneurysms? Stroke 1996; 27:10-12.
44. Coselli JS, LeMaire SA, Bücket S. Marfan syndrome: The variability and outcome of operative management. J Vasc Surg 1995; 21:432-443.
45. Crawford ES. Marfan's syndrome. Broad spectral surgical cardiovascular manifestations. Ann Surg 1983; 198:487-505.
46. Becker AE, van Mantgem J-P. The coronary arteries in Marfan's syndrome. A morphologic study. Am J Cardiol 1975; 36:315-321.
47. Medley TL, Cole TJ, Gatzka CD et al. Fibrillin-1 genotype is associated with aortic stiffness and disease severity in patients with coronary artery disease. Circulation 2002; 105:810-815.
48. Mozes G, Gloviczki P, Park WM et al. Spontaneous dissection of the infrarenal abdominal aorta. Semin Vasc Surg 2002; 15:128-136.
49. Pruzinsky MS, Katz NM, Green CE et al. Isolated descending thoracic aortic aneurysm in Marfan's syndrome. Am J Cardiol 1988; 61:1159-1160.
50. Pan CW, Chen CC, Wang SP et al. Echocardiographic study of cardiac abnormalities in families of patients with Marfan syndrome. J Am Coll Cardiol 1985; 6:1016-1020.
51. Hirata K, Triposkiadis F, Sparks E et al. The Marfan syndrome: cardiovascular physical findings and diagnostic correlates. Am Heart J 1992; 123:743-752.
52. Carr NJ, Cullen SA. Acute aortic valve prolapse in Marfan's syndrome. Postgrad Med J 1991; 67:208-209.
53. MacMahon SW, Roberts JK, Kramer-Fox R et al. Mitral valve prolapse and infective endocarditis. Am Heart J 1987; 113:1291-1298.
54. Soman VR, Breton G, Hershkowitz M et al. Bacterial endocarditis of mitral valve in Marfan syndrome. Brit Heart J 1974; 36:1247-1250.
55. Hetzer R, Gehle P, Ennker J. Results of cardiovascular surgery for Marfan syndrome in Berlin. In: Hetzer R, Gehle P, Ennker J, eds. Cardiovascular aspects of Marfan syndrome 1995; Steinkopff Verlag; Darmstadt: 109-118.
56. Siracusa LD, McGrath R, Ma Q et al. A tandem duplication within the fibrillin 1 gene is associated with the mouse tight skin mutation. Genome Res 1996; 6:300-313.
57. Porciani MC, Giurlani L, Chelucci A at al. Diastolic subclinical primary alterations in Marfan syndrome and Marfan related disorders. Clin Cardiol 2002; 25:416-420.
58. Gross DM, Robinson LK, Smith LT et al. Severe perinatal Marfan syndrome. Pediatrics 1989; 84:83- 89.
59. Raghunath M, Superti-Furga A, Godfrey M et al. Decreased deposition of fibrillin and decorin in neonatal Marfan syndrome fibroblasts. Hum Genet 1993; 90:511-515.
60. Geva T, Sanders SP, Diogenes MS et al. Two-dimensional and Doppler-echocardiographic and pathologic characteristics of the infantile Marfan syndrome. Am J Cardiol 1990; 65:1230-1237.
61. Basso C, Frescura C, Corrado D et al. Congenital heart disease and sudden death in the young. Hum Pathol 1995; 26:1065-1072.
62. Maron BJ, Shirani J, Poliac LC et al. Sudden death in young competitive athletics. Clinical, and pathologic profiles. JAMA 1996; 276:199-204.
63. Maron BJ, Roberts WC, McAllister HA et al. Sudden death in young athletes. Circulation 1980; 62:218-229.
64. Kligfield P, Levy D, Devereux RB et al. Arrhythmias and sudden death in mitral valve prolapse. Am Heart J 1987; 113:1298-1307.
65. Davies MJ, Moore BP, Braimbridge MV. The floppy mitral valve: study of incidence, pathology, and complications in surgical, necropsy, and forensic material. Br Heart J 1978; 40:373-481.
66. Barnett HJ, Boughner DR, Taylor DW et al. Further evidence relating mitral-valve prolapse to cerebral ischemic events. N Engl J Med 1980; 302:139-144.
67. Chen S, Fagan LF, Nouri S et al. Ventricular dysrhythmias in children with the Marfan's syndrome. Am J Dis Child 1985; 139:273-276.

68. Yetman AT, Bornemeier RA, McCrindle BW. Long-term outcome in patients with Marfan syndrome: is aortic dissection the only cause of sudden death? J Am Coll Cardiol 2003; 41:329-332.
69. Konishi Y, Tatsuta N, Kumada K et al. Dissecting aneurysm during pregnancy and the puerperium. Japan Circ J 1980; 44:726-733.
70. Immer FF, Bansi AG, Immer-Bansi AS et al. Aortic dissection in pregnancy: analysis of risk factors and outcome. Ann Thorac Surg 2003; 76:309-314.
71. Brice G, Treasure T, Pumphrey C et al. Serial echocardiography is of limited value in predicting aortic dissection in pregnant Marfan patients. Eur J Pediatr 1996; 155:745-746.
72. Pyeritz RE. Maternal and fetal complications of pregnancy in the Marfan syndrome. Am J Med Genet 1981; 71:784-790.
73. Lind J, Wallenburg HCS. Pregnancy and the Marfan syndrome-A Dutch study. 155. 1996; Eur J Pediatr:745.
74. Rossiter JP, Repke JT, Morales AJ et al. A prospective longitudinal evaluation of pregnancy in the Marfan syndrome. Am J Obstet Gynecol 1995; 173:1599-1606.
75. Lind J, Wallenburg HC. The Marfan syndrome and pregnancy: a retrospective study in a Dutch population. Eur J Obstet Gynecol Reprod Biol 2001; 98:28-35.
76. Lipscomb KJ, Smith JC, Clarke B et al. Outcome of pregnancy in women with Marfan syndrome. Br J Obstet Gynaecol 1997; 104:201-206.
77. Elkayam U, Ostrzega E, Shotan A et al. Cardiovascular problems in pregnant women with the Marfan syndrome. Ann Intern Med 1995; 123:117-122.
78. Detrano R, Moodie DS, Gill CC et al. Intravenous digital subtraction aortography in the preoperative and postoperative evaluation of Marfan's aortic disease. Chest 1985; 88:249-253.
79. Yin FCP, Brin KP, Ting C et al. Arterial hemodynamic indexes in Marfan's syndrome. Circulation 1989; 79:854-862.
80. Jeremy RW, Huang H, Hwa J et al. Relation between age, arterial distensibility, and aortic dilatation in the Marfan syndrome. Am J Cardiol 1994; 74:369-373.
81. Franke A, Mühler EG, Klues HG et al. Detection of abnormal aortic elastic properties in asymptomatic patients with Marfan syndrome by combined transesophageal echocardiography and acoustic quantification. Heart 1996; 75:307-311.
82. Recchia D, Sharkey AM, Bosner MS et al. Sensitive detection of abnormal aortic architecture in Marfan syndrome with high-frequency ultrasonic tissue characterization. Circulation 1995; 91:1036-1043.
83. American Academy of Pediatrics. Health supervision for children with the Marfan syndrome. Pediatrics 1996; 98:978-982.
84. Braverman AC. Exercise and the Marfan syndrome. Med Sci Sports Exerc 1998; 30:387-395.
85. Peters KF, Horne R, Kong F et al. Living with the Marfan syndrome II. Medication adherence and physical activity modification. Clin Genet 2001; 60:283-292.
86. Nienaber CA, von Kodolitsch Y. Meta-analysis of changing mortality pattern in thoracic aortic dissection. Herz 1992; 17:398-416.
87. Leach SD, Toole AL, Stern H et al. Effect of beta-adrenergic blockade on the growth of abdominal aortic aneurysms. Arch Surg 1988; 123.
88. Shores J, Berger KR, Murphy EA et al. Progression of aortic dilatation and benefit of long-term ß-adrenergic blockade in Marfan's syndrome. N Engl J Med 1994; 330:1335-1341.
89. Haouzi A, Berglund H, Pelikan PCD et al. Heterogeneous aortic response to acute ß-adrenergic blockade in Marfan syndrome. Am Heart J 1997; 133:60-63.
90. Hirata K, Triposkiadis F, Sparks E et al. The Marfan syndrome: abnormal aortic elastic properties. J Am Coll Cardiol 1991; 18:57-63.
91. Rios AS, Silber EN, Bavishi N et al. Effect of long-term ß-blockade on aortic root compliance in patients with Marfan syndrome. Am Heart J 1999; 137:1057-1061.
92. Salim MA, Alpert BS, Ward JC et al. Effect of beta-adrenergic blockade on aortic root rate dilation in the Marfan syndrome. Am J Cardiol 1994; 74:629-633.
93. Tahernia AC. Cardiovascular abnormalities in Marfan's syndrome: the role of echocardiography and ß-blockers. South Med J 1993; 86:305-310.
94. Rossi-Foulkes R, Roman MJ, Rosen SE et al. Phenotypic features and impact of beta blocker or calcium antagonist therapy on aortic lumen size in the Marfan syndrome. Am J Cardiol 1999; 83:1364-1368.
95. Hirata K, Triposkiadis F, Bowen J et al. The Marfan syndrome: rate of aortic root dilatation. Circulation 1989; 80(suppl II):II-529.
96. Reed CM, Fox ME, Alpert BS. Aortic biochemical properties in pediatric patients with the Marfan syndrome and the effects of atenolol. Am J Cardiol 1993; 71:606-608.

97. Miller DC. Valve-sparing aortic root replacement in patients with the Marfan syndrome. J Thorac Cardiovasc Surg 2003; 125:773-778.
98. Chen TO. Caution in use of beta blockers during pregnancy. Cathet Cardiovasc Diagn 1995; 34:186.
99. Nagashima H, Sakomura Y, Aoka Y et al. Angiotensin II type 2 receptor mediates vascular smooth muscle cell apoptosis in cystic medial degeneration associated with Marfan syndrome. Circulation 2001; 104(suppl I):I282-I287.
100. Yetman AT, Bornemeier RA, McCrindle BW. A comparison of beta-blockers and angiotensinogen converting enzyme inhibitors on elastic properties, aortic growth, and left ventricular size and function in Marfan patients. J Am Coll Cardiol 2003; 41(suppl B):492 (A).
101. Dajani AS, Bisno AL, Chung KJ. Prevention of bacterial endocarditis: Recommendations by the American Heart Association. JAMA 1990; 264:2919-2922.
102. Treasure T. Cardiovascular surgery for Marfan syndrome. Heart 2000; 84:674-678.
103. Savolainen A, Savolainen H, Savunen T et al. Results of cardiovascular surgery in the Marfan syndrome. A retrospective study of 49 patients. Scand J Thor Cardiovasc Surg 1995; 29:11-15.
104. Sisk HE, Zahka KG, Pyeritz RE et al. The Marfan syndrome in early childhood: analysis of 15 patients diagnosed at less than 4 years of age. Am J Cardiol 1983; 52:353-358.
105. Tsang VT, Pawade A, Karl TR et al. Surgical management of Marfan syndrome in children. J Card Surg 1994; 9:50-54.
106. Nienaber CA, Fattori R, Lund G et al. Nonsurgical reconstruction of thoracic aortic dissection by stent-graft placement. N Engl J Med 1999; 340:1539-1545.
107. Heinemann MH, Bühner B, Jurmann MJ et al. Aortic disease in Marfan syndrome: surgery, results and special aspects. In: Hetzer R, Gehle P, Ennker J, eds. Cardiovascular aspects of Marfan Syndrome 1995; Darmstadt; Steinkopff: 101.
108. Crawford ES, Coselli JS. Marfan's syndrome: Combined composite graft replacement of the aortic root and transaortic mitral valve replacement. Ann Thorac Surg 1988; 45:296-302.
109. Geva T, Hegesh J, Frand M. The clinical course and echocardiographic features of Marfan's syndrome in childhood. Am J Dis Child 1987; 141:1179-1182.
110. Carrel T, Pasic M, Jenni R et al. Reoperations after operation on the thoracic aorta: etiology, surgical techniques, and prevention. Ann Thorac Surg 1993; 56:259-269.
111. Soulen RL, Fishman EK, Pyeritz RE et al. Marfan syndrome: Evaluation with MR imaging versus CT. Radiology 1987; 165:697-701.
112. von Kodolitsch Y, Simic O, Nienaber CA. Aneurysms of the ascending aorta: diagnostic features and prognosis in patients with the Marfan syndrome versus hypertension. Clin Cardiol 1998; 21:817-824.
113. Smith JA, Fann JI, Miller C et al. Surgical management of aortic dissection in patients with the Marfan syndrome. Circulation 1994; 90(part II):II-235-II-242.

CHAPTER 5

Cardiovascular Surgery:
Surgical Management of the Marfan Patient at the Johns Hopkins Hospital

Duke E. Cameron and Vincent L. Gott

O ver the past 25 years, there has been remarkable progress in surgical management of the Marfan patient with aortic root aneurysm. The introduction in 1966 of the composite graft procedure by Mr. Hugh Bentall in London was a major step forward in the successful treatment of these aneurysms and was complemented later by use of homograft aortic roots. More recently, valve-sparing procedures introduced by Sir Magdi Yacoub in London and Tirone David in Toronto, Canada have provided new options for Marfan patients that eliminate life-long anticoagulation.

It is generally agreed that all adults with a Marfan aneurysm should have operative intervention when the aneurysm reaches 5.5 cm in diameter; if there is a history of aortic dissection in the family, surgical intervention is recommended when the aortic root reaches 5.0 cm. The optimal timing of surgery in children with Marfan aortic root aneurysm is more problematic, but adolescents should be managed as adults. Guidelines for operative intervention in children under 12 include enlargement of 1.0 cm or more in 12 months, worsening aortic insufficiency and the development of an "adult size" root aneurysm. Also, the presence of worsening mitral regurgitation may mandate earlier surgical intervention in children.

Although the Bentall composite graft procedure has provided excellent long-term results in both adults and children, it is our feeling that the current procedure of choice is a valve-sparing operation utilizing the David I reimplantation technique. A newer Dacron graft with built-in "sinuses of Valsalva" may offer significant advantages over the straight tubular Dacron graft originally used by David.

Since there is a potential for late aneurysmal dilatation in the residual thoracic aorta after root replacement, followup CT or MRI studies should be performed on a yearly basis after root replacement. If the residual aorta remains stable for 5 to 8 years after aortic root surgery, CT/MRI monitoring may be extended to longer intervals. Patients who have had the valve-sparing procedure should be followed with annual echocardiograms to monitor aortic valve function.

Fortunately, both the Bentall composite graft procedure and valve-sparing procedures have provided excellent long-term results with minimal late morbidity and mortality.

Introduction

Before the Bentall composite graft operation for aortic root replacement, surgical management of Marfan aortic aneurysms was unsatisfactory. Between 1965 and 1976, one of the authors (VLG) performed aortic root replacement in five Marfan patients. Four patients presented with an acute ascending aortic dissection and underwent emergent surgery. A fifth patient had severe left ventricular failure secondary to aortic and mitral insufficiency and under-

Marfan Syndrome: A Primer for Clinicians and Scientists, edited by Peter N. Robinson and Maurice Godfrey. ©2004 Eurekah.com and Kluwer Academic / Plenum Publishers.

went elective surgery. At that time, the only operation available consisted of aortic valve replacement with a mechanical prosthesis and replacement of the supracoronary aneurysmal section of the ascending aorta with a Dacron fabric sleeve graft. This operation was extremely difficult because of the thin, friable aortic wall and the size discrepancy between the Dacron graft and the aneurysmal aortic root. Only 2 of these 5 patients survived; a hospital mortality rate of 60%.

This dismal picture of surgical management of Marfan aneurysm of the aortic root changed overnight in 1968 with the introduction of a remarkable new operative procedure developed by Mr. Hugh Bentall of London.[1] At the operating table, Mr. Bentall fabricated a composite graft consisting of a mechanical valve placed within a tubular Dacron graft. With this, he was able to completely replace the aneurysmal root of patients with large Marfan aneurysms; in doing so, he performed the first total aortic root replacement.

With the new Bentall composite graft procedure, outlook for these Marfan patients improved dramatically. We reported our 24 year experience with aortic root replacement in 271 Marfan patients in 2002.[2] In this series, 235 Marfan patients underwent elective aortic root replacement with no 30 day mortality. The only 2 early deaths in this series occurred among the 36 patients who underwent urgent or emergent operations. Both of these early deaths occurred in patients arriving in the operating room with rupture of their aortic root.

Excellent results with aortic root replacement in Marfan patients have also been obtained in many other centers. In a paper published in the *New England Journal of Medicine* in 1999, results of aortic root replacement were evaluated in 675 patients from 10 patient centers worldwide.[3] Three centers were in Europe, one was in Canada and six were in the United States. Thirty-day mortality rate was 1.5% among the 455 patients having elective repair, 2.6% among 170 patients with urgent repair (within seven days after surgical consultation), and 11.7% among 103 patients with emergent repair (within 24 hours after surgical consultation). This paper emphasized the low operative risk in patients undergoing elective repair, in sharp contrast to the high operative mortality in the setting of emergent repair, usually for acute aortic dissection.

Indications for Cardiovascular Surgery

It is generally agreed that elective aortic root replacement is indicated for any Marfan patient with a root diameter of 5.5 cm or larger. Because the operative mortality for elective root replacement is low, many Marfan centers even recommend operative intervention when the aortic root reaches 5.0 cm. In our view, if there is a family history of aortic dissection, elective aortic root replacement should proceed when the aneurysm reaches a diameter of 5.0 cm.

Approximately one fifth of Marfan patients undergoing aortic root replacement will have sufficient mitral regurgitation to warrant mitral valve repair or replacement at the time of the aortic root surgery. If a Marfan patient has significant mitral regurgitation and modest aortic root dilatation (4 to 5 cm), we feel it is also prudent to resect the aortic root concomitantly.

Optimal timing of aortic root replacement in children with the Marfan aneurysm is not well defined. In children with severe mitral insufficiency and heart failure, timing of intervention is straight-forward. In the absence of mitral insufficiency, it is our opinion that an aortic root diameter increase of 1 cm or more in 12 months is an indication for root replacement. Progressive aortic insufficiency is also an indication, particularly if a valve-sparing procedure is contemplated. Finally, an acute or chronic dissection of the ascending aorta in the Marfan patient is an absolute indication for urgent operative intervention.

There are two classifications of aortic dissection, the DeBakey classification and the Stanford classification (Fig. 1). The DeBakey Type I dissection commences in the ascending aorta and involves the arch and descending aorta, whereas a DeBakey Type II dissection is limited to the

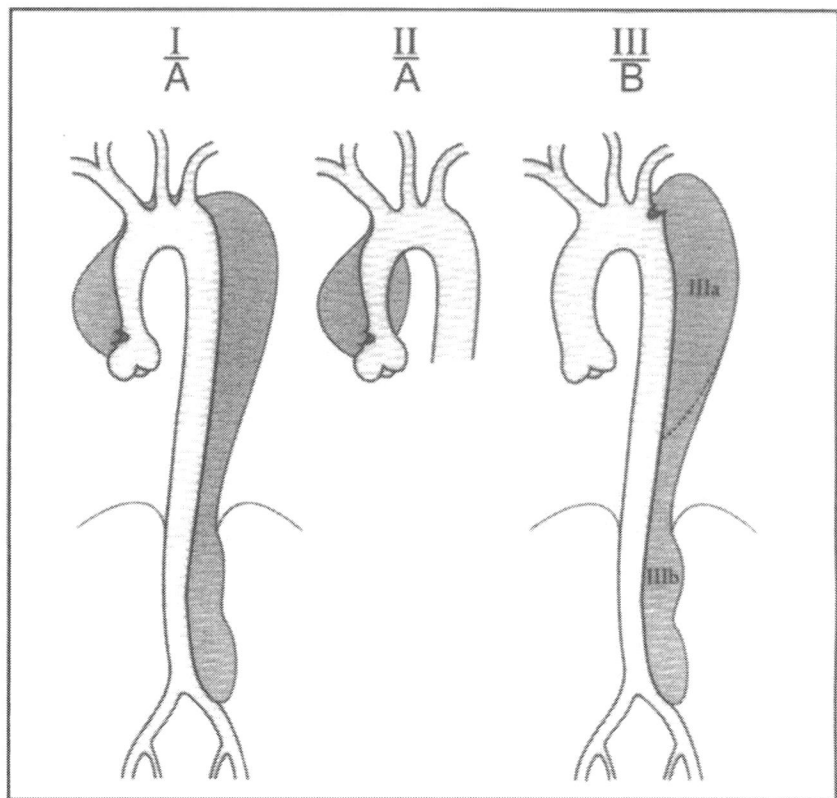

Figure 1. Commonly used classifications of aortic dissection. DeBakey: Types I, II and III. Stanford: Types A and B. Reprinted from Borst HG, Heinemann MK, Stone CD. Surgical treatment of aortic dissection. with permission from Elsevier, Inc. © 1991.

ascending aorta. Marfan patients with either acute DeBakey Type I or DeBakey Type II dissection require urgent operative intervention. The DeBakey Type III dissection commences in the descending thoracic aorta and may involve a limited segment of the descending thoracic aorta or all of this segment plus the abdominal aorta and even the iliac vessels. Most cardiac surgeons and cardiologists feel that it is not as imperative to routinely operate on patients with DeBakey Type III dissections. These dissections however, must be monitored carefully with serial CT scans or MRI studies. As long as the diameter of the descending thoracic aortic remains under 5 cm, these patients can be managed primarily by beta blocker medications and observation. If the overall diameter of the descending thoracic aorta approaches 5.0 cm, surgery should be considered. Also, patients with Type III dissection and visceral ischemia should undergo early operative intervention regardless of aortic diameter.

The Stanford classification of aortic dissection is preferred by some surgical groups. It should be noted that the Stanford Type A dissection includes both the DeBakey Type I and DeBakey Type II dissections (Fig. 1). The Type B Stanford dissection is identical to the DeBakey Type III dissection. Again, ascending aortic dissection requires immediate surgical correction, whereas dissection involving the descending thoracic aorta is best treated medically and may never require surgical intervention. In addition to monitoring these descending thoracic aortic dissections, beta blockers are given to lower blood pressure and reduce dp/dt.

Surgical Procedures

Bentall Composite Graft

As mentioned above, a significant advance in the surgical management of Marfan aortic root aneurysms occurred with the introduction of the Bentall composite graft procedure.[1] In the original Bentall procedure, the left and right coronary artery ostia were anastomosed side-to-side to "buttonhole" openings in the Dacron graft. In the early 1990s, surgeons abandoned the side-to-side coronary anastomosis in favor of wide mobilization and reimplantation of the two coronary arteries to the side of the Dacron graft (Fig. 2). With this modification, low-lying coronary ostium can be anastomosed without difficulty and late coronary dehiscence should not occur.

Valve-Sparing Root Replacement

Another advance in the treatment of Marfan aortic root aneurysms was the valve-sparing operation by Yacoub in 1994 (Fig. 3). In this procedure, the aneurysmal aortic root is resected and replaced with a Dacron tubular graft. The aortic valve is retained and the proximal end of the Dacron graft is trimmed to produce three separate tongues that serve as neo-sinuses and are sutured to the aortic valve annulus.

Advantages of the valve-sparing procedure are that a prosthetic valve is avoided, thereby eliminating the need for life-time anticoagulation (mechanical prosthesis) or reoperation for valve degeneration (bioprosthetic valve).

The results of the Yacoub or remodeling procedure have been good, but a small percentage of patients develop late aortic regurgitation because of splaying out of the three Dacron fabric tongues due to annular dilatation. Because of the potential development of aortic root dilatation with the Yacoub remodeling procedure, Tirone David and associates in Toronto introduced a procedure in 1992, in which the Dacron sleeve graft is brought down externally around the aortic valve and secured below the annulus.[5] This repair prevents late dilatation of the aortic root and is the so-called David I reimplantation technique (Fig. 4). This procedure provided excellent early results for Marfan patients, but some other surgical groups reported damage to the aortic leaflets from impingement by the non-compliant Dacron graft that lacked sinuses. Because of this rare complication, David introduced an operation that in many ways, resembles the Yacoub remodeling procedure, the David II procedure.[6] In recent years there have also been reports of aortic root dilatation with the David II remodeling procedure resulting in late aortic insufficiency. Because of this, David returned to the reimplantation procedure and to avoid the potential problem of leaflet damage, modified the sleeve graft by creating aortic sinuses in the graft.[7] This is achieved by plicating the graft at the levels of the aortic annulus and the sino-tubular junction. In a recent paper,[8] David reported that the eight year freedom from moderate/severe AI was 55% ± 6% with the remodeling procedure and 90% ± 6% with the reimplantation procedure. He also reported exclusive use of the reimplantation technique (with neo-aortic sinuses in the sleeve graft) over the last two years.

DePaulis of Rome recently introduced a commercial Dacron graft with "built-in" aortic sinuses that appears suitable for the David I reimplantation technique.[9]

Homograft Root Replacement

Before the availability of the valve-sparing procedure, cryopreserved homograft aortic roots were briefly employed in young children requiring aortic root replacement. Homografts had the advantage of avoiding anticoagulants. A paper from our institution, currently in press,[10] reports our experience in 10 children who received a cryopreserved homograft aortic root. Five of these patients required early operation because of severe mitral insufficiency; in fact, three of these children had simultaneous mitral valve repair at the time of homograft root replacement, and two had had an earlier separate mitral repair. Two of the five children having mitral valve surgery are currently 13 and 15 years after homograft root surgery and three children died 0.3,

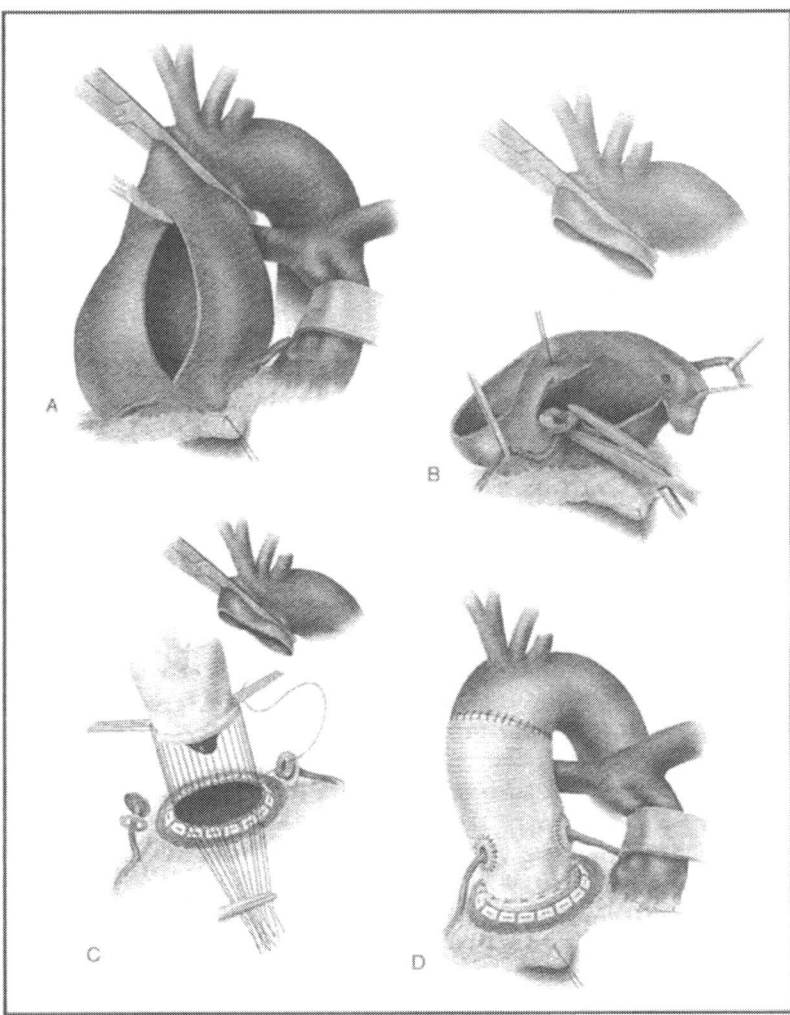

Figure 2. Technique of composite graft insertion. A) Aorta is cross-clamped and aneurysm opened. B) Coronary arteries are excised from the aneurysm wall. C) Sutures placed through aortic annulus and base of prosthesis. D) Completed repair after coronary arteries implanted to side of graft. Reprinted from Cameron DE. Surgical techniques-ascending aorta. Card Clin North Am 1999; 17:739-50 with permission from Elsevier, Inc. © 1999.

4 and 5 years after homograft replacement and mitral valve surgery. All five children in our series who received a homograft root but did not require mitral valve surgery, were NYHA Class I at study closure. Four of the children receiving a homograft root have required replacement of the homografts 5.7 to 7.2 years post-operatively (mean 6.4 years). All four excised homograft roots had moderate to heavy calcification of the aortic root. Thus, homograft aortic root replacement was effective for young children ages 3 to 7, but lacked durability, as all homografts were replaced within 7.2 years of the original surgery. With the availability of the valve-sparing procedures, use of homografts has diminished markedly.

Figure 3. A schematic drawing of the remodelling procedure. A Dacron graft is tailored to conform to the shape of the 3 aortic sinuses and then anastomosed to the aortic root. (From Schäfers, et al. Valve-preserving replacement of the ascending aorta: remodelling versus reimplantation. J Thorac Cardiovasc Surg 1998; 116:990-6. Reprinted with permission.)

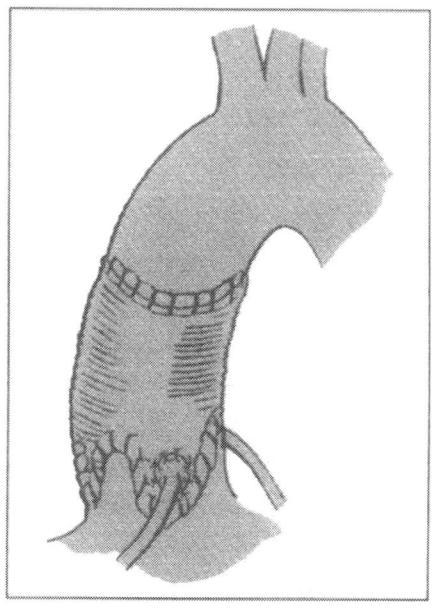

Figure 4. A schematic drawing of the reimplantation procedure. After mobilization of the aortic root, a Dacron graft is anchored to the aortoventricular junction. The native aortic valve is then resuspended within the vascular graft. (From Schäfers et al. Valve-sparing replacement of the ascending aorta: remodelling versus reimplantation. J Thorac Cardiovasc Surg 1998; 116:990-6. Reprinted with permission.)

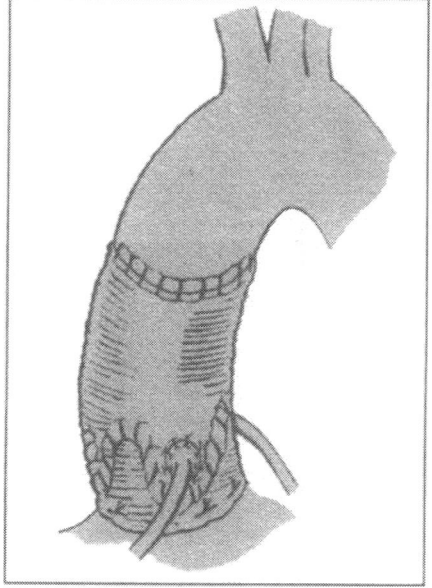

At the present time, we believe the valve-sparing procedure is the operation of choice for most Marfan patients requiring aortic root replacement. For those patients who are not candidates for the valve-sparing procedure because of significant aortic insufficiency, or those who wish to minimize need for reoperation surgery and are accepting of anticoagulants, the composite graft can be used with low early and late morbidity and mortality.

Table 1. Late mortality among 269 hospital survivors of aortic root replacement

Cause of Death	No. of Patients	Cause of Death	No. of Patients
Arrhythmia	8	Drug overdose*	2
Dissection/rupture of residual aorta	8	Dehiscence of coronary anastomoses	1
Endocarditis	3	Congestive heart failure	1
Systemic infection	3	Motor vehicle accident	1
Intracerebral /spinal hemorrhage	2	Unknown	12
Cancer	2		
		TOTAL	43(16%)

*This patient developed endocarditis 6 years post op and had a successful root replacement; patient died 18 months later of on overdose of illicit drugs. The homograft was free of infection. (From Gott VL, Cameron DE, Alejo DE. Aortic root replacement in 271 Marfan patients: a 24-year experience. Ann Thor Surg 2002; 72:438-43. Reprinted with permission of the Society of Thoracic Surgeons.)

Mitral Valve Surgery

In our 24 year experience with aortic root replacement in 271 Marfan patients, 57 or 21% required a mitral valve procedure.[2] Twenty-six of these patients had a concomitant mitral valve repair and 14 patients (all operated early in this series), had a concomitant mitral valve replacement. Four patients had a mitral repair as a procedure separate from aortic root replacement and 13 had mitral valve replacement at a time different than their aortic root surgery. The majority of our patients undergoing mitral valve replacement were operated on during the early years of this surgical series. At the current time, virtually all Marfan patients requiring mitral valve surgery for mitral insufficiency, may be managed with a mitral repair procedure as opposed to replacement. Virtually all patients with mitral insufficiency have annular dilatation and respond well to simple ring annuloplasty.

Long Term Results of Surgery

In the authors' personal series of 271 operated Marfan patients, 232 received a composite graft, 15 received a homograft, and 24 had a valve-sparing procedure.[2] Eighty-three percent of the patients in this series are currently alive. Table 1 depicts causes of late mortality in 43 deaths among the 269 hospital survivors. The two most common causes of late death were arrhythmia (8 patients) and dissections or rupture of the residual aorta (8 patients). Figure 5 shows the actuarial survival of the entire group of 271 patients.

Multiple patient and procedure-related variables were screened by univariate and subsequent multivariate analysis as risk factors for early or late death (Table 2). These variables included preoperative New York Heart Association Functional Class, urgent/elective operation, presence of aortic dissection, age, gender, and mitral valve operation. In multivariate analysis, only New York Heart Association Class and urgent operation emerged as significant independent predictors of early or late death. Age, gender and mitral valve surgery were not significant independent predictors of early or late death.

Major complications among the 269 hospital survivors are listed in Table 3. Thromboembolism was the most common late complication after aortic root replacement. One patient thrombosed a Bjork-Shiley valve 10 years postoperatively and had successful re-replacement. Fourteen patients developed emboli (13 cerebral and 1 peripheral); nine had complete recovery, but five had mild persistent incapacity. Freedom from thromboembolism was 96% at 5 years and 93% at 10 and 20 years.

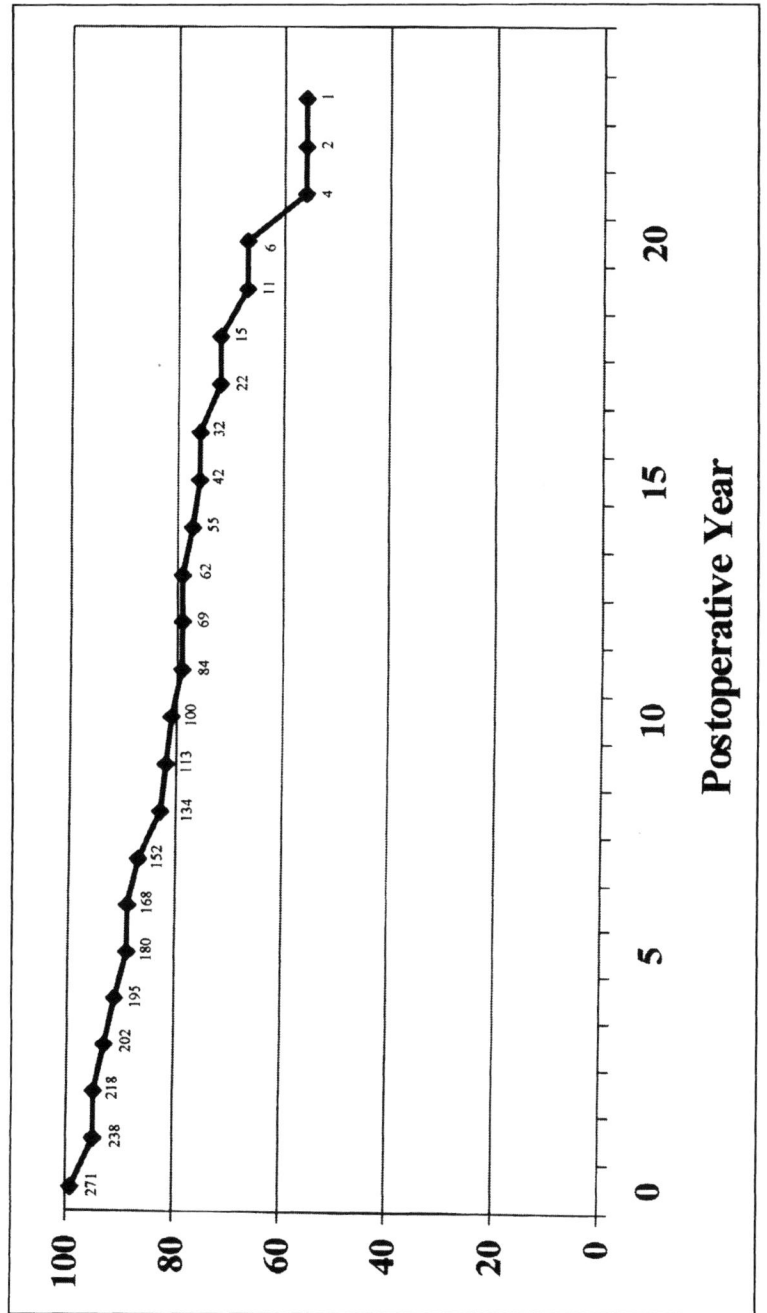

Figure 5. Actuarial survival of 271 Marfan patients after aortic root replacement. Survival was 89% at 5 years, 18% at 10 years, 76% at 15 years, and 67% at 20 years. Patients at risk are depicted at yearly intervals. (From Gott VL, Cameron DE, Alejo DE et al. Aortic root replacement in 271 Marfan patients: A 24 year experience. Ann Thorac Surg 2002; 72:438-43. Reprinted with permission of the Society of Thoracic Surgeons.)

Eleven of the 269 patients discharged from the hospital developed late endocarditis. Five were treated successfully with antibiotics and three were successfully operated on by re-replacement of the prosthetic valve with a homograft root. There were three late deaths from endocarditis that resulted without benefit of a homograft root. Freedom from endocarditis was 97% at 5 years, 94% at 10 years, 94% at 15 years and 90% at 20 years.

Table 2. Univariate and multivariate predictors of mortality

Risk factors	P value	
	Univariate	Multivariate
NYHA Class	0.022	0.017
Urgent surgery	0.001	0.007
Preop dissection	0.001	NS
Age	NS	NS
Gender	NS	NS
Mitral valve surgery	NS	NS

(NS= Not Significant). (From Gott VL, Cameron DE, Alejo DE. Aortic root replacement in 271 Marfan patients: a 24-year experience. Ann Thor Surg 2002; 72:438-43. Reprinted with permission of the Society of Thoracic Surgeons.)

Table 3. Late morbidity and mortality among 269 survivors of aortic root replacement

Complications and Outcome	No. of Patients	
Thromboembolism	15	
Cerebral emboli – complete recovery		8
Cerebral emboli – mild incapacity		5
Peripheral embolus – complete recovery		1
Thrombosis of B. S. valve*		1
Endocarditis	11	
Successful treatment		
Antibiotics		5
Homograft aortic root replacement		3**
Demise		
Antibiotics		3
Coronary dehiscense	3	
Successful repair		2
Demise		1
Total	29	

*Successfully replaced. B.S.= Bjork-Shiley **One of these patients died of drug overdose 18 months later. (From Gott VL, Cameron DE, Alejo DE. Aortic root replacement in 271 Marfan patients: a 24-year experience. Ann Thor Surg 2002; 72:438-43. Reprinted with permission of the Society of Thoracic Surgeons.)

Three patients developed late coronary dehiscence. Two were successfully repaired at one month and 2.2 years respectively. In one patient, there was sudden death seven years post-operatively. Interestingly all three patients with late coronary dehiscence were operated on before 1984. Since 1990, we have used the coronary-button technique and therefore do not anticipate further late coronary dehiscences.

In our series of 271 Marfan patients from the Johns Hopkins Hospital, 24 had a valve-sparing procedure. Figure 6 depicts the long-term results with this procedure in 14 adults and

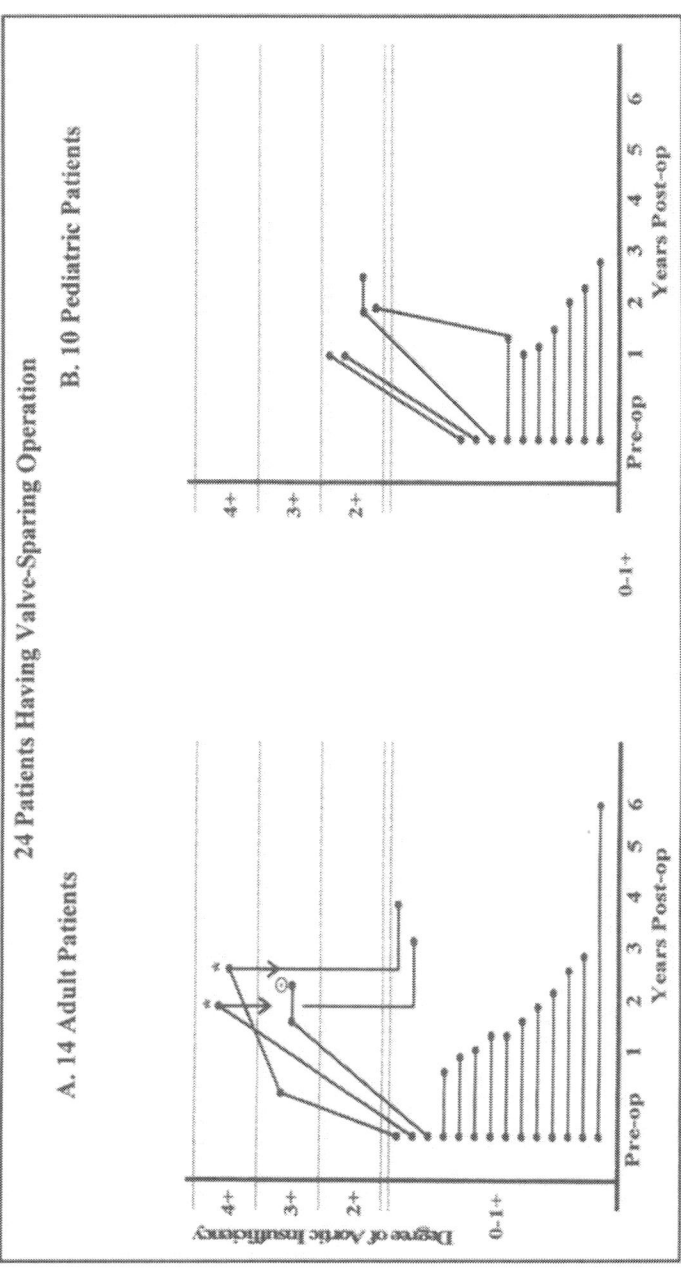

Figure 6. A and B) Postoperative echocardiography results in 24 patients having the valve-sparing operation. (*These two patients received a prosthetic valve at this time in their postoperative course. ⊙This patient is in New York Heart Association functional class II. All four children with late 2+ aortic insufficiency are New York Heart Association functional class I or II.) (From Gott VL, Cameron DE, Alejo DE et al. Aortic root experience in 271 Marfan patients: a 24 year experience. Ann Thorac Surg 2002; 73:438-43. Reprinted with permission of the Society of Thoracic Surgeons.)

10 children. Generally, late results of this operation have been good; 11 of the 14 adults have had excellent long-term outcomes. Three adults developed significant late aortic insufficiency; two have required valve replacement with a mechanical prosthesis. Six of the 10 children undergoing a valve-sparing have no progressive aortic insufficiency since operation. Four patients developed late mild to moderate aortic insufficiency. Subsequent to the publication of this paper, two of these children have required aortic valve replacement.

Late results in the series of 675 Marfan patients operated on in 10 centers worldwide, have similarly been encouraging.[3] There were 114 late deaths (> 30 days postop) among the 675 Marfan patients operated upon. Principal causes of late death were dissection or rupture of the residual aorta (22 patients) and arrhythmia (21 patients). Risk of death was greatest within the first 60 days after surgery, and then decreased rapidly to a constant level by the end of the first year.

Long Term Management Following Aortic Root Replacement

Following aortic root replacement, all patients should have CT or MRI studies. This is particularly true for patients who came to aortic root surgery with a DeBakey Type I or Type III dissection. For adults or children who have had a valve-sparing operation, echocardiograms are important to evaluate aortic root and aortic valve function every 6 months after surgery; if the aortic root appears stable, echocardiograms can be obtained every 12 months. For patients with a composite graft, INR monitoring is critical.

Another potential late complication after aortic root replacement is endocarditis. In our series of 271 patients 11 with a composite graft developed late endocarditis. Eight were treated successfully, five with antibiotics alone, and three with homograft aortic root replacement. Three of the 11 patients with endocarditis who were managed only with antibiotics died. If a patient with endocarditis does not improve within two to three weeks on antibiotics, homograft aortic root replacement is indicated.

In summary, elective aortic root replacement for Marfan patients can be performed with a low operative risk. Elective operative repair is recommended before the aortic root reaches 6 cms in diameter to minimize the risk of dissection and rupture. For those patients without significant aortic insufficiency, the operation of choice may be the David I reimplantation procedure with sinuses of Valsalva created in the tubular Dacron graft. All other patients can anticipate good long-term results with a composite graft.

Acknowledgements

The authors wish to thank Ms. Eileen Wright, Ms. Barbara Dobbs and Ms. Diane Alejo for their assistance in preparing this manuscript. This study was supported by the Broccoli Aortic Center at Johns Hopkins Medical Institution.

References

1. Bentall HH, DeBono A. A technique for complete replacement of the ascending aorta. Thorax 1968; 23:338-9.
2. Gott VL, Cameron DE, Alejo DE et al. Aortic root replacement in 271 Marfan patients: A 24 year experience. Ann Thorac Surg 2002; 73:438-43.
3. Gott VL, Greene PJ, Alejo DE et al. Replacement of the aortic root in patients with Marfan's syndrome. N Eng J Med 1999; 340:1207-13.
4. Sarsan MAI, Yacoub M. Remodeling of the aortic valve annulus. J Thorac Cardiovasc Surg 1993; 105:435-8.
5. David TE, Feindel CM. An aortic valve-sparing operation for patients with aortic incompetence and aneurysm of the ascending aorta. J Thorac Cardiovasc Surg 1992; 103:617-22.
6. David TE. Remodeling the aortic root and preservation of the native aortic valve. Oper Tech Cardiol Thorac Surg 1996; 1:44-50.
7. David TE. Aortic valve-sparing operations. Ann Thorac Surg 2002; 73:1029-30.
8. David TE, Ivonov J, Armstrong S et al. Aortic valve-sparing operations in patients with aneurysms of the aortic root and ascending aorta. Ann Thor Surg 2002; 74:S1758-61.
9. De Paulis R, De Matteis GM, Nordi P et al. One year appraisal of a new aortic root conduit with sinuses of Valsalva. J Thorac Cardioivasc Surg 2002; 123:33-39.
10. Cattaneo SM, Bethea BT, Alejo DE et al. Surgery for aortic root aneurysms in children: A 21 year experience in 50 patients. Ann Thorac Surg 2004; 77(1):168-76.

CHAPTER 6

Surgery for Cardiovascular Disorders in Marfan Syndrome:
The Atrioventricular Valves, Distal Aortic Segments and Myocardium

Roland Hetzer, Reinhard Pregla and Frank Barthel

The life expectancy of patients with Marfan syndrome is primarily limited by severe complications of the malformations of the aortic root and the ascending aorta typical of the disease. The characteristic structural defects of the aortic wall give rise to acute type A dissection, i.e., disruption of the intima and part of the media at the site of greatest distension in the area of the aortic root. This may lead to primary aortic rupture, which accounts for most of the lethal cases. Occasionally, the weakened though intact aortic wall or, more frequently, the site of "healed" local dissection may lead to direct rupture without dissection. Thus, the average life expectancy of patients with untreated Marfan syndrome is around 32 years, almost exclusively determined by such ascending aorta catastrophes.[1] Cardiovascular surgery for Marfan-associated disorders of the aortic valve and ascending aorta has been covered in Chapter 5 by Drs. Cameron and Gott.

Dr. Vincent Gott has acquired immense experience with Marfan aortic root surgery through his admirable efforts over several decades to follow these patients after operation. In this book he has shared with us his vast expertise. This leaves to us the less frequent lesions of the heart and the aorta which, nevertheless, are quite typical for Marfan patients and, although less dramatic, still have quite significant consequences. Our chapter will primarily cover Marfan-type pathology and surgical treatment of the atrioventricular valves, of the distal aortic segments, of the myocardium and the coronary arteries, of the specific features in childhood and of some chest wall deformities and other noncardiac abnormalities that may have relevance to cardiac operations.

The basis of our findings and the ensuing concepts has been our clinical experience with Marfan patients who were treated and followed during the period between 1991 and 2001 at the German Heart Institute Berlin. There were 353 Marfan patients with ages ranging from 6 to 69 years (mean 39.5 years); 246 were male, resulting in a male-to-female ratio of 2.3 to 1. Of these 353 patients we have operated on 252 (192 men, 60 women) with a total of 293 operations.

There were 157 cases of replacement of the ascending aorta, including 68 Bentall–DeBono procedures and 44 Cabrol procedures, 6 aortic-valve-sparing operations, and 6 homograft implantations. Thirty-nine patients had surgery on distal segments of the aorta (between arch and infrarenal segment). Eighty-two patients had surgery for mitral and twenty for tricuspid incompetence.

Marfan Syndrome: A Primer for Clinicians and Scientists, edited by Peter N. Robinson and Maurice Godfrey. ©2004 Eurekah.com and Kluwer Academic / Plenum Publishers.

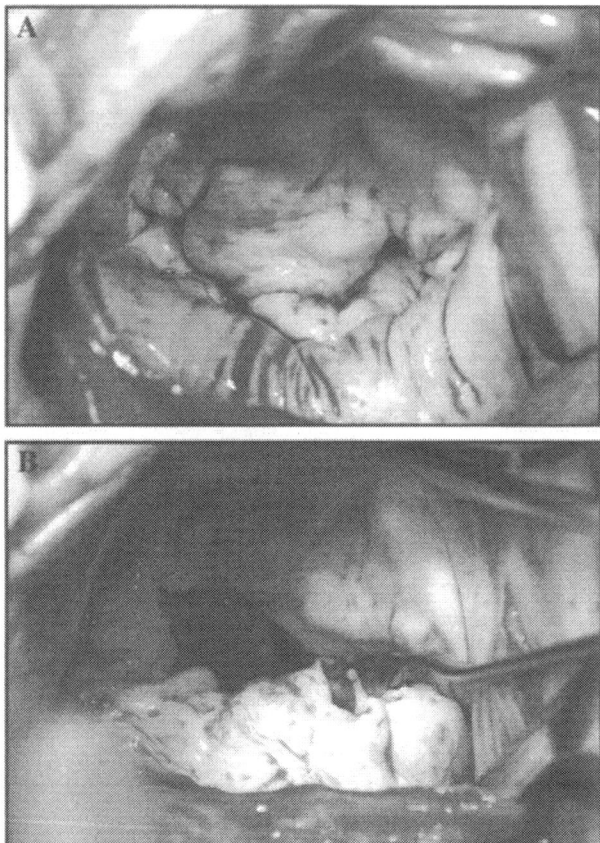

Figure 1. Intra-operative photographs of a Marfan-type mitral valve. The annulus is enormously distended and has a diameter of 6 cm (A). The leaflets are enlarged and show "billowing" (A). The chordae are elongated and there is rupture of chordae along the central scallop of the posterior leaflet (B). The valve is grossly incompetent.

We performed 9 heart transplantations in Marfan patients, 4 without and 5 with preceding mechanical support. Sixty-three patients had more than one operation, with up to six operations in two patients.

Atrioventricular Valves in Marfan Patients

Both atrioventricular (AV) valves show macroscopic and histological alterations characteristic of Marfan syndrome. Since it is the mitral valve which more frequently comes to clinical attention, the patho-anatomical terms of mitral malformation have also been applied to the Marfan valves. Typically the AV valves in Marfan patients have an enormously enlarged annulus with mitral valve orifice areas of 14, 16 and more cm^2 (with normal values being in the range of 9 cm^2) and mitral valve circumferences of as much as 17 cm or even more in the adult (normal ca. 9 cm). The leaflets are grossly enlarged, the posterior leaflet shows exaggerated scalloping and the chordae are thin and elongated. Thus, leaflet "billowing" and prolapse are the consequences, and the coaptation area between the leaflets is small (Fig. 1). This abnormal condition has found many names in clinical terminology, such as "billowing mitral valve",

Table 1. Frequency of mitral incompetence in 353 persons with Marfan syndrome followed at the German Heart Institute Berlin from 1991 to 2001

Patient Group		MI All Grades	MI I	MI II	MI III	MI IV
Marfan syndrome	353	135	45	13	62	15
		38.2%	12.7%	3.7%	17.6%	4.2%
Patients with MFS who underwent cardiovascular surgery	252	86	4	8	60	14
		34.1%	1.6%	3.2%	23.8%	5.6%
Patients with MFS who underwent mitral valve surgery	82	82	2	7	59	13
		100%	2.4%	8.5%	72%	15.9%

MFS = Marfan syndrome

"mitral valve prolapse syndrome", "floppy valve syndrome", and "Barlow's disease", among others.

The mitral leaflet tissue displays a characteristic histological appearance of so-called myxoid degeneration, with severe abnormalities of the extracellular matrix constituting the central fibrous layer of the leaflets. Large areas show a loss of fibrous tissue with fragmentation, coiling, and disruption of collagen bundles. Most destruction is centered around the sites of chordal insertion, and these destroyed areas usually contain pools of glycosaminoglycans and residual collagen strands.[2]

Quite frequently the annulus is calcified, most often along the central portion of the posterior annulus.[3] On echocardiography and in angiography, severe forms of leaflet prolapse may be seen. We have observed clinically significant mitral incompetence in 25.5% of our Marfan patients (grade II or more), and 38.5% of our patients have grade I or higher mitral incompetence (Table 1).

Amongst all Marfan patients who underwent any form of cardiovascular surgery at the German Heart Institute Berlin, the incidence of grade III or IV mitral insufficiency was somewhat higher than in the entire Marfan group including Marfan patients without surgery (Table 1). As expected, the incidence of grade III or IV mitral insufficiency was highest in those patients who received mitral valve surgery (Table 1).

In Marfan patients, mitral insufficiency is caused either by the dilatation of the annulus and resulting inability of the leaflet to co-apt, or as a consequence of chordal rupture. Although chordal rupture most likely occurs along the affected central scallop of the posterior leaflet, any part of the mitral apparatus may be subject to chordal rupture. In our group of 82 Marfan patients who underwent mitral valve surgery, we observed chordal rupture in 46% of cases (Table 2).

Less attention has been paid to the tricuspid valve in Marfan syndrome, presumably due to the less prominent clinical features of imperfect leaflet coaptation and extremely rare occurrence of chordal rupture on this valve. Twenty percent of our Marfan patients (72/353) had tricuspid insufficiency. However, only 5.7% (20/353) required tricuspid valve surgery due to clinically relevant tricuspid insufficiency. In all cases, tricuspid reconstruction was performed. We did not observe even one case of tricuspid chordal rupture.

Indications for Atrioventricular Valve Surgery

The indications for operations on the mitral valve in Marfan patients do not differ from those in other types of mitral diseases,[4] although there is still some controversy concerning the optimal surgical management of mitral valve lesions in patients with Marfan Syndrome.[5] Mitral incompetence either in the chronically progressive form or with acute chordal rupture

Table 2. *Chordal rupture in MFS patients receiving mitral valve surgery at the German Heart Institute Berlin from 1991 to 2001. Thirty seven out of 82 patients had some form of chordal rupture; the table shows a breakdown of anatomic location of chordal rupture among these patients.*

PML	AML	Both Leaflets	Anterior Commissure	Posterior Commissure
28/37	3/37	1/37	3/37	2/37
76%	8%	2%	8%	5%

PML = posterior mitral leaflet; AML = anterior mitral leaflet

constitutes the predominant clinical setting. Surgery is indicated when signs of heart failure, illustrated by decreased work capacity, tachycardia or even tachyarrhythmia become apparent and are accompanied by mitral insufficiency significantly higher than grade II by echocardiography. In this situation the left atrium is generally enlarged, except for cases of acute mitral incompetence. When mitral incompetence persists for longer periods of time, the left ventricle will also dilate, and in the later course of the disease chronic pulmonary hypertension will result.

The relative importance of mitral incompetence must be seen in context with aortic valve incompetence, the two being a frequent combination in Marfan patients. This raises the question of concomitant operation of a mild form of mitral incompetence when aortic root surgery is performed and vice versa. This question has not been conclusively answered.[6-8] In our own experience, 47.9% [169/353] of our Marfan patients received surgical treatment of the aortic root, of whom 75.7% [128/169] simultaneously received replacement of the aortic valve and the ascending aorta. A total of 28 patients underwent ascending aortic surgery only; 13 had only aortic valve surgery. Of the patients who received any aortic root surgery, 18.9% [32/169] had mitral insufficiency (grade I-III). In 56% of these cases [18/32], mitral surgery was performed simultaneously (MI III in 12 cases, MI II in 4 cases, MI I and floppy leaflets in one case each). The patients who did not receive simultaneous mitral surgery all had grade I mitral incompetence (except for one patient with grade II mitral incompetence). None of these patients have developed progression of MI requiring surgical treatment over a period of 1 to 13 years (mean 6.2 years).

Since, however, most of the Marfan patients are in a younger age group we have recently adopted a more liberal attitude towards combined surgery when either the mitral valve or aortic root displays a clear indication for surgery and the other only mild dysfunction.

Techniques of Surgery

Both classical approaches to mitral valve surgery, i.e., prosthetic replacement and valve repair, may be indicated for the Marfan patient.[9,10] The choice of the individual procedure is determined principally by the patient's own preference after being informed of the individual pathological findings, the specific long-term consequences of different prothesis types and of repair, and the experience of the surgeon with the procedures. Replacement with a prosthetic valve minimizes the risk of later re-operation, but involves the need of life-long anticoagulant treatment, which may be hazardous in the face of the probability of subsequent complications and surgery on various parts of the aorta. Replacement with xenograft valves may avoid these hazards; however, since most of the Marfan patients undergo operation at a relatively young age, xenograft degeneration and re-operation must be expected within 8-15 years at most.

Various types of valve repair have become an attractive alternative, but, later re-operation is not ruled out, in particular when the primary repair has a suboptimal result.

In the past, Marfan mitral valves were considered surgically difficult due to the friability of the leaflet and annulus tissue and the often enormous discrepancy between the natural annulus size and the size of the available prostheses or size of the repaired valve. At present, however, the experience acquired with Marfan valves allows safe surgery, avoiding any leaks or rupture. Mitral valve replacement nowadays is uniformly performed with preservation of the subvalvular apparatus as this has been recognized to be highly important for left ventricular function.[9,11,12]

Apart from the easier technique of preserving the chordal integrity of the posterior leaflet alone, several techniques have been introduced to preserve the entire apparatus, including attachment of the anterior leaflet chordae to the trigona,[13] attachment of the anterior leaflet chordae to the posterior annulus,[14] attachment of the anterior leaflet chordae to the anterior annulus,[15,16] radial incisions at the commissures and mid-leaflets,[17] and radial incisions and reefing of the leaflets to the annulus.[17] In Marfan mitral valves, great care should be taken not to allow retained leaflet tissue to impair prosthetic valve function or to create subaortic obstruction. A generous number of pledgeted sutures should be applied in order to ensure solid prosthetic seating when reducing the large natural annulus to the size of the prosthetic sewing ring.

Valve repair follows the principle of re-establishing a large enough coaptation area between the anterior and the posterior leaflet. With pure annular dilatation and otherwise intact chordae, this may be achieved by shortening the posterior annulus with either suture techniques or by implantation of a prosthetic ring. Great care must be taken in the excessively large Marfan leaflets not to shorten the annulus excessively in order to avoid systolic anterior movement (SAM) of the anterior leaflet, thus creating subaortic stenosis.

With concomitant chordal rupture, the respective segment may be resected and the resection edges brought together with sutures or the segment may be plicated towards the ventricle. Additional annular shortening with a suture or ring is obligatory. Our approach has been the continuous suture annular shortening system, modified after the original "Paneth" technique which includes anchoring of a double running suture at both trigona and shortening towards the center of the posterior annulus to a degree that the remaining valve orifice has a 2.3 to 2.5 cm diameter and the anterior leaflet fills the opening completely (Fig. 2).[18]

In the case of posterior leaflet chordal rupture we have continued to apply a modified "Gerbode" plasty,[18] plicating the posterior leaflet towards the ventricular cavity by a line of re-enforced sutures perpendicular to the annulus. Additional annulus stabilization is required (Fig. 3).

We fasten a strip of autologous pericardium to the shortened annulus to improve stability. In very large valves with excessive tissue, some parts of the leaflets may be plicated or resected individually in order to obtain an acceptable valve. As a rule, whenever no ideal repair results can be obtained, we prefer valve replacement.

In the rare instance of tricuspid incompetence of a significant degree, repair should be performed. Twenty patients with signs of Marfan syndrome and isolated tricuspid insufficiency were operated on in our institution. Invariably the pathology displayed an extremely dilated annulus but intact leaflets and chordae, some with signs of mild myxoid degeneration. Tricuspid repair was performed in all cases with our modification of the DeVega type plasty (Fig. 4A,B).[19]

Surgery of the Distal Aortic Segments

Non-Dissecting Aneurysms

Aneurysm formation without dissection may occur at any site of the aorta distal to the ascending part. However, the incidence of such distal aneurysms is much lower than in the

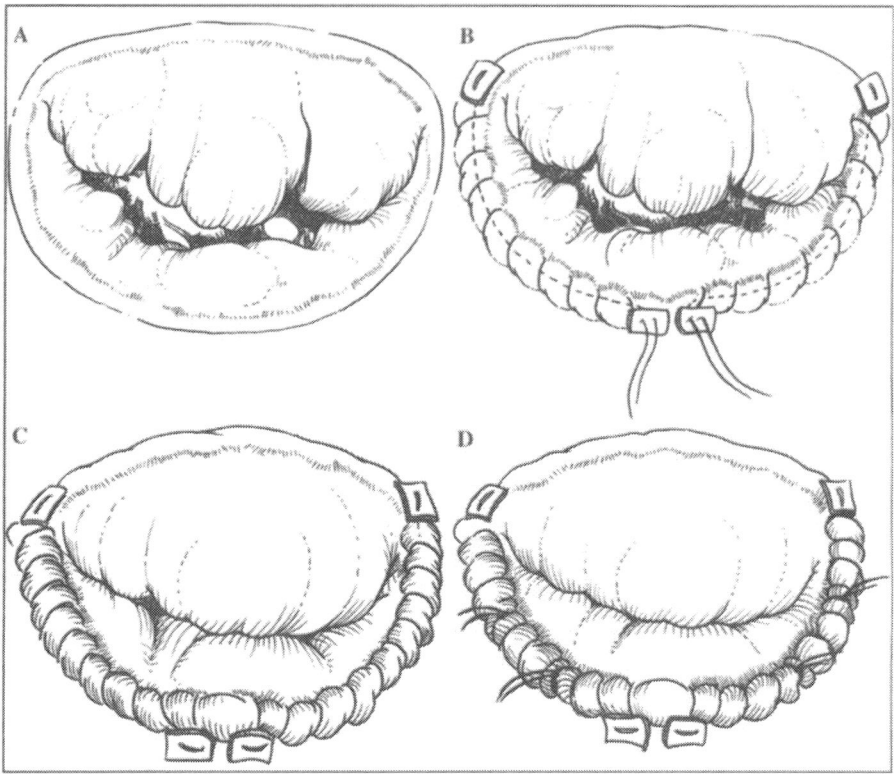

Figure 2. The modified "Paneth" plasty is applied in cases of mitral incompetence with annular dilatation and billowing leaflets, but without chordal rupture (A). Two running sutures are anchored with pledgets at both trigona and the posterior annulus is shortened by the sutures whereby the first is run along the annulus and the second around the first and the annular tissue (B). The sutures are tied over pledgets in order to reduce the orifice to an appropriate size, resulting in a competent valve without systolic anterior movement of the anterior leaflet (C). Additional securing sutures may be added to the running suture (D) or even a strip of autologous pericardium may be sewn onto the shortened annulus.

ascending aorta.[20] In individual cases of multilocular non-dissecting aortic aneurysms repeat interventions on several aortic segments may become necessary[21] (Fig. 5).

Second in frequency to the aortic root aneurysm we have observed infra-renal aneurysms that reach from directly distal to the origin of the renal arteries to the bifurcation and show either a fusiform or a reverse pear-like shape (Fig. 6). They may grow to considerable diameter without rupture, but at operation we have observed horizontal intimal tears which could have become the origin of dissection or could have propagated to rupture. Such tears were accompanied by new-onset abdominal pain. Considering the extremely thin wall and friable tissue such pain must be considered an indication for immediate surgery. The technique of surgery follows the standard of infra-renal aortic prosthetic replacement via a transperitoneal approach and using the graft inclusion technique.[22] Whereas now we treat many of the atherosclerotic infra-renal aneurysms with the transfemoral stent graft replacement technique, there has been hesitation about application of this technique in Marfan patients because of concern as to the limited stability of the Marfan aortic wall for solid anchoring of the stents at both ends.

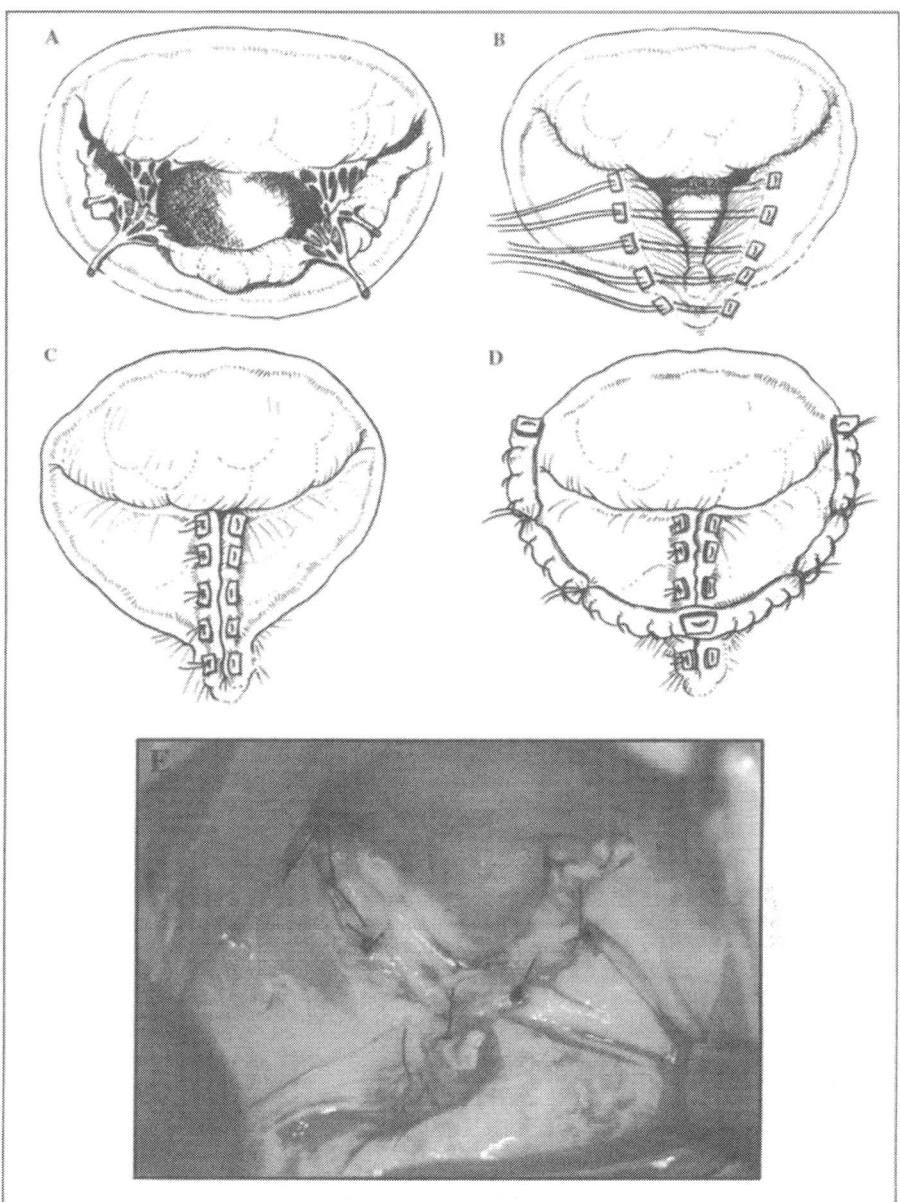

Figure 3. A modified "Gerbode" plasty is applied in cases of chordal rupture along the central scallop of the posterior leaflet which is the most frequent rupture site (A). The segment where the chordae are ruptured is plicated towards the ventricular cavity by a line of mattress sutures pledgeted with pieces of autologous pericardium (B). This procedure eliminates both the prolapse of the "flail" segment and at the same time reduces the posterior annulus and the valve opening area (C). A strip of autologous pericardium is sutured onto the posterior annulus and anchored at both trigona in order to re-enforce the shortened posterior annulus and to neutralize strain (D). Photograph (E) shows completed repair (same case as shown in Fig. 1).

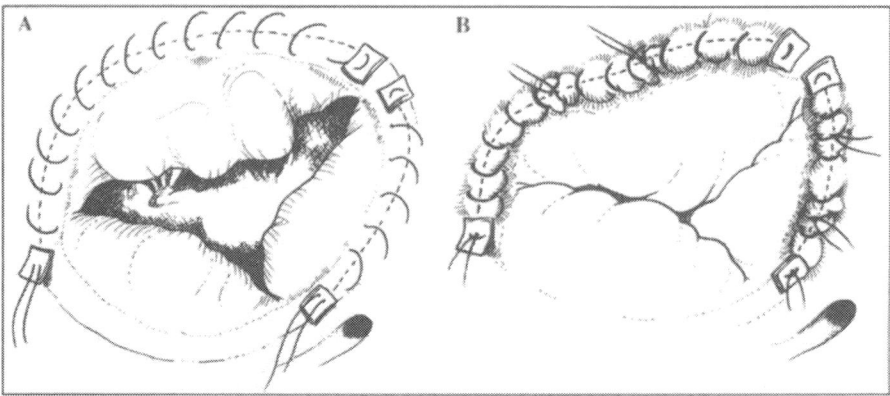

Figure 4. The preferred technique of tricuspid valve repair in all cases of annular dilatation and intact leaflets and chordae has been a modification of the DeVega principle. Two continuous sutures are anchored at the commissure between anterior and posterior leaflets with pledgets. One suture is run along the annulus and the second is stitched in an encircling mode around the first suture (A). Both the posterior and the anterior annulus are shortened by tying those sutures in a separate way, thus allowing individual modulation of the final valve orifice (B).

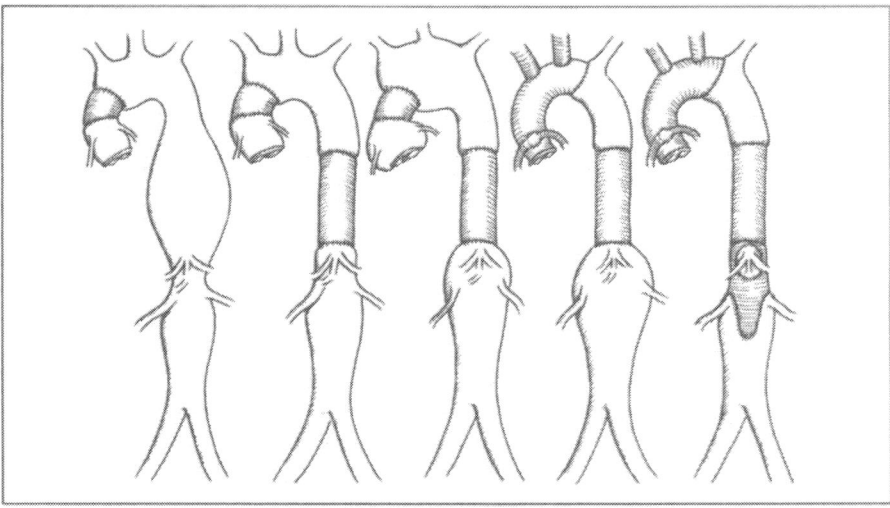

Figure 5. Case of a now 40 year old Marfan patient who has undergone six operations. Closure of persistent ductus arteriosus was performed at age one year and separate aortic valve replacement (Bjork-Shiley) and ascending replacement had been performed elsewhere at age 23. At age 28 he presented with a fusiform distal descending aneurysm which was replaced. Two years later he showed increasing aneurysms of the aortic root and the entire aortic arch. Composite aortic valve (St. Jude Medical), ascending and arch replacement according to Cabrol and implantation of prosthetic side arms to the supra-aortic branches was performed in deep hypothermia. One year later, aneurysmal dilatation of the infradiaphragmatic aorta had occurred which required the next step of aortic segment replacement. During the later years the patient showed signs of progressive heart failure with dilatated cardiomyopathy and in 2001 he underwent orthotopic heart transplantation at our institution with an uneventful course since then.

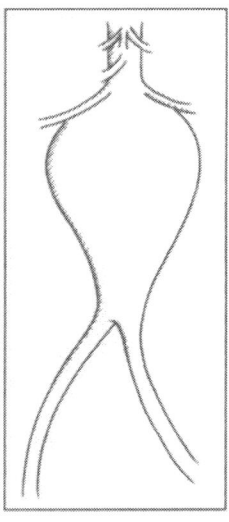

Figure 6. Typical shape of infrarenal aortic aneurysm in the Marfan aorta, either fusiform or "reverse pear"-like. Characteristically the neighboring aortic and iliac segments are rather thin in caliber and the aortic wall is extremely thin and friable.

Figure 7. Typical configuration of aneurysms of the false lumen in persistent dissection distal to the replaced ascending aorta in erstwhile acute type A dissection. The largest aneurysm diameter is most often seen at the junction between arch and descending aorta. In general, however, large aneurysms may develop at any site. The true lumen is compressed and thin in caliber. Intestinal, and in particular renal arteries may arise from either lumen. The surgical concept includes consecutive segmental aortic replacement in the order of the most prominent and most threatening aneurysm at a given time. As a rule, distal perfusion should be allowed to both lumina.

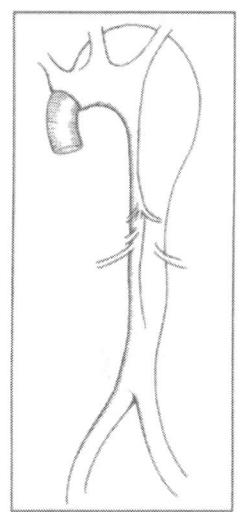

Distal (Type B) Dissections and Aneurysm Formation in the Distal Aorta Following Ascending Replacement in Type A Dissection

Whereas type A aortic dissection with the entry just above the coronary ostia is the most frequent cardiovascular pathology in Marfan patients, primary aortic dissection beginning distal to the ascending aorta (type B) is rare in these patients.[23]

The most frequent involvement of the distal aorta is observed in the type A dissections where the ascending aorta has been replaced and the distal aorta remains dissected. These distal segments, i.e., the false lumen, may become aneurysmal, whereas often the true lumen becomes rather compressed and thin. The preferred site of aneurysm growth is the very proximal part of the descending aorta (Fig. 7); however, any other segment of the dissected aorta may become aneurysmal and may rupture.

In our experience, 63 Marfan patients (25% of all Marfan patients operated on) had more than one operation on the aorta, 54 had two, six had three, one had four and two had more than four.

Close surveillance of Marfan patients after successful surgical treatment of the ascending aortic segment thus becomes very important in the follow-up period.

Therefore, potentially every Marfan patient who has suffered acute dissection and had ascending aorta replacement may face further operations on the distal aortic segments, and in some patients more or less the entire aorta may need to be replaced by prosthetic graft in several steps. The sequence of intervention is dictated by the size and growth of the most threatening aneurysm formation. If possible, the steps should be performed going from proximal to distal, always performing the distal graft anastomosis on the entire aortic circumference, thus leaving both lumina perfused in downstream dissection. In this way all arterial branches will be perfused regardless of whether they arise from the true or the false lumen.

In the case of arch and proximal descending aorta replacement and in any case of long segment replacement we prefer deep hypothermia and circulatory arrest, which allows the best view of the intra-luminal pathology of the aorta, construction of an optimal replacement and inspection of the large arterial branches. With this technique neurological damage to the brain or spinal cord has been avoided in all of our Marfan patients so far.

Cardiomyopathy in Marfan Patients

A small number of Marfan patients have developed myocardial failure[21,24,25] with dilated cardiomyopathy, in some primarily and in a certain proportion in those who had undergone one or several operations on the aorta. It was assumed that myocardial ischemia and possible afterload increase during operation may be less well tolerated by the Marfan hearts and may lead to myopathy later on, as in the case of the patient shown in Figure 5. Structural abnormalities in the myocardium of Marfan children were described by Savolainen in 1994.[26]

This assumption was also underlined by the finding of very thin-caliber and loosely embedded coronary arteries typical for Marfan hearts and a highly abnormal capacity of the coronary arteries to respond to variations in blood flow and wall sheer stress. We found a strongly diminished flow-dependent vascular dilatation which was not accompanied by an increase in cyclic-GMP concentration in the vascular smooth muscle cells which normally occurs (Siegel, personal communication, 2003).

There are further indications of abnormalities of the coronary arteries in Marfan syndrome such as the finding of a separate ostium of the conus artery from the right aortic sinus just left of the right coronary ostium. This abnormality has been seen in 9% of our patients who underwent any kind of surgery on the aortic root.

So far, 12 patients developed dilated cardiomyopathy primarily and 3 following previous aortic operations. Mechanical circulatory support as a bridge-to-transplantation was required for five patients, and nine patients were transplanted.

Heart transplantation in Marfan patients has remained a topic of dispute since it was assumed that the immunosuppressive treatment, in particular with steroids, may promote aneurysm formation in the already diseased or dissected aortic segments. This has not been our experience so far and we do not hesitate to transplant Marfan patients in advanced heart failure, if needed.

Cardiovascular Disease in Children with Marfan Syndrome

The typical manifestations of Marfan syndrome in the heart and the aorta have been observed as early as in infancy.[27,28] Most of our pediatric cases are young adolescents, mostly in Marfan families where up to four children were affected by aortic root dilatation, aortic dissection and mitral incompetence.

While medical treatment with beta blockers is routine[29] the decision when to replace the aortic root has been disputed. The indication for surgery is clear when significant aortic and/or mitral incompetence is present. In competent aortic valves we would recommend surgery on the aortic root when this is dilated to twice the diameter of the distal ascending aorta or the descending aorta or when a rapid progression of aortic root diameter is observed.

In the children, valve-sparing operations were performed as a preference, still assuming that prosthetic replacement will most likely become necessary later on, some time in early adult life.[30]

The preponderance of mitral incompetence as the primary finding of cardiovascular disease is more frequent in the pediatric group.[31] Mitral repair is the preferred technique, also to allow for growth and to avoid the side effects of prosthesis which are a greater burden in childhood than in adults.

Our own experience with Marfan children includes 22 patients who underwent surgery at ages between 4 and 17 years, with a mean of 13 years. There were 14 patients who received ascending aortic replacement, 7 with mechanical valve conduits, and 7 with valve-sparing operations. Four children underwent mitral valve repair and one of these had also composite ascending aortic replacement.

Noncardiac Marfan Anomalies Affecting Cardiac Surgery Operations

Extensive chest wall deformities frequently associated with Marfan Syndrome such as pectus excavatum or pectus carinatum may become important when surgery on cardiac structures must be performed.

Simultaneous correction of such deformities may be unavoidable in order to allow sufficient exposure of the heart and is often requested by patients for cosmetic reasons. It has been our policy to perform such chest wall corrections at the time of cardiovascular surgery.

Spontaneous pneumothorax is not uncommon in Marfan syndrome[32,33] and may occur following cardiovascular surgery and even short-term mechanical ventilation, which may prolong post-operative care, and extended periods of chest drainage may be required. In our clinic we have seen 8 cases of spontaneous pneumothorax during the period 1991-2001. Four of these patients had Marfan syndrome. Two other Marfan patients had severe bullous pulmonary emphysema.

References

1. Murdoch JL, Walker BA, Halpern BL et al. Life expectancy and causes of death in the Marfan syndrome. N Engl J Med 1972; 286(15):804-808.
2. Becker AE, Davies MJ. Pathomorphology of mitral valve prolapse. In: Boudoulas H, Wooley CF, eds. Mitral Valve: Floppy Mitral Valve, Mitral Valve Prolapse, Mitral Valve Regurgitation. Armonk: Futura Publishing Company; 2000:91-114.
3. Roberts WC, Honig HS. The spectrum of cardiovascular disease in the Marfan syndrome: A clinico-morphologic study of 18 necropsy patients and comparison to 151 previously reported necropsy patients. Am Heart J 1982; 104(1):115-135.
4. Gillinov AM, Hulyalkar A, Cameron DE et al. Mitral valve operation in patients with the Marfan syndrome. J Thorac Cardiovasc Surg 1994; 107(3):724-731.
5. Fuzellier JF, Chauvaud SM, Fornes P et al. Surgical management of mitral regurgitation associated with Marfan's syndrome. Ann Thorac Surg 1998; 66(1):68-72.
6. Gott VL, Gillinov AM, Pyeritz RE et al. Aortic root replacement. Risk factor analysis of a seventeen-year experience with 270 patients. J Thorac Cardiovasc Surg 1995; 109(3):536-544; discussion 544-535.
7. Baumgartner WA, Cameron DE, Redmond JM et al. Operative management of Marfan syndrome: The Johns Hopkins experience. Ann Thorac Surg 1999; 67(6):1859-1860; discussion 1868-1870.
8. Pasic M, von Segesser L, Carrel T et al. Surgical treatment of cardiovascular complications in Marfan syndrome: a 27-year experience. Eur J Cardiothorac Surg 1992; 6(3):149-154; discussion 155.
9. Hetzer R, Bougioukas G, Franz M et al. Mitral valve replacement with preservation of papillary muscles and chordae tendineae — revival of a seemingly forgotten concept. I. Preliminary clinical report. Thorac Cardiovasc Surg 1983; 31(5):291-296.

10. Hetzer R, Drews T. Mitral valve replacement. In: Franco KL, Verrier ED, eds. Advanced Therapy in Cardiac Surgery. Hamilton, London, St. Louis: B.C.Decker; 1999:232-244.
11. David TE, Uden DE, Strauss HD. The importance of the mitral apparatus in left ventricular function after correction of mitral regurgitation. Circulation 1983; 68(3 Pt 2):II76-82.
12. Lillehei CW, Morris MD, Levy MJ et al. Mitral valve replacement with preservation of papillary muscles and chordae tendineae. J Thorac Cardiovasc Surg 1964; 532-543.
13. Miki S, Kusuhara K, Ueda Y et al. Mitral valve replacement with preservation of chordae tendineae and papillary muscles. Ann Thorac Surg 1988; 45(1):28-34.
14. Feikes HL, Daugharthy JB, Perry JE et al. Preservation of all chordae tendineae and papillary muscle during mitral valve replacement with a tilting disc valve. J Card Surg 1990; 5(2):81-85.
15. Rose EA, Oz MC. Preservation of anterior leaflet chordae tendineae during mitral valve replacement. Ann Thorac Surg 1994; 57(3):768-769.
16. Straub U, Feindt P, Huwer H et al. Mitral valve replacement with preservation of the subvalvular structures where possible: an echocardiographic and clinical comparison with cases where preservation was not possible. Surgical technique and early postoperative course. Thorac Cardiovasc Surg 1994; 42(1):2-8.
17. Hetzer R, Drews T, Siniawski H et al. Preservation of papillary muscles and chordae during mitral valve replacement: possibilities and limitations. J Heart Valve Dis 1995; 4 Suppl 2:S115-123.
18. Komoda T, Hubler M, Siniawski H et al. Annular stabilization in mitral repair without a prosthetic ring. J Heart Valve Dis 2000; 9(6):776-782.
19. Hetzer R. Herzchirurgie, Kirschnersche allgemeine und spezielle Operationslehre. Vol VI, Teil 2. zweite Auflage ed. Heidelberg / New York: Springer; 1991.
20. Coselli JS, Büket S. Marfan Syndrome: The variability of operative management. In: Hetzer R, Gehle P, Ennker J, eds. Cardiovascular Aspects of Marfan Syndrome. Darmstadt: Steinkopff; 1995.
21. Hetzer R, Gehle P, Ennker J. Results of cardiovascular surgery for Marfan syndrome in Berlin. In: Hetzer R, Gehle P, Ennker J, eds. Cardiovascular Aspects of Marfan Syndrome. Darmstadt: Steinkopff; 1995:109-118.
22. Crawford ES. Thoraco-abdominal and abdominal aortic aneurysms involving renal, superior mesenteric, celiac arteries. Ann Surg 1974; 179(5):763-772.
23. Pruzinsky MS, Katz NM, Green CE et al. Isolated descending thoracic aortic aneurysm in Marfan's syndrome. Am J Cardiol 1988; 61(13):1159-1160.
24. Kesler KA, Hanosh JJ, O'Donnell J et al. Heart transplantation in patients with Marfan's syndrome: a survey of attitudes and results. J Heart Lung Transplant 1994; 13(5):899-904.
25. Mullen JC, Lemermeyer G, Bentley MJ. Recurrent aortic dissection after orthotopic heart transplantation. Ann Thorac Surg 1996; 62(6):1830-1831.
26. Savolainen A, Nisula L, Keto P et al. Left ventricular function in children with the Marfan syndrome. Eur Heart J 1994; 15(5):625-630.
27. Geva T, Hegesh J, Frand M et al. The clinical course and echocardiographic features of Marfan's syndrome in childhood. Cardiac manifestations of Marfan syndrome in infancy and childhood. Natural history of cardiovascular manifestations in Marfan syndrome. Am J Dis Child 1987; 141(11):1179-1182.
28. van Karnebeek CD, Naeff MS, Mulder BJ et al. Natural history of cardiovascular manifestations in Marfan syndrome. Arch Dis Child 2001; 84(2):129-137.
29. Shores J, Berger KR, Murphy EA et al. Progression of aortic dilatation and the benefit of long-term beta-adrenergic blockade in Marfan's syndrome. N Engl J Med 1994; 330(19):1335-1341.
30. Carrel T, Berdat P, Pavlovic M et al. Surgery of the dilated aortic root and ascending aorta in pediatric patients: techniques and results. Eur J Cardiothorac Surg 2003; 24(2):249-254.
31. Phornphutkul C, Rosenthal A, Nadas AS et al. Cardiac manifestations of Marfan syndrome in infancy and childhood. Natural history of cardiovascular manifestations in Marfan syndrome. Circulation 1973; 47(3):587-596.
32. Sensenig DM, LaMarche P. Marfan's syndrome and spontaneous pneumothorax. Am J Surg 1980; 139(4):602-604.
33. Hall JR, Pyeritz RE, Dudgeon DL et al. Pneumothorax in the Marfan syndrome: prevalence and therapy. Ann Thorac Surg 1984; 37(6):500-504.

CHAPTER 7

Mutation Analysis of the *FBN1* Gene in Individuals with Marfan Syndrome:
Sensitivity, Methods, Clinical Indications

Anne De Paepe, Bart Loeys and Paul Coucke

Since the discovery of the *FBN1* gene as the gene responsible for the Marfan syndrome (MFS), molecular testing for this condition has become possible. Although MFS is a clinical diagnosis, certain situations may occur in which molecular analysis of the *FBN1* gene is wanted, either for diagnostic, management or genetic counseling purposes.

The clinical diagnosis can be challenging in children not yet fulfilling the diagnostic criteria or in instances of variable expressivity of the condition. In addition, molecular testing allows one to offer prenatal and preimplantation genetic diagnosis, an option for which demands are steadily increasing.

Nevertheless, molecular analysis is at present a laborious and expensive analysis which cannot be offered on a routine basis. The efficiency of mutation detection in the *FBN1* gene depends on a number of factors, such as the type of mutation or substrate, the technique used and —above all— the accuracy of the clinical diagnosis. Today, mutation screening techniques allow a mutation detection rate approaching 90%, providing the clinical selection is made appropriately.

Diagnosis of the Marfan Syndrome: Interpretation and Limitations of the Ghent Nosology

One of the principal characteristics of the Marfan syndrome (MFS) is the pleiotropic nature of its clinical manifestations and its great phenotypic variability, both within and between families. These are important aspects to consider in the diagnostic workup of a potential Marfan patient. The diagnosis of MFS is based primarily on clinical criteria and requires a combination of "major" and "minor" clinical manifestations in different organ systems. The four major manifestations of the MFS are ectopia lentis, dissection of the ascending aorta, dural ectasia and presence of at least four of eight manifestations in the skeletal system. The simultaneous occurrence of at least two of these major manifestations is highly specific for the condition, although each one can be found separately as a significant feature in certain other conditions. Minor manifestations are numerous but less specific, because they are commonly found in many other conditions as well as in the general population. The presence of minor criteria only is insufficient to establish the clinical diagnosis of MFS, but indicates that a particular body system is "involved". According to the current nosological criteria (Ghent nosology, 1996)[1] the diagnosis of MFS in a sporadic patient requires the presence of a major manifestation in at least two different organ systems and the involvement of a third organ system. In familial instances, a positive family history, which means the presence of a relative who independently fulfils the diagnosis of MFS, is considered to be a major diagnostic criterion. The diagnosis of MFS in a

Marfan Syndrome: A Primer for Clinicians and Scientists, edited by Peter N. Robinson and Maurice Godfrey. ©2004 Eurekah.com and Kluwer Academic / Plenum Publishers.

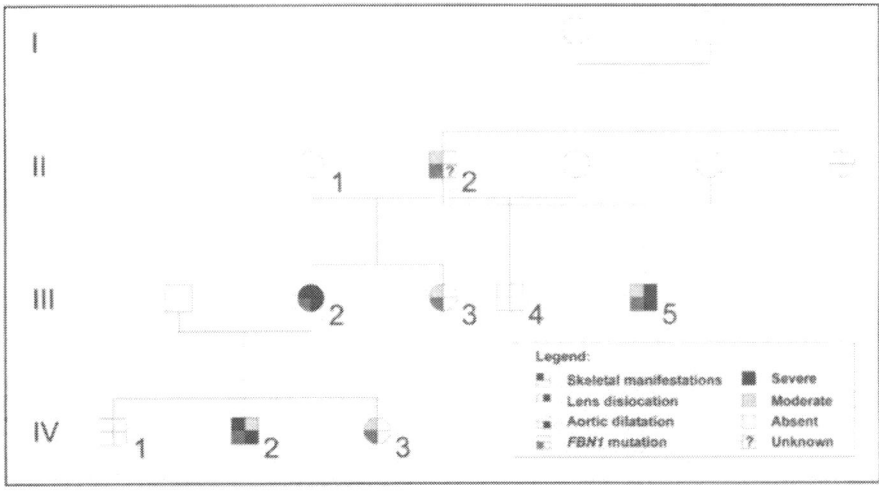

Figure 1. A 28 year old mother (III2) visited the genetic clinic because of suspected MFS in her 4 year old son (IV2), who presented with tall stature, slender asthenic build and mild developmental delay. She had been treated in childhood for bilateral lens dislocation but was never evaluated for MFS. Upon physical examination, she presented several skeletal and facial characteristics of MFS. Additional investigations revealed a dilated aortic root in both mother and son and a subluxation of the lenses in the boy. The 3 year old daughter (IV3) of the proband showed some mild skeletal characteristics but otherwise the family history was reportly negative for MFS. Mutational studies identified a premature truncation mutation (PTM) in the *FBN1* gene (R1596X)) in mother and both her children. Moreover, the mutation was detected in the mother's younger sister (III3), who showed only mild facial and minimal skeletal characteristics, in a half-brother (III5) who was subsequently diagnosed with lens subluxation, aortic root dilatation and prominent dural ectasia, and in the proband's father (II2).

relative then requires the presence of one additional major manifestation in one organ system and involvement of a second organ system.

A correct and timely diagnosis of MFS is important in view of the great progress in clinical management of MFS patients over the last decade, especially with respect to the cardiovascular complications. The medical and surgical treatments which are now available have significantly improved the life expectancy in MFS. This should encourage an early diagnosis of the condition in order to start timely preventive and therapeutic measures.

In most instances, the diagnostic criteria as defined in the Ghent nosology allow the establishment or exclusion of the diagnosis of MFS on clinical grounds. The interpretation of these criteria requires some flexibility. This is not always obvious for a number of reasons. Firstly, the condition is known for its extensive phenotypic variability both within and between families (see example of intrafamilial variability in Fig. 1). This may cause underdiagnosis of the condition.[2] Secondly, establishing a diagnosis of MFS in children can be difficult because several manifestations of MFS are age-dependent and may not yet be present in childhood. Thirdly, clinical overlap exists between the MFS and other, so-called "Marfan-like" conditions, which share some of the features of MFS but do not necessarily have the same outcome. Differentiation between one of those conditions and plain variability in expression of the true MFS in a sporadic individual can be difficult at first consultation and may only become obvious over time.

Methodology for Mutational Analysis of the *FBN1* Gene

In 1986, Sakai and coworkers first identified the protein fibrillin as a new component of the extracellular matrix microfibrils.[3] In 1990, Hollister and coworkers showed that patients with Marfan syndrome display abnormalities of the microfibrillar system.[4] They developed a immunohistochemical technique to demonstrate reduced expression of fibrillin and abnormal extracellular microfibrils in MFS. In the same year Kainulainen et al, 1990 mapped the gene for the MFS to chromosome 15q15-21.1.[5] Shortly thereafter, the *FBN1* gene was cloned and identified as the gene for the MFS.[6]

The discovery of the *FBN1* gene has paved the way for introducing molecular testing in the diagnostic work up of the MFS. With a cumulative LOD-score of more than 150 for linkage between the *FBN1* gene and classical MFS[7] and the identification of over 500 *FBN1* mutations in individuals with MFS,[8] the causal relationship between the *FBN1* gene and the MFS is now firmly established. Molecular analysis of the *FBN1* gene therefore seems a logical aid to the clinical diagnosis of MFS especially in situations of clinical uncertainty. In practice, however, the large and complex structure of the *FBN1* gene and the wide scope of *FBN1* mutations have hampered the clinical implementation of *FBN1* testing. At present, only a limited number of laboratories offer molecular testing for MFS in a clinical setting.[9] Moreover, literature data show great variation in detection rates of *FBN1* mutations and methodologies used. This can be accounted for by a variety of factors such as the type of mutational analysis method, the substrate (genomic versus cDNA) and, most importantly, the accuracy of the clinical diagnosis.

The studies which have looked at sensitivity and specificity of *FBN1* mutation analysis show substantial differences in one or several of these factors. For example, on the one hand, there are studies which have looked prospectively in a cohort of MFS and/or Marfan related patients of whom no prior molecular information was available, while other studies were designed to evaluate a new method in a group of MFS patients in which the mutation was already known. In Table 1 an overview is provided of the different studies and the mutation detection rates observed according to the type of method, substrate and clinical diagnosis. This overview only covers those studies in which the complete coding sequence of the fibrillin-1 gene was screened for mutations and the total number of patients screened was indicated.

The first mutational studies were performed on cDNA sequences derived from *FBN1* transcripts. The results of these studies were rather disappointing with regard to the low efficiency of *FBN1* mutation detection, yielding mutation rates of 9%[10] to 23%.[11] These studies used SSCP (Single Stranded Conformation Polymorphism), a technique based on detecting conformational differences between single strand DNA sequences. Subsequent mutational studies, used genomic DNA as template for analysis of the 65 individual exons of the *FBN1* gene (average DNA fragment size of 250 bp). With this approach, higher detection rates were obtained that varied however according to the type of screening method used : SSCP, HA (Heteroduplex Analysis) in MDE (Mutation Detection Enhancement) gels, long range PCR, CSGE (Conformation Sensitive Gel Electrophoresis) , EMD (Enzymatic Mutation Detection), TGGE (Temperature Gradient Gel Electrophoresis), direct sequencing and DHPLC (Denaturing High Performance Liquid Chromatography) (see Table 1).

Mutation detection with SSCP based methods was performed in several studies, yielding mutation rates that varied from 16%,[12] 50%,[13] 59%,[14] to 80% in a study of 10 British patients with MFS.[15] Nijbroek et al[16] used HA in MDE gels as screening method of *FBN1* genomic DNA and found mutations in seven of nine MFS patients (78%). Liu et al[17] advocated the use of long range PCR to detect exon skipping and genomic deletion mutations in 10% of MFS patients. Using a combination of this method with SSCP and HA, they obtained detection rates of 30%. One Japanese study reports the detection of 9 disease-causing mutations in a panel of 20 MFS patients (45%), using a direct sequencing strategy of all 65 exons.[18] Several

Table 1. *Overview of published studies on FBN1 analysis in patients with marfan syndrome or a marfan-related phenotype (adapted from Katzke et al[23])*

Substrate	Method	Number of Mutations			Reference
		True MFS + Marfan-Related	True MFS	Marfan-Related	
cDNA	SSCP	4/44 (9%)	4/32 (13%)	0/12	10
cDNA	SSCP	10/40 (25%)	9/39 (23%)	1/1	11
gDNA	HA in MDE gel	7/9 (78%)	7/9 (78%)	-	16
cDNA	Long range RT-PCR	6/60 (10%)	6/55 (11%)	0/5	17
gDNA	SSCP and HA	17/60 (28%)	10/20 (50%)		13
gDNA	DHPLC	39/93 (42%)	34/61 (56%)	5/32	24
gDNA	SSCP	8/34 (24%)	8/34 (24%)	-	26
	HA	4/34 (12%)	4/34 (12%)		
gDNA	CSGE	8/14 (57%)	7/10 (70%)	1/4	20
gDNA	EMD	3/6 (50%)	3/4 (75%)	0/2 0/2	28
	HA	1/6 (17%)	1/4 (25%)		
gDNA	SSCP	6/38 (16%)	6/38 (16%)	-	12
gDNA	Direct sequencing	9/20 (45%)	9/20 (45%)	-	18
gDNA	SSCP and HA	8/10 (80%)	8/10 (80%)	-	15
g+cDNA	SSCP and CSGE	71/171 (42%)	62/94 (66%)	3/65	30
gDNA	CSGE	23/31 (74%)	18/20 (90%)	2/11	22
gDNA	TGGE	53/126 (42%)	40/78 (51%)	4/32	23
gDNA	DHPLC	-		29	
gDNA	SSCP	14/76 (18%)	16/27 (59%)	3/58	14,35
gDNA	CSGE and/or DHPLC	21/35 (60%)	17/22 (77%)	4/13	25

other groups used CSGE[19] with good results as shown in Halliday et al[20] and Pepe et al,[21] who obtained mutation detection rates of 57% and 73%, respectively. These studies show that CSGE is a sensitive and specific method for *FBN1* mutational analysis. Their findings were further supported by Korkko et al,[22] who tested 20 MFS patients and identified 18 *FBN1* mutations. In addition a *FBN1* mutation has been identified in a family with EL (Ectopia Lentis), and in one multi-generation family diagnosed with AAE (Annulo Aortic Ectasia) characterized by aortic dilatation and skeletal features of MFS but no lens dislocation. However, dural ectasia was not investigated in this family. In the presence of dural ectasia, the diagnostic criteria for the MFS would have been met in this family. Katzke et al[23] used TGGE to screen the *FBN1* coding sequence in a cohort of 78 patients with MFS and obtained an overall mutation detection rate of 51%. Liu et al[24] performed exon-by-exon screening of *FBN1* using DHPLC, a technique reported to be very sensitive for mutation detection in human disease genes, and obtained a mutation detection rate of 56%. Other studies have confirmed the high efficiency of mutation detection of DHPLC, such as Halliday et al [25] who identified 17 *FBN1* mutations in 22 MFS patients. Overall, these studies all show variable mutation detection rates, even when the same technique is used.

Some groups have compared different techniques for a panel of MFS patients. Perez et al[26] compared the efficiency of SSCP versus HA in a panel of 34 MFS patients and found that SSCP was the more efficient method (23.5 vs. 17.6%). Yuan et al[27] compared HA, direct sequencing and enzyme mismatch cleavage and found EMC as the most sensitive but SSCP with heteroduplex as the most cost effective technique. In a small cohort of 6 MFS patients,

Youil et al[28] tested the EMD method, an improved version of the original mismatch cleavage method and found better results with the EMD as compared to an HA based approach. Matyas et al[29] reported a very high sensitivity for DHPLC with a detection rate approaching 100% in a cohort of 64 patients with known *FBN1* sequence variants. However they found a significant percentage of false positives which lower specificity. In our opinion, DHPLC is the most efficient approach for mutation detection, also because of its potential for automation when combined with robotic PCR.

There are different reasons to explain the variability of the mutation detection rate. The type of mutation and substrate can influence this phenomenon to some extent. Screening at the cDNA level has the advantage of being quicker as compared to the genomic DNA analysis since only 24 instead of 65 PCR fragments need to be analyzed. However, the mutation detection rate is significantly lower. Mutations resulting in the generation of a PTC and nonsense-mediated decay are missed with cDNA analysis. Adding cycloheximide to the fibroblast cultures in order to stabilize mRNA transcripts may prevent them from being destroyed by nonsense mediated decay. On the other hand, genomic analysis, although more "patient-friendly" since a skin biopsy is not required, has the disadvantage that it is labor intensive, that large genomic deletions can be overlooked and that the effect of splicing mutations can not always be assessed.

The most important factor influencing the mutation detection rate however, appears to be the clinical homogeneity or heterogeneity of the patient population. Indeed several studies have shown that the incidence of *FBN1* mutations is significantly higher in patients fulfilling the diagnostic criteria than in patients who do not. Hayward et al[13] screened all 65 exons of the *FBN1* genomic DNA using both SSCP and HA in 60 unrelated MFS patients, 50 of which had evidence for familial MFS and 10 of which were sporadic patients. They found that the mutation detection rates in large, well-characterized families was considerably higher (78%) than either in families with only one affected family member available for analysis (17%) or in sporadic cases, suggesting (or showing) that reliability of clinical diagnosis is an important factor influencing the mutation detection rates. In a large study including a cohort of 94 MFS patients and 77 patients with a MFS related phenotype, Loeys et al[30] found *FBN1* detection rates of 66% versus 5% respectively. They demonstrated that fulfilling the clinical diagnosis of MFS in itself is a good predictor of the outcome of *FBN1* mutation analysis. Katzke et al[23] supported these findings by their study which showed a much higher incidence of mutations in a group of MFS patients versus those with a MFS related condition and concluded that clinical overdiagnosis is the most important overall explanation for low *FBN1* mutation detection rate. Rommel et al[14] also found significant differences in mutation detection rate between patients fulfilling versus not fulfilling the clinical criteria for MFS. Finally, in a study by Halliday et al[25] it was shown that the majority of patients in which an *FBN1* mutation was found, met the clinical diagnosis of MFS. This study further supported the opinion that the robustness of selection criteria is the most important determinant of the outcome of mutational studies.

Clinical Indications for Mutation Analysis of the *FBN1* Gene

Several situations can occur in which molecular studies of the *FBN1* gene may be helpful. For example, to resolve a clinical dilemma in a patient in whom the diagnosis of MFS is suspected but can not be confirmed on clinical grounds.

In adult patients with lens dislocation, MFS is always the first diagnosis to consider. If they have MFS, then patients usually have major manifestations also in the cardiovascular or skeletal system and involvement of a third organ system. The diagnosis is then straightforward on clinical grounds. Therefore, neither MRI studies to detect dural ectasia nor molecular studies are needed to confirm the diagnosis. In patients who present with skeletal, cardiovascular and/ or possibly other manifestations of MFS, but have no involvement of the ocular system, it can be difficult to establish the diagnosis strictly on clinical grounds, particularly in absence of a positive family history. Here however, MRI studies can reveal the presence of dural ectasia in which case the diagnostic criteria may still be met. Dural ectasia is a highly sensitive clinical manifestation of MFS and can be detected by MRI or CT studies of the lumbosacral spine. Because there are no good standard measurements for children, this examination is not normally performed in children with suspected MFS. Moreover, no good data exist about the age at which dural ectasia develops in patients with MFS, although some studies suggest that this can be an early sign.[31] Several mutational studies which report *FBN1* mutations in Marfan-like patients have not verified the presence or absence of dural ectasia, so that the possibility remains that the diagnostic criteria of the Ghent nosology are in fact met. In cases where no definitive conclusion can be reached with the clinical data, molecular analysis of the *FBN1* gene is an alternative option.

Because of the evolving nature of the phenotype - particularly so for the cardiovascular and skeletal manifestations - children with suspected MFS may not yet fulfil the diagnostic criteria. In those instances, it is better to postpone a final diagnosis rather than to establish the diagnosis prematurely. This is important to avoid stigmatization or anxiety. On the other hand, careful follow up until well into adulthood is important before definitively excluding the diagnosis. Children presenting with only one or two major manifestations could still develop additional features later in life. The identification of an *FBN1* mutation in children or young adults not (yet) fulfilling the diagnostic criteria can help to identify those who need to be clinically surveyed with particular attention. Kainulainen et al[32] reported two patients with ectopia lentis, mild skeletal involvement and an *FBN1* mutation, who later developed cardiac manifestations of the MFS.[33] This clearly illustrates the importance of rigorous medical follow-up in children and adults, in particular in those in which an *FBN1* mutation is found despite the fact that they do not meet the Ghent criteria at their initial examination.

The availability of a molecular test also allows prenatal or preimplantation diagnosis to prospective parents. This is one option for which requests appear to be steadily increasing.[34] Although in familial instances linkage analysis can sometimes be used to perform prenatal diagnosis, there are many instances where this is not possible. For example, noninformativeness of the DNA markers, too small family size, or technical difficulties may make identification of the causal mutation necessary to achieve the quality and validity needed for the best possible informativeness. In couples, in which one of the prospective parents is affected with MFS, preimplantation diagnosis may be psychologically easier to accept because the prospect of termination of pregnancy for a condition by which they are personally affected is unacceptable.

Conclusion

There is no single protocol for analysis of the *FBN1* gene and the strategy of choice is dependent of the goals that must be reached. If *FBN1* mutational studies are implemented as an aid to clinical diagnosis, screening at the genomic DNA level, using a sensitive technique such as DHPLC or CSGE, is likely to be the strategy of choice, since it is the most practical and patient-friendly approach. On the other hand, in cases in which an exhaustive *FBN1* mutation analysis may be required, such as for example in the instance of prenatal diagnosis, screening of the *FBN1* gene at the cDNA level can be done as a second step, using either DHPLC, CSGE or direct sequencing. In those instances however, a skin biopsy is required as a source for the *FBN1* mRNA. At present, no method exists which has a 100% mutation detection efficiency. However, it can be expected that ongoing advances in biotechnology will gradually overcome technical shortcomings and enhance the use of molecular approaches in the diagnosis of MFS. In all instances however, it is imperative to assemble good clinical information and to perform a stringent clinical selection prior to molecular analysis.

References

1. De Paepe A, Devereux RB, Dietz HC et al. Revised diagnostic criteria for the Marfan syndrome. Am J Med Genet 1996; 62:417-426.
2. Pereira L, Levran O, Ramirez F et al. A molecular approach to the stratification of cardiovascular risk in families with Marfan's syndrome. N Engl J Med 1994; 331:148-153.
3. Sakai LY, Keene DR, Engvall E. Fibrillin, a new 350-kD glycoprotein, is a component of extracellular microfibrils. J Cell Biol 1986; 103:2499-2509.
4. Hollister DW, Godfrey M, Sakai LY et al. Immunohistologic abnormalities of the microfibrillar-fiber system in the Marfan syndrome. N Engl J Med 1990; 323:152-159.
5. Kainulainen K, Pulkkinen L, Savolainen et al. Location on chromosome 15 of the gene defect causing Marfan syndrome. N Engl J Med 1990; 323:935-939.
6. Dietz HC, Cutting GR, Pyeritz RE et al. Marfan syndrome caused by a recurrent de novo missense mutation in the fibrillin gene. Nature 1991; 352:337-339.
7. Ramirez F. Fibrillin mutations in Marfan syndrome and related phenotypes. Curr Opin Genet Dev 1996; 6:309-315.
8. Le Bourdelles S, Dehaupas I, Grey S. More than 500 FBN1 mutations associated with Marfan syndrome and overlapping disorders in the extensive update of the FBN1 mutation database. Am J Hum Genet 2002; suppl vol 71:525 abstract 2084.
9. Geneclinics website: http://www.geneclinics.org/.
10. Tynan K, Comeau K, Pearson M et al. Mutation screening of complete fibrillin-1 coding sequence: Report of five new mutations, including two in 8-cysteine domains. Hum Molec Genet 1993; 11:1813-1821.
11. Kainulainen K, Karttunen L, Puhakka L et al. Mutations in the fibrillin gene responsible for dominant ectopia lentis and neonatal Marfan syndrome. Nat Genet 1994; 6:64-69.
12. Oh MR, Kim JS, Beck NS et al. Six novel mutations of the fibrillin-1 gene in Korean patients with Marfan syndrome. Ped Int 2000; 42:488-491.
13. Hayward C, Porteous ME, Brock DJH. Mutation screening of all 65 exons of the fibrillin-1 gene in 60 patients with Marfan syndrome: Report of 12 novel mutations. Hum Mutat 1997; 10:280-289.
14. Rommel K, Karck M, Haverich A et al. Mutation screening of the fibrillin-1 (FBN1) gene in 76 unrelated patients with Marfan syndrome or Marfanoid features leads to the identification of 11 novel and three previously reported mutations. Hum Mutat 2002; 20:406-407. Mutation in Brief #546 Online.
15. Comeglio P, Evans AL, Brice GW et al. Detection of six novel FBN1 mutations in British patients affected by Marfan syndrome. Hum Mutat 2001; 18:251. Mutation in Brief #438 Online.
16. Nijbroek G, Sood S, McIntosh I et al. Fifteen novel FBN1 mutations causing Marfan syndrome detected by heteroduplex analysis of genomic amplicons. Am J Hum Genet 1995; 57:8-21.
17. Liu W, Qian C, Comeau K et al. Mutant fibrillin-1 monomers lacking EGF-like domains disrupt microfibril assembly and cause severe Marfan syndrome. Hum Molec Genet 1996; 5:1581-1587.
18. Matsukawa R, Iida K, Nakayama M et al. Eight novel mutations of the FBN1 gene found in Japanese patients with Marfan syndrome. Hum Mutat 2001; 17:71-72. Mutation in Brief #383 Online.
19. Ganguly A, Rock MJ, Prockop DJ. Conformation-sensitive gel electrophoresis for rapid detection of single-base differences in double-stranded PCR products and DNA fragments: Evidence for solvent-induced bends in DNA heteroduplexes. Proc Natl Acad Sci USA 1993; 90:10325-10329.
20. Halliday D, Hutchinson S, Kettle S et al. Molecular analysis of eight mutations in FBN1. Hum Genet 1999; 105:587-597.
21. Pepe G, Giusti B, Evangelisti L et al. Fibrillin-1 (FBN1) gene frameshift mutations in Marfan patients: Genotype-phenotype correlation. Clin Genet 2001; 59:444-450.
22. Körkkö J, Kaitila I, Lönnqvist L et al. Sensitivity of conformation sensitive gel electrophoresis in detecting mutations in Marfan syndrome and related conditions. J Med Genet 2002; 39:34-41.
23. Katzke S, Booms P, Tiecke F et al. TGGE screening of the entire FBN1 coding sequence in 126 individuals with Marfan syndrome and related fibrillinopathies. Hum Mutat 2002; 20:197-208.
24. Liu W, Oefner P, Qian C et al. Denaturing HPLC-identified novel FBN 1 mutations, polymorphisms, and sequence variants in Marfan Syndrome and related connective tissue disorders. Genetic Testing 1997; 1:237-242.
25. Halliday DJ, Hutchinson S, Lonie L et al. Twelve novel FBN1 mutations in Marfan syndrome and Marfan related phenotypes test the feasibility of FBN1 mutation testing in clinical practice. J Med Genet 2002; 39:589-593.
26. Perez ABA, Pereira LV, Brunoni D et al. Identification of 8 new mutations in Brazilian families with Marfan syndrome. Hum Mutat 1999; 13:84. Mutation in Brief #211 Online.

27. Yuan B, Thomas JP, von Kodolitsch Y et al. Comparison of heteroduplex analysis, direct sequencing, and enzyme mismatch cleavage for detecting mutations in a large gene, FBN1. Hum Mutat 1999; 14:440-446.
28. Youil R, Toner TJ, Bull E et al. Enzymatic mutation detection (EMD™) of novel mutations (R565X and R1523X) in the FBN1 gene of patients with Marfan syndrome using T4 endonuclease VII. Hum Mutat 2000; 16:92-93. Mutation in Brief #345 Online.
29. Matyas G, De Paepe A, Halliday D et al. Evaluation and application of denaturing HPLC for mutation detection in Marfan syndrome: Identification of 20 novel mutations and two novel polymorphisms in the FBN1 gene. Hum Mutat 2002; 19:443-456.
30. Loeys B, Nuytinck L, Delvaux I et al. Genotype and phenotype analysis of 171 patients referred for molecular study of the fibrillin-1 gene (FBN1) because of suspected Marfan syndrome (MFS). Arch Int Med 2001; 161:2447-2454.
31. Fattori R, Nienaber CA, Descovich B et al. Importance of dural ectasia in phenotypic assessment of Marfan's syndrome. Lancet 1999; 354:910-913.
32. Kainulainen K, Karttunen L, Puhakka L et al. Mutations in the fibrillin gene responsible for dominant ectopia lentis and neonatal Marfan syndrome. Nat Genet 1994; 6:64-69.
33. Black C, Withers A, Gray J et al. Correlation of a recurrent FBN1 mutation (R122C) with an atypical familial Marfan syndrome phenotype. Hum Mutat 1998; Suppl 1:S198-S200.
34. Loeys B, Nuytinck L, Van Acker P et al. Strategies for prenatal and preimplantation genetic diagnosis in Marfan syndrome (MFS). Prenat Diagn 2002; 22:22-28.
35. El-Aleem AA, Karck M, Haverich A et al. Identification of 9 novel FBN1 mutations in German patients with Marfan syndrome. Hum Mutat 1999; 14:181.

CHAPTER 8

The Marfan Mutation Database

Gwenaëlle Collod-Béroud and Catherine Boileau

M utations in the fibrillin gene (*FBN1*) were described at first in Marfan syndrome (MFS) patients. Subsequently, the gene was shown to harbor mutations related to a spectrum of conditions clinically related to MFS, the "type-1 fibrillinopathies". In an effort to standardize the information regarding these mutations, to facilitate their mutational analysis and the identification of structure/function and phenotype/genotype relationships, we created, in 1995, a mutation database named "UMD-*FBN1*". The database follows the guidelines on mutation databases of the Hugo Mutation Database Initiative and gives access to a software package that provides specific routines and optimized multicriteria research and sorting tools. Recently, we have also developed a *FBN1* polymorphism database in order to facilitate diagnosis.

Mutations in the *FBN1* gene are associated with a wide range of phenotypes that show considerable variation in tissue distribution, timing of onset, and severity. The severe end of this broad spectrum of phenotypes is neonatal MFS. Conditions at the mild end include the MASS syndrome (Mitral valve prolapse, Aortic dilatation, and Skin and Skeletal manifestations) (OMIM#604308), mitral valve prolapse syndrome (OMIM#157700), isolated skeletal features, familial ascending aortic aneurysm and dissection (OMIM#132900), and ectopia lentis (OMIM#129600) with relatively mild skeletal features. Many of these conditions show significant overlap with MFS and are quite common in the general population. Finally, two syndromes [dominant Weill-Marchesani[1] and Shprintzen-Goldberg OMIM#182212] have also been associated with mutations in the *FBN1* gene.

In an effort to standardize the information regarding *FBN1* mutations, we created a computerized database in 1995 using the UMD (Universal Mutation Database) software. This generic software was devised by Cariello et al[2] to create locus-specific databases (LSDBs) with the 4th Dimension® package from ACI.[2,3] The software includes an optimized structure to assist and secure data entry and to allow the input of clinical information. The software package contains routines for the analysis of the database. The use of the 4th Dimension® (4D) SGDB gives access to optimized multicriteria research and sorting tools to select records from any field. The software has been successfully adapted to many different genes either involved in cancer (*p53*,[2,4] *APC*,[5,6] *VHL*,[7,8] and *WT1*[9]) or in genetic diseases (*LDLR*,[10-12] *VLCAD* (very long-chain acyl-CoA dehydrogenase) (unpublished) and *MCAD* (medium-chain acyl-CoA dehydrogenase) (unpublished)). Routines were specifically developed for the *FBN1* database and are described below.[3,13-15]

Database

The *FBN1* database contains mutations either published or only reported in meetings proceedings, or contributed by the co-authors of published papers.[13-16] Recently, the database was modified to follow the guidelines on mutation databases of the Hugo Mutation Database Initia-

The Marfan mutation database can be found at http://www.umd.be

Marfan Syndrome: A Primer for Clinicians and Scientists, edited by Peter N. Robinson and Maurice Godfrey. ©2004 Eurekah.com and Kluwer Academic / Plenum Publishers.

tive including the new nomenclature.[17] Mutation names are numbered with respect to the *FBN1* gene cDNA sequence obtained from the Genbank database (Genbank database accession number L13923; Complete coding sequence of HUM-FIBRILLIN Homo sapiens fibrillin mRNA). Intron-exon boundaries are as defined by Pereira et al[18] and module organization is as in SWISS-PROT (accession number P35555).

The mutation records of the database include point mutations, large and small deletions, insertions, and splice mutations in the *FBN1* gene. The database cannot accommodate complex mutations or two mutations on the same allele. These are entered as two different records linked by the same sample ID. Each record contains the molecular and the clinical data for a given mutation in a standardized, easily accessible and summary form. Several levels of information are provided: at the gene level (exon and codon number, wild type and mutant codon, mutational event, mutation name), at the protein level (wild type and mutant amino acid, affected domain) and at the clinical level (absence or presence of skeletal, ocular, cardiovascular, central nervous system and other various manifestations). Data on fibrillin protein biosynthesis classification groups have been added when available as described in Aoyama et al.[19] In 2003, the database was updated to 563 mutations, of which over 400 are new entries.

Routines

Initially, 6 routines were specifically developed.[13] The database now contains 13 routines that are described in the order in which they were implemented:[14,15]

1. **"Position"** studies the distribution of mutations at the nucleotide level to identify preferential mutation sites.

2. **"Statistical evaluation of mutational events"** is comparable to "Position" but also indicates the mutational event. The result can either be displayed as a table or in a graphic representation.

3. **"Frequency of mutations"** allows one to study the relative distribution of mutations at all sites and to sort them according to their frequency. A graphic representation is also available and displays a cumulative chart of mutation distributions.

4. **"Stat exons"** studies the distribution of mutations in the different exons. It enables detection of a statistically significant difference between observed and expected mutations.

5. **"Protein"** studies the distribution of mutational events in the various protein domains and repeated motifs (NH2 unique region, EGF-like motifs, cbEGF-like motifs, 8-cysteine motifs, hybrid motifs, proline rich region and COOH unique region).

6. **"Insertions and deletion analysis"** searches for repeated sequences flanking the mutation and possibly involved in the mutational mechanism.

7. **"Restriction enzyme"** appears on the first page of the mutation file. If the mutation modifies a restriction site, the program shows a restriction map displaying the new or abolished site and the enzyme of interest.

8. **"Amino acid type search"** studies the mutations with respect to phylogenetic conservation. In effect, the fibrillin gene has been identified and sequenced in two mammalian species [complete coding sequence of mouse fibrillin (*Fbn-1*) mRNA and complete coding sequence of bovine fibrillin (*BovFib*) mRNA, Genbank database accession numbers L29454 and L28748 respectively]. The identity at the amino level between the human and bovine sequences is 97.8% and between the human and the mouse sequences of 96.2%. Therefore, the routine lists the mutations affecting non-conserved amino acids in the bovine, in the mouse, or in both sequences.

9. **"Binary comparison"** compares two mutation groups, each group being defined by distinct research criteria chosen from the database (molecular, clinical, age of onset, sex...). The result can be displayed as either of several graphic representations (by amino acids, by exon, or by protein domain) of the distribution of the sorted mutations. Furthermore the sorted mutations can appear by groups or in a detailed format (insertion, deletion, missense, nonsense).

10. **"Amino acid changes"** lists for each of the 20 amino acids the observed substitutions throughout the protein.
11. **"Base modification"** lists the observed mutations with respect to their position within the codon for each of the 4 bases.
12. **"CpG"** studies the distribution of mutations occurring at CpG sites throughout the coding sequence. The result is displayed in a graphic representation.
13. **"Distribution of mutations"** lists the proportion of each of the mutational events observed in a selected group of mutation records.

Finally, several routines can be simultaneously studied and multicriteria searches can be performed.

Mutation Analyses

To date, 563 *FBN1* mutations have been identified and reported. Since the software cannot accommodate complex mutational events in a given individual, several mutations are not included in the current version of the database (3901_3904del; 3908_3909del;[20] 1642del3ins20pb and 1888delAAinsC[21] and 1882_1884delinsAAA.[22] The double mutant 3212T>G;3219A>G (I1071S;E1073D) reported by Wang et al[23] is reported in two different records linked by the same sample ID, as well as for double mutant 3797A>T; 5746T>A (Y1266F;C1916S) found in a French proband (unpublished data). Other mutations are spread throughout almost the entire gene without obvious predilection for any given region. So far, mutations were thought to be private and generally non recurrent. However, the recent database update shows that approximately 12% of mutations are recurrent (56 recurrent mutations representing 156 events). Most of these mutations have been found in 2 to 3 probands. However, 8 are found in more than 4 subjects (Table 1). Interestingly, almost all these mutational events affect a CpG, a mutation hot spot, and could truly be recurrent rather than associated with an unique haplotype. The case of mutation I2585T is baffling since it does not affect a CpG, it is absent in over 300 control chromosomes and has been found only in Marfan patients. Although these subjects were identified in different countries, the existence of a founder effect cannot be ruled out and further molecular and functional data need to be collected to investigate this finding.

Information on transmission is available for 398 mutations. Among these, there is a surprising number of de novo mutations compared to transmitted mutations (188 de novo vs. 210 familial cases). This finding could imply that sporadic cases are far more frequent than the 25% usually reported. However, the observed figure may reflect a bias related to the fact that clinical diagnosis of MFS is often difficult and that the yield in mutation identification is low in the incomplete forms of the syndrome. Thus, molecular studies may be biased in favor of the study of probands with a complete and severe form of the disease. In effect, screening for *FBN1*

Table 1. Recurrent mutations found in the FBN1 gene

Events	Position (AA)	WT Codon	Mutant Codon	Mutation Name	Number of Records
184C>T	62	CGT	TGT	R62C	5
IVS2+1G>A	83	CCC	spl+1	IVS2+1G>A	5
364C>T	122	CGC	TGC	R122C	4
1633C>T	545	CGC	TGC	R545C	4
3037G>A	1013	GGA	AGA	G1013R	6
4930C>T	1644	CGA	TGA	R1644X	4
IVS46+5G>A	1930	GAT	spl-2	IVS46+5G>A	8
7754T>C	2585	ATT	ACT	I2585T	6

Table 2. FBN1 exons displaying an abnormal number of mutations

Exon	Domain Encoded	Expected Mutations	Observed Mutations	p
1	NH2-ter	8.0	1	<0.02
7	cb EGF-like # 1	6.0	1	<0.05
13	cb EGF-like # 4	7.2	17	<0.001
26	cb EGF-like # 12	7.6	18	<0.001
27	cb EGF-like # 13	6.9	16	<0.001
28	cb EGF-like # 14	7.0	13	<0.05
43	cb EGF-like # 25	7.1	13	<0.05
45	cb EGF-like # 27	7.5	2	<0.05
57	TGF β BP-like # 7	11.8	5	<0.05
65	NH2-ter	20.0	7	<0.01

The "Stat exons" routine studies the distribution of exonic mutations and enables detection of a statistically significant difference between observed and expected mutations. The algorithm takes into account the mutability of each base from an exon. The mutability is defined as follows: for each base, the significance of a mutation is defined by its ability to produce a new amino acid. In these conditions, the specific position of the base within the codon has a major incidence. If any substitution result in a new amino acid, its individual mutability is 3. If only two substitutions result in a new amino acid, mutability is 2... The exon's mutability is defined by the addition of all mutabilities for each base. The expected value is calculated by the formula:

(exon mutability / by the CDNA mutability)*observed mutations

The p value is calculated using the usual Chi square formula.

mutations has remained laborious, expensive, and limited to the 65 exons and their intronic flanking sequences. Thus, the mutations located within the large non coding sequence as well as in the promoter and the 5' and 3' regulatory regions go undetected. Finally, genetic heterogeneity could also partly explain the low number of mutations reported for familial forms of MFS.

The 492 mutations located within the coding sequence are generally distributed in all exons. However, when comparing the number of mutational events expected in a given exon to the number of mutations identified in the given exon, statistically significant differences appear in a few exons: there is under-representation of mutations in exons 45 (cbEGF-like#27) and 57 (8-cystein#7) while over-representation is observed in exons 13 (cbEGF-like#4), 26 (cbEGF-like#12), 27 (cbEGF-like#13), 28 (cbEGF-like#14) and 43 (cbEGF-like#25) (Table 2). Over-representation in exons 26 to 28 can be explained by the fact that almost all the mutations identified in neonatal cases of MFS are located within this area (exons 24-32). Furthermore, mutations in this region are more likely to be associated with a severe clinical phenotype.[24] Severity probably leads to a bias in the types of patients selected for mutation detection.

The fibrillin gene has been identified and sequenced in vertebrates species: bovine (L28748), mouse (L29454), rat (AF135059), dog (partial cDNA, AF29080), pig (AF073800) and invertebrates: medusa (partial cDNA, L39930). The identity at the amino acid level between mammals is so high (for example 97.8% human-bovine, 98% human-rat, and 96.2% human-mouse) that very often phylogenic conservation should be observed at the amino acid position affected by a given missense mutation. In fact, all reported mutations in the *FBN1* gene affect a conserved amino acid with respect to the bovine sequence. It is interesting to note that only 6 mutational events affect amino acids that are not conserved between mouse and man: three deletions (4179_4187del (Lesley Adès, Katherine Holman, personal communication 2000), 4177_4177delG and 7965_7977del,[21] a duplication (6409_6411dup (Collod-Béroud, in preparation), a nonsense mutation (6339T>G),[25] for which the causality is obvious, and a missense mutation (3382G>A corresponding to V1128I).[26] In the medusa (*Podocoryne carnea*) the primary amino acid sequence (> 40% sequence identity with mammalians), the highly repetitive multidomain structure, as well as the beaded microfibril appearance of fibrillin are highly conserved in man.[27] Fibrillin, as well as collagen, is thought to be a very early invention of metazoans in evolution.[28] Reber-Müller et al[27] suggest that with the invention of fibrillin, resilience and elasticity might have been added to the characteristics of ECM, thus providing the biomechanical basis for the development of a free-swimming medusa life stage.

Large Rearrangements

In the *FBN1* gene, no major deletions have been reported except for the exon 60-62 genomic deletion[29] and two other multi-exon deletions.[30] Deletions of contiguous EGF-like domains have different effects depending on their location within the fibrillin-1 molecule. Deletion of the three contiguous cbEGF-like domains encoded by exons 44-46 resulted in a severe phenotype with onset in infancy and a rapidly progressing clinical course.[30] In frame deletion of exons 42-43 was characterized in a patient presenting with bilateral ectopia lentis and Marfanoid skeletal features[30] and finally, deletion of the cbEGF-like domains encoded by exons 60-62 in the C-terminal domain of fibrillin-1 results in a much less severe phenotype characterized by a moderate Marfanoid habitus.[29] In the patient presenting with inframe deletion of exons 42-43, two sets of identical pentamers (cagta and ggaaa) were identified near the breakpoints in intron 41 and 43. In the deletion of exons 44-46, the exchange occurred within an identical pentamer (atttt). None of these sequences are known to predispose to genomic rearrangements. For the exons 60-62 genomic deletion, data are not available.

Presently, the Human Gene Mutation Database (HGMD at www.hgmd.org) which is the largest general mutation database, contains 33 252 mutations in 1338 genes (public dataset numbers available online). Of these mutations, 1827 (5.49%) are gross deletions, 274 (0.86%) gross insertions and duplications, 57 (0.17%) are repeat variations, and 340 (1.0%) are complex rearrangements. Thus, although generally less frequent than point mutations, major rearrangements represent a mutational mechanism found in many disease genes. Therefore, it is surprising that so few have been reported in the *FBN1* gene. However, it is unclear if this mutation type has always been searched for since Southern blotting is a long and time consuming technique that is no longer performed in many diagnostic laboratories.

Small Insertion/Deletion Mutations (< 20nt), Duplications

Among the 80 small insertion/deletion mutations, 72 create a premature termination codon (PTC). These account for 12.9% of the total mutations (small insertions: 22 cases and small deletions: 50 cases). They act as dominant negatives but display a highly variable clinical phenotype, from severe to mild. The severity of the phenotype may be related to the quantitative expression of the mutant allele and to the percentage of truncated proteins incorporated in the microfibrils.[20,31] In the 55 small deletions reported (Table 3), 18 single base pair deletions can be the result of a mechanism of slipped mispairing, 4 small deletions (5791_5793delGTT, 8525_8529delTTAAC, 755_762del and 3603_3668del) are flanked by direct repeats and 3 mutations are deletions of a repeated sequence (635_636delCA, 3355_3358delAGAG and 4920_4923delTGAA). For the 30 other mutations, the mechanisms have yet to be determined by the search, among others, for the presence of quasi-palindromic sequences, inverted repeats or symmetric elements which facilitate the formation of secondary-structure intermediates.[32] Twenty-seven insertions have been reported so far (Table 4). Six are insertions within runs of identical bases and can be explained by slippage mispairings at the replication fork. Eleven single base pair insertions correspond to the duplication of an existing base. Seven insertions are small duplications of existing sequences. For three mutations, the mechanisms, similar to those described above, have yet to be determined.

Table 3. Small deletions identified in the FBN1 gene involving repeated sequences

Mutation Name	File #	Exon	Mutation Position	Codon	WT Codon	Mutant Codon	Event	Repeat Type	WT AA	Mutant AA
7965_7977del	594	63	7963	2655	GCG	del13c	Stop at 2677	Deletion flanked by direct repeats	Ala	Fr.
5791_5793delGTT	467	47	5791	1931	GTT	del3a	In frame del	Deletion flanked by direct repeats	Val	InF
8525_8529delTTAAC	328	65	8524	2842	CTT	del5b	Stop at 2848	Deletion flanked by direct repeats	Leu	Fr.
755_762del	225	7	754	252	GCC	del8b	Stop at 261	Deletion flanked by direct repeats	Ala	Fr.
4178_4178delA	586	33	4177	1393	GAA	del1b	Stop at 1412	Deletion in a repetition of A	Glu	Fr.
3767_3767delA	585	30	3766	1256	AAT	del1b	Stop at 1275	Deletion in a repetition of A	Asn	Fr.
7577_7577delA	435	61	7576	2526	AAT	del1b	Stop at 2681	Deletion in a repetition of A	Asn	Fr.
4704_4704delA	208	37	4702	1568	AAA	del1c	Stop at 1580	Deletion in a repetition of A	Lys	Fr.
7291_7291delA	207	58	7291	2431	ACT	del1a	Stop at 2437	Deletion in a repetition of A	Thr	Fr.
3192_3192delA	82	25	3190	1064	GAA	del1c	Stop at 1087	Deletion in a repetition of A	Glu	Fr.
1836_1836delA	102	14	1834	612	AAA	del1c	Stop at 624	Deletion in a repetition of A	Lys	Fr.
3238_3238delC	500	26	3238	1080	CTC	del1a	Stop at 1087	Deletion in a repetition of C	Leu	Fr.
526_526delC	492	5	526	176	CAG	del1a	Stop at 189	Deletion in a repetition of C	Gln	Fr.
4485_4485delC	286	36	4483	1495	ACC	del1c	Stop at 1519	Deletion in a repetition of C	Thr	Fr.
2399_2399delC	230	19	2398	800	CCT	del1b	Stop at 802	Deletion in a repetition of C	Pro	Fr.
5311_5311delC	205	43	5311	1771	CGG	del1a	Stop at 1892	Deletion in a repetition of C	Arg	Fr.
4020_4020delC	52	32	4018	1340	ACC	del1c	Stop at 1412	Deletion in a repetition of C	Thr	Fr.
124_124delG	436	1	124	42	GCC	del1a	Stop at 107	Deletion in a repetition of G	Ala	Fr.
6996_6996delT	522	56	6994	2332	CTT	del1c	Stop at 2397	Deletion in a repetition of T	Leu	Fr.
4356_4356delT	495	35	4354	1452	CTT	del1c	Stop at 1474	Deletion in a repetition of T	Leu	Fr.
6018_6018delT	206	48	6016	2006	CTT	del1c	Stop at 2058	Deletion in a repetition of T	Leu	Fr.
1604_1604delT	51	13	1603	535	TTA	del1b	Stop at 578	Deletion in a repetition of T	Leu	Fr.
635_636delCA	354	6	634	212	ACA	del2b	Stop at 221	Deletion of a repeated sequence	Thr	Fr.
4920_4923delTGAA	193	39	4918	1640	AAT	del4c	Stop at 1648	Deletion of a repeated sequence	Asn	Fr.
3355_3358delAGAG	583	27	3355	1119	AGA	del4a	Stop at 1160	Deletion of a repeated sequence	Arg	Fr.
6497_6616del	555	52-53	6496	2166	GAT	del120b	In frame del	gDNA defect not found yet	Asp	InF

Reference sequence on which the +1 nt residue is based is L13923. File #= indicates the file record in the database; Mutant codon= "del" followed by the number of deleted bases then by the position of the deletion "a"= first base of the codon deleted, "b"= second base of the codon deleted, "c"= third base of the codon deleted. InF= in frame; Fr= Frameshift. (Adapted from Collod-Béroud et al, ref. 16)

Table 4. Insertions and duplications reported in the FBN1 gene and involving repeated sequences

Mutation Name	File #	Exon	Mutation Position	Codon	WT Codon	Mutant Codon	Event	Mutation Type	WT AA	Mutant AA
4699_4721dup	519	37	4720	1574	TGT	ins23c	Stop at 1588	Duplication	Cys	Fr.
5470_5484dup	518	44	5485	1829	GGC	ins15a	In frame ins	Duplication	Gly	InF
6409_6411dup	513	52	6412	2138	AAA	ins3a	In frame ins	Duplication	Lys	InF
3228_3232dup	554	26	3232	1078	CCT	ins5b	Stop at 1089	Duplication	Pro	Fr.
5470_5484dup	481	44	5485	1829	GGC	ins15a	In frame ins	Duplication	Gly	InF
5236_5240dup	248	42	5239	1747	CTC	ins5c	Stop at 1894	Duplication	Leu	Fr.
5134_5137dup	9	41	5137	1713	AAC	ins4b	Stop at 1735	Duplication	Asn	Fr.
4703_4704dup	423	37	4705	1569	GCC	ins2a	Stop at 1581	Duplication = Insertion after an A repeat	Ala	Fr.
3444_3444dup	237	27	3445	1149	AAC	ins1a	Stop at 1158	Duplication = Insertion after a C repeat, Consensus sequence for N-glycosylation	Asn	Fr.
6285_6285dup	595	50	6285	2096	TGC	Ins1a	Stop at 2104	Duplication = Insertion after an A repeat	Cys	Fr.
5065_5065dup	424	40-41	5065	1689	GAT	ins1b	Stop at 1702	Duplication = Insertion after a G repeat	Asp	Fr.
7818_7818dup	575	62-63	7819	2607	GAT	ins1a	Stop at 2607	Duplication = Insertion after a T repeat	Asp	Fr.
3525_3525dup	584	28	3526	1176	GGG	ins1a	Stop at 1192	One base duplication	Gly	Fr.
3525_3525dup	584	28	3526	1176	GGG	ins1a	Stop at 1192	One base duplication	Gly	Fr.
7116_7116dup	553	57	7117	2373	CCC	ins1a	Stop at 2376	One base duplication	Pro	Fr.
623_623dup	469	6	622	208	CTC	ins1c	Stop at 222	One base duplication	Leu	Fr.
2586_2586dup	434	21	2587	863	GAG	ins1a	Stop at 863	One base duplication	Glu	Fr.
5499_5499dup	433	44	5500	1834	GAC	ins1a	Stop at 1834	One base duplication	Asp	Fr.
959_959dup	432	8	958	320	TAC	ins1c	Stop at 320	One base duplication	Tyr	Fr.
1378_1378dup	306	11	1378	460	TGT	ins1b	Stop at 475	One base duplication	Cys	Fr.
7790_7790dup	298	62	7789	2597	CTC	ins1c	Stop at 2607	One base duplication	Leu	Fr.
2570_2570dup	297	21	2569	857	GTC	ins1c	Stop at 859	One base duplication	Val	Fr.
6185_6185dup	254	50	6184	2062	TAT	ins1c	Stop at 2062	One base duplication	Tyr	Fr.

Reference sequence on which the +1 nt residue is based is L13923. File #= indicates the file record in the database; Mutant codon="ins" followed by the number of inserted bases then by the position of the insertion "a"= insertion before the first base of the codon, "b"= insertion before the second base of the codon, "c"= insertion before the third base of the codon. InF= in frame; Fr= Frameshift. (Adapted from Collod-Béroud et al, ref. 16)

Splice Mutations

The pre-mRNA splicing machinery recognizes exons and joins them together to form mRNAs with intact translational reading frames. Splicing requires canonical sequences at the intron/exon boundary. Three categories of mutations can be identified. The first one corresponds to mutations in canonical sequences and represents 60/73 splice mutations found in the *FBN1* gene. They cause abnormal splicing patterns by the use of the nearest and strongest consensus splice site. The second category (10 mutations) corresponds to mutations not located in canonical sequences (Table 5). These last few years, different studies have indicated that distinct sequence elements that are distant from the splice sites are also needed for normal splicing. These elements can affect splice-site recognition during constitutive splicing and also play important roles in directing alternative splicing[33] [Cooper, 1997]. They can be auxiliary splicing elements (ASEs) required for cell-specific modulation of alternative splicing within introns that flank alternative exons, or exonic splicing enhancers (ESEs) within both coding and noncoding exons that direct specific recognition of splice sites during constitutive and alternative splicing. Two exonic mutations, a nonsense mutation 6339T>G (Y2113X)[34] and a silent exonic mutation 6354C>T (I2118I)[35] have been reported as inducing in-frame skipping of the entire exon 51 and demonstrate the existence of an ESE sequence in exon 51.[36,37] Thirteen other mutations could belong to this category with mutations up to 53 bp away from the canonical sequence. In most cases, cDNA analysis is not available and abnormal splicing has not been demonstrated. Therefore, causality is still uncertain. Finally, the third category of mutations is provided by single base pair changes that introduce novel splice sites that substitute for the wild-type sites. A single recurrent mutation possibly creating a potential donor splice site has been reported but an abnormal splicing pattern has not been demonstrated (3294C>T). In the majority of splice site mutations, exon skipping results in an in-frame mRNA and produces a mutant fibrillin-1 missing a whole domain. The mutant allele produces abnormal monomers that considerably interfere with the assembly of fibrillin molecules in the microfibrils network. In a small number of patients (9 cases), the skipping of an exon causes a frameshift, a premature termination codon (PTC) and reduced mutant RNA levels through nonsense-mediated decay of the mutant transcript.[38] Furthermore, MFS patients have been reported in whom the donor splice site mutation results in the incorporation of intronic sequence (IVS46+1G>A, IVS27+1G>A) into the transcript or in the use of a cryptic splice site inducing partial exon deletion (IVS18+2T>C, IVS37+5G>T).

Nonsense/Missense

Nonsense (61 cases) and missense (337 cases) mutations represent 10.9% and 60.3% of mutations, respectively. Among missense mutations, more than three quarters (263/337) are located in calcium binding modules. These mutations either create (20/263, 7.6%) or substitute (129/263, 49%) cysteine residues potentially implicated in disulfide bonding. Pulse-chase studies on fibrillin-1 secretion from MFS patient fibroblasts have shown that these mutations often result in a delay in secretion/intracellular retention of profibrillin.[19,39-41] As three disulfide bonds are required to maintain the native cbEGF-like module fold, suppression or addition of cysteine residues would result in cbEGF-like module misfolding, which impairs trafficking.[42-44] The majority of the remaining mutations in these modules affects residues of the calcium consensus sequence and results in reduced calcium affinity, which may in turn destabilize the interface between two consecutive cbEGF-like modules. Calcium binding would rigidify the interdomain region between two cbEGF-like modules and allow multiple tandem cbEGF-like modules to take on a rigid, rod-like conformation.[45-47] Increased protease susceptibility due to reduced calcium affinity is a mechanism also reported for missense mutations. This pathological mechanism emphasizes the importance of calcium binding for the structural integrity of fibrillin-1. Mutations which do not belong to one of these subclasses may likely be involved in protein-protein interactions. Other modules are carriers of one quarter of missense mutations and pathological mechanisms have yet to be clearly demonstrated.

Table 5. Unusual splice mutations

Mutation Name	File #	Exon	Mutation Position	Codon	WT Codon	Mutant Codon	Event	Probable Effect on cDNA	WT AA	Mutant AA
I. Mutations not located in canonical donor or acceptor splice site sequence										
IVS28+15del3	216	28-29	3589	1197	GAC	spl+15	del3	ND	Asp	Spl.
IVS56+17delG	479	56-57	6997	2333	GAC	spl+17	delG	ND	Asp	Spl.
IVS30+28C>A	217	30-31	3838	1280	GAT	spl+28	C>A	ND	Asp	Spl.
IVS12+21G>A	189	12-13	1588	530	GAC	spl+21	G>A	ND	Asp	Spl.
IVS52+50C>T	155	52-53	6496	2166	GAT	spl+50	C>T	ND	Asp	Spl.
IVS20+53T>A	213	20-21	2539	847	GAA	spl+53	T>A	ND	Glu	Spl.
IVS2-7T>G	314	2-3	247	83	CCC	spl-7	T>G	ND	Pro	Spl.
IVS35-8G>A	440	35-36	4459	1487	GAT	spl-8	G>A	ND	Asp	Spl.
6339T>G	57	51	6339	2113	TAT	TAG	T>G	Nonsense mutation inducing skipping of exon 51, in frame	Tyr	Stop
6354C>T	145	51	6354	2118	ATC	ATT	C>T	Silent mutation inducing skipping of exon 51, in frame	Ile	Ile
II. Mutations creating potential donor or acceptor splice site										
3294C>T	215	26	3294	1098	GAC	GAT	C>T	ND	Asp	Asp
3294C>T	359	26	3294	1098	GAC	GAT	C>T	ND	Asp	Asp
3294C>T	329	26	3294	1098	GAC	GAT	C>T	ND	Asp	Asp

Reference sequence on which the +1 nt residue is based is L13923. File #= indicates the file record in the database. ND= Skipping of exon not demonstrated by cDNA analysis. Spl= splice; PTC= Premature Termination Codon. (Adapted from Collod-Béroud et al, ref. 16)

The global molecular analysis of *FBN1* mutations reveals 2 classes of mutations. The first one, which represents more than one third of the mutations (38.6%), contains mutations predicted to result in shortened fibrillin-1 molecules: 61 nonsense mutations, 71 splicing errors, 23 insertions and duplications, 51 deletions and 10 inframe deletions. They act as dominant negatives but display a highly variable clinical phenotype, of which severity is directly related to the quantitative expression of the mutant allele and to the percentage of truncated or shortened proteins incorporated in the microfibrils.[20,31] The second one represents less than two thirds (60.3%) of the mutations and contains missense mutations, mostly located in cbEGF-like modules (78%). They can be subclassified in (a) mutations creating or substituting cysteine residues potentially implicated in disulfide bonding and consequently in the correct folding of the monomer and (b) amino acids implicated in calcium binding and subsequently in interdomain linkage, rigidification of monomer and in protease susceptibility.

FBN1 Polymorphism Database

Recently, we created an independent database for *FBN1* polymorphisms. Its goal is to make available a complete set of *FBN1* gene variations (mutations and polymorphisms) so that causative mutations may be quickly identified. The polymorphism database contains molecular as well as population data: size and ethnicity of all the populations in which a given variation was identified, the number of tested chromosomes, and the frequency of each allele. In the future, patients in which these polymorphisms have been found should be added. This information can be helpful to determine if certain *FBN1* genotypes are associated with more severe phenotypes. These data should provide tools to start to interpret the phenotypic variability associated with a mutation in different probands or in the same family.

Conclusion

Elucidating the molecular basis of MFS and related fibrillinopathies is the major goal of the teams working on this subject.[16,48] The extreme clinical variability, the difficulties associated with clinical diagnosis and the low detection rate of mutations in this large gene all conspire to negatively impact on progress. At present it is not possible to predict the phenotype for a given *FBN1* mutation. On the one hand, mutations affecting different positions within a given module may be associated with quite different phenotypes. On the other hand, mutations affecting an analogous residue within two different modules may also be associated with differing phenotypes. Therefore, it is apparent that neither the location of the affected structural module in the protein nor the position of the altered residue is, in itself, sufficient to predict potential genotype-phenotype correlations.[50] The high degree of intrafamilial variability suggests that environmental and perhaps stochastic or epigenetic factors are important for the phenotypic expression of disease. The effects of unknown modifier (enhancing or protecting) genes on the clinical expression as well as conjugation of different alleles of the multiple fibrillin-interacting proteins are likely to constitute the foundation of an enhanced susceptibility for disease severity. All these hypotheses are starting points for future research.

References

1. Faivre L, Gorlin R, Wirtz M et al. In-frame fibrillin-1 gene deletion in autosomal dominant Weill-Marchesani syndrome. J Med Genet 2003; 40:34-36.
2. Cariello N, Cui L, Béroud C et al. Database and software for the analysis of mutations in the human p53 gene. Cancer Res 1994; 54: 4454–4460.
3. Béroud C, Collod-Béroud G, Boileau C et al. UMD (Universal mutation database): a generic software to build and analyze locus specific databases. Hum Mutat 2000; 15:86-94.
4. Béroud C, Soussi T. p53 gene mutation: Software and data-base. Nucl Acids Res 1998b; 26:200-204.
5. Béroud C, Soussi T. APC gene: database of germline and somatic mutations in human tumors and cell lines. Nucl Acids Res 1996a; 24:121–124.
6. Laurent-Puig P, Béroud C, Soussi T. APC gene: Database of germline and somatic mutations in human tumors and cell lines. Nucl Acids Res 1998; 26:269–270.

7. Béroud C, Joly D, Gallou C et al. Software and database for the analysis of mutations in the VHL gene. Nucl Acids Res 1998a; 26:256–258.
8. Gallou C, Joly D, Méjean A et al. Mutations of the VHL gene is sporadic renal cell carcinoma definition of a risk fac-tor for VHL patients to develop an RCC. Hum Mutat 1999; 13:464–475.
9. Jeanpierre C, Béroud C, Niaudet P et al. Software anddatabase for the analysis of mutations in the human WT gene. Nucl Acids Res 1998; 26:271–274.
10. Varret M, Rabes J, Collod-Béroud G et al. Software and database for the analysis of mutations in the human LDL receptor gene. Nucl Acids Res 1997; 25:172-80.
11. Varret M, Rabes J, Thiart R et al. LDLR Database, 2nd ed. New additions to the database and the soft-ware, and results of the first molecular analysis. Nucl Acids Res 1998; 26:248–252.
12. Villeger L, Abifadel M, Allard D et al. The UMD-LDLR database: Additions to the software and 490 new entries to the database. Hum Mutat 2002; 20:81-87.
13. Collod G, Béroud C, Soussi T et al. Software and database for the analysis of mutations in the human FBN1 gene. Nucl Acids Res 1996; 24(1):137-140.
14. Collod-Béroud G, Béroud C, Ades L et al. Marfan Database (second edition): Software and data-base for the analysis of mutations in the human FBN1 gene. Nucl Acids Res 1997; 25(1):147-50.
15. Collod-Béroud G, Béroud C, Ades L et al. Marfan Database (third edition): New mutations and new routines for the software. Nucl Acids Res 1998; 26(1):229-3.
16. Collod-Béroud G, Le Bourdelles S, Ades L et al. New update of the UMD-FBN1 mutation data-base and creation of a FBN1 polymorphism database. Hum Mutat 2003; 22:199-208.
17. den Dunnen J, Antonarakis S. Nomenclature for the description of human sequence variations. Hum Genet 2001; 109:121-4.
18. Pereira L, D'Alessio M, Ramirez F et al. Genomic organization of the sequence coding for fibrillin, the defective gene product in Marfan Syndrome. Hum Mol Genet 1993; 2:961-968.
19. Aoyama T, Francke U, Dietz H et al. Quantitative differences in biosynthesis and extracellular deposition of fibrillin in cultured fibroblasts distinguish five groups of Marfan syndrome patients and suggest distinct pathogenetic mechanisms. J Clin Invest 1994; 94:130-137.
20. Nijbroek G, Sood S, McIntosh I et al. Fifteen novel FBN1 mutations causing Marfan syndrome detected by heteroduplex analysis of genomic amplicons. Am J Hum Genet 1995; 57:8-21.
21. Schrijver I, Liu W, Odom R et al. Premature termination mutations in FBN1: Distinct effects on differential allelic expression and on protein and clinical phenotypes. Am J Hum Genet 2002; 71:223-237.
22. Rommel K, Karck M, Haverich A et al. Mutation screening of the fibrillin-1 (FBN1) gene in 76 unrelated patients with Marfan syndrome or Marfanoid features leads to the identification of 11 novel and three previously reported mutations. Hum Mutat 2002; 20:406-407.
23. Wang M, Kishnani P, Decker-Phillips M et al. Double mutant fibrillin-1 (FBN1) allele in a pa-tient with neonatal Marfan syndrome. J Med Genet 1996; 33(9):760-3.
24. Tiecke F, Robinson P, Booms P et al. Classic, early-onset severe and neonatal Marfan syndrome: Twelve mutations and genotype-phenotype correlations in FBN1 exons 24-40. Eur J Hum Genet 2001; 9:13-21.
25. Dietz H, Valle D, Francomano C et al. The skipping of constitutive exons in vivo induced by nonsense mutations. Science 1993b; 259:680-683.
26. Loeys B, Nuytinck L, Delvaux I et al. Genotype and phenotype analysis of 171 patients referred for molecular study of the fibrillin-1 gene FBN1 because of suspected Marfan syndrome. Arch Intern Med 2001; 161(20):2447-54.
27. Reber-Müller S, Spissinger T, Schuchert P et al. An extracellular matrix protein of jellyfish ho-mologous to mammalian fibrillins forms different fibrils depending on the life stage of the animal. Dev Biol 1995; 169:662-672.
28. Doolittle R. Reconstructing history with amino acid sequences. Protein Sci 1992; 1:191-200.
29. Kainulainen K, Sakai L, Child A et al. Two mutations in Marfan syndrome resulting in truncated fibrillin polypeptides. Proc Natl Acad Sci USA 1992; 89:5917-5921.
30. Liu W, Schrijver I, Brenn T et al. Multi-exon deletions of the FBN1 gene in Marfan syndrome. BMC Med Genet 2001; 2(1):1.
31. Karttunen L, Ukkonen T, Kainulainen K et al. Two novel fibrillin-1 mutations resulting in prema-ture termination codons in different mutant transcript levels and clinical phenotypes. Hum Mutat 1998; supplement 1:S34-S37.
32. Krawczak M, Cooper D. Gene deletions causing human genetic disease: mechanisms of mutagen-esis and the role of the local DNA sequence environment. Hum Genet 1991; 86:425-441.
33. Cooper T, Mattox W. Gene regulation'97. The regulation of splice-site selection, and its role in Human disease. Am J Hum Genet 1997; 61:259-266.

34. Dietz H, McIntosh I, Sakai L et al. Four novel FBN1 mutations: Significance for mutant transcript level and EGF-like domain calcium binding in the pathogenesis of Marfan syndrome. Genomics 1993a; 17:468-475.

35. Liu W, Qian C, Francke U. Silent mutation induces exon skipping of fibrillin-1 gene in Marfan syndrome. Nat Genet 1997; 16(4):328-9.

36. Dietz H, Kendzior RJ. Maintenance of an open reading frame as an additional level of scrutiny during splice site selection. Nat Genet 1994; 8(2):183-8.

37. Caputi M, Kendzior RJ, Beemon K. A nonsense mutation in the fibrillin-1 gene of a Marfan syndrome patient induces NMD and disrupts an exonic splicing enhancer. Genes Dev 2002; 16(14):1754-9.

38. Frischmeyer PA, Dietz HC. Nonsense-mediated mRNA decay in health and disease. Hum Mol Genet 1999; 8(10):1893-1900.

39. Aoyama T, Tynan K, Dietz HC et al. Missense Mutations Impair Intracellular Processing of Fibrillin and Microfibril Assembly in Marfan Syndrome. Hum Mol Genet 1993; 2(12):2135-2140.

40. Halliday D, Hutchinson S, Kettle S et al. Molecular analysis of eight mutations in FBN1. Hum Genet 1999; 105:587-597.

41. Schrijver I, Liu W, Brenn T et al. Cysteine substitutions in epidermal growth factor-like domains of fibrillin-1: distinct effects on biochemical and clinical phenotypes. Am J Hum Genet 1999; 65:1007-20.

42. Johnson A, Haigh N. The ER translocon and retrotranslocation: is the shift into reverse manual or automatic? Cell 2000; 102:709-712.

43. Lippincott-Schwartz J, Roberts T, Hirshberg K. Secretory protein trafficking and organelle dynamics in living cells. Ann Rev Cell Dev Biol 2000; 16:557-589.

44. Lord J, Davey J, Frigerio L et al. Endoplasmic reticulum-associated protein degradation. Cell Dev Biol 2000; 11:159-164.

45. Downing A, Knott V, Werner J et al. Solution structure of a pair of calcium-binding epidermal growth factor-like domains: implications for the Marfan syndrome and other genetic disorders. Cell 1996; 85:597-605.

46. Knott V, Downing AK, Cardy CM et al. Calcium binding properties of an epidermal growth factor-like domain pair from human fibrillin-1. J Mol Biol 1996; 255(1):22-7.

47. Cardy CM, Handford PA. Metal ion dependency of microfibrils supports a rod-like conformation for fibrillin-1 calcium-binding epidermal growth factor-like domains. J Mol Biol 1998; 276(5):855-60.

48. Collod-Béroud G, Boileau C. Marfan syndrome in the third millenium. Eur J Hum Genet 2003; 10:673-681.

49. Robinson P, Booms P, Katzke S et al. Mutations of FBN1 and genotype-phenotype correlations in Marfan syndrome and related fibrillinopathies. Hum Mutat 2002; 20:153-61.

50. Palz M, Tiecke F, Booms P et al. Clustering of mutations associated with mild Marfan-like phenotypes in the 3' region of FBN1 suggests a potential genotype-phenotype correlation. Am J Med Genet 2000; 91:212-21.

CHAPTER 9

Familial Thoracic Aortic Aneurysms and Dissections

Sumera N. Hasham and Dianna M. Milewicz

Introduction

Ascending aortic aneurysms leading to type A aortic dissections are the major cardiovascular complication of the Marfan syndrome (MFS). MFS is a major genetic syndrome predisposing individuals to these aortic conditions but other genetic syndromes also have similar aortic problems. In addition, ascending thoracic aortic aneurysms and dissections can be inherited in an autosomal dominant manner with decreased penetrance and variable expression, and the locations of the genes contributing to familial thoracic aortic aneurysms and dissections are beginning to be mapped in the human genome.

Aortic Aneurysms and Dissections

An aortic aneurysm is defined as the permanent localized dilatation of the aorta having a diameter at least 1.5 times that of the expected normal diameter of that given segment. Aortic aneurysms can develop anywhere along the length of the aorta and are most commonly classified based on the anatomical location. Abdominal aortic aneurysms (AAAs) occur in the infrarenal abdominal aorta, which is the most common location for aneurysm formation. The next most common aneurysm involves the ascending thoracic aortic aorta above the aortic valve. Less common are aneurysms involving the arch, descending or thoracoabdominal aortic aneurysms (TAAAs), which originate in the descending aorta and extend to the abdominal aorta. Morphologically, aortic aneurysms may be fusiform or, less commonly, saccular. Fusiform aneurysms are characterized by circumferential widening of the aorta, whereas saccular aneurysms represent localized outpocketings of the aortic wall. Nomograms, which demonstrate the range of normal aortic size within the context of body size and age, are used as a reference point for defining aortic aneurysms and in the decision-making process for treatment of aortic enlargement.[1] Nomograms have been established for the ascending thoracic aorta but not for other anatomic regions of the aorta.

Aortic dissection, a term coined by Leannec, is defined as the separation of aortic media, due to the flow of blood within the layers of aortic wall, which creates an extraluminal channel called the false lumen. Following a tear in the intimal layer, the diseased medial layer is exposed to the pulse pressure of the intraluminal blood, which then penetrates the diseased medial layer dissecting the aortic wall. The ascending aorta within several centimeters of the sinuses of Valsalva is the most favored area for the primary intimal tear and dissections involving this segment are termed type A dissections (Stanford classification) or Type I or II dissections (Debakey classification).[2] The next most likely site for origin of a dissection is the descending thoracic aorta just distal to the branch of the subclavian artery (Type B or Type III dissection). Once a false lumen is formed, there can be a reentry into the true lumen by secondary tear or there can be an external rupture, such as into the pericardial space.[3]

Marfan Syndrome: A Primer for Clinicians and Scientists, edited by Peter N. Robinson and Maurice Godfrey. ©2004 Eurekah.com and Kluwer Academic / Plenum Publishers

A type A aortic dissection can occur without a predisposing ascending thoracic aneurysm or can occur following the progressive dilatation of an aortic aneurysm. Therefore, these conditions are interrelated; with both disorders reflecting an underlying weakening of the arterial wall. Recent data indicates that it is the active remodeling of the aortic wall involving a complex milieu of ECM proteins, proteases and their inhibitors that manifest in the form of these aortic diseases.[4]

In contrast to AAA in which atherosclerosis is the pathology involved, aneurysms involving the ascending thoracic aorta, including in Marfan syndrome (MFS), are associated with a pathologic lesion defined as medial degeneration, originally termed cystic medial necrosis (CMN) by Erdheim.[5] Medial degeneration is not a single disease entity but is commonly seen in aging aorta and in patients with hypertension, and hypertension is a risk factor for both aneurysms and dissections.[6,7] However, the changes are qualitatively and quantitatively seen to be much greater in patients with thoracic aortic aneurysms and dissections (TAAD).

Cystic medial necrosis is neither cystic nor necrotic, but rather associated with the loss of elastic fibers and smooth muscle cells in the medial layer of the aorta, and subsequent accumulation of mucoid material. Increased immunoreactivity for metalloproteinases (MMPs) and tissue inhibitors of MMPs (TIMPs) in SMCs is evident in vascular specimens from MFS patients.[8] This imbalance in the proteases and their inhibitors may promote both fragmentation of medial elastic layers and elastolysis. However, the detailed mechanism of the remodeling involved in this pathological process is not completely understood. Studies have implicated an altered phenotype of the smooth muscle cells (SMCs) in the aortic wall of patients with aneurysms that may contribute to the elastic fiber loss observed in medial degeneration. In the aortas of patients with TAAD, ultrastructural analysis of the SMCs was characteristic of the synthetic phenotype with enlarged endoplasmic reticulum, and decreased amount of myofilaments, and increased expression of osteopontin, MMP 1, 2, and 9 and tissue inhibitor of MMP 1 and 2.[9] The detailed mechanism of smooth muscle cells loss is not completely understood, but apoptosis is emerging as a process of smooth muscle cell loss in diseased aorta.[10,11] Peroxisome proliferator activated receptor-γ (PPAR-γ) expression has been seen to be upregulated in the SMCs in MFS and parallels CMN, suggesting its involvement in the pathogenesis of CMN and aortic disease progression.[12] The cause of the transition between the altered SMC phenotype to SMC apoptosis has not been determined.

Thoracic Aortic Aneurysms and Dissection Associated with Genetic Syndromes

Medial degeneration of the proximal aorta leading to aneurysms and dissections is a major feature of Marfan syndrome (MFS), and is less commonly associated with other genetic syndromes such as Ehler-Danlos syndrome,[13] polycystic kidney disease, congenital contractural arachnodactyly (CCA or Beals syndrome),[14] bicuspid aortic valve,[15] as well as Turner,[16] and Noonan[17] syndrome.

A second locus for a MFS-like disorder was identified at 3p24.2-p25 (*MFS2*, OMIM #154705) based on linkage analysis of a large French family but the findings in this family have been controversial.[18] Although the affected members of the family did not meet the diagnostic criteria for classical MFS, the phenotype was similar to MFS with complications in the skeletal and cardiovascular systems, but no ocular features. The characterization of the clinical phenotype and the mapping of the locus were controversial and discounted by many investigators.[18-21] It was suggested that intrafamilial variability for MFS was present in the French family rather than genetic heterogeneity for MFS. No subsequent families with MFS have demonstrated linkage to *MFS2*, hence raising further concerns about both the validity of the *MFS2* locus and the possibility of genetic heterogeneity for MFS. More recently, a locus for familial TAAD has been mapped to this region, as described below.

Mutations in the gene for fibrillin-2, *FBN2*, cause congenital contractural arachnodactyl (CCA, OMIM #121050), a syndrome that is closely related to MFS.[22] Patients with CCA have congenital contractures of joints, primarily the fingers, toes, elbows and knees, an abnormal, crumpled appearance to the helix of the ears, and skeletal manifestations of MFS, in particular scoliosis. Although CCA patients were initially reported not to have aortic root dilatation, it has been shown that a minority of CCA patients with characterized *FBN2* mutations have aortic root dilatation.[14,23] It has not been determined if the aortic root dilatation observed in CCA patients will progress to aortic dissection or rupture, as occurs in patients with MFS.

Ehlers Danlos syndrome (EDS) is a group of heterogeneous disorders characterized by joint laxity, skin fragility, and abnormal scarring, and classified based primarily on the molecular defect for each form. EDS, vascular type, resulting from mutations in the *COL3A1* gene (which encodes the peptides of type III collagen), is an autosomal dominant disorder characterized by spontaneous rupture and dissection of the aorta and large to medium sized arteries.[24] The arterial complications of EDS, vascular type, typically involve rupture of medium size arteries but three families have been described with primarily aortic disease.[25,26] Furthermore, an analysis of type III collagen in patients with aortic disease indicated that COL3A1 mutations were a rare cause of aortic disease.[27]

EDS, classic type, is associated with joint laxity, skin fragility and abnormal scarring, while EDS, hypermobility type, has hypermobility as the primary feature.[28] EDS, classic type, results from mutations in type V collagen however there is genetic heterogeneity with many cases due to mutations in unidentified genes.[29,30] A mutation in COL3A1 has been reported in one case of EDS, hypermobility type, but there have been no subsequent reports of the involvement of this gene.[31] Wenstrup et al. found that dilatation of the ascending aorta is common finding (28%) in individuals with both EDS, classic type, and EDS, hypermobility type.[32,33]

Turner syndrome is a sex aneuploidy syndrome in which the most frequent chromosome constitution is 45, X. The cardiovascular problems include bicuspid aortic valve, coarctation of the aorta, hypertension, and TAAD.[16,34] Aortic root dilatation is present in approximately 40% of Turner patients. Aortic dissection has been reported in Turner patients as young as 4 years of age, but more commonly occurs in patients above 20 years of age.[34]

Familial Thoracic Ascending Aortic Aneurysms and Dissections

Although TAADs occur as a major manifestation of connective tissue disorders, like Marfan syndrome, and aneuploidy syndromes, such as Turner syndrome, the vast majority of TAAD occurs in patients without an identified genetic syndrome. More than 60% of patients with an ascending aortic aneurysm are asymptomatic, and the aneurysms usually are detected on chest radiography obtained for another reason. Among symptomatic patients, chest pain (17% of cases) and back pain are the most common complaints, and generally are associated with ascending aortic aneurysms and descending aneurysms, respectively. Other symptoms related to thoracic aneurysms like dyspnea and syncope usually arise from compression of adjacent structures or from aneurysm rupture. Ascending aortic aneurysms predispose an individual to a type A dissection; in addition, a type A dissection can occur without predisposing aneurysms. Acute chest pain is the most common presentation of aortic dissection and occurs in 90% to 95% of patients.[35] About 70% of the patients have a radiating pattern of pain, based on the extent and progression of the dissection. Patients who do not have pain may have other associated symptoms, such as dyspnea, syncope and unconsciousness. The asymptomatic nature of the aortic aneurysms and life threatening dissections that result from undiagnosed aneurysms make it important to identify the genes that predispose individuals to these life-threatening aortic conditions.

To clarify the genetic contribution in these patients with nonsyndromic TAAD, Biddinger and colleagues compared first-degree relatives of patients referred for surgical repair of ascending aortic aneurysms or type A dissections with a control group without aortic disease for differences in occurrence of these disorders.[26] Eleven percent of the 158 individuals interviewed had a first degree relative with TAAD. Additionally, the study also demonstrated that the first-degree relatives of the probands had a higher prevalence of sudden, unexplained death.

A subsequent study interviewed one hundred and thirty five patients, of which 19.3% of the patients had a first degree relative with TAAD. The mean age of onset was found to be younger for the familial cases than sporadic cases but older than cases with MFS.[36] Analysis of the TAAD families in these two familial aggregation studies indicated more than one mode of transmission was apparent in the families, though an autosomal dominant mode of inheritance was found in majority of the families. These studies confirmed that the first-degree relatives of patients with TAAD are at a higher risk of TAADs compared with a control group. In addition, these studies support the hypothesis that genetic factors play a role in the etiology of TAADs in patients who do not have an identified syndrome causing aortic disease.

Families with multiple members presenting with nonsyndromic aortic disease have been reported in the literature.[37,38] Nicod and colleagues described a family in which nine members over two generations had an aortic dissecting aneurysm or aortic or arterial dilation at a young age.[37] None of the patients had Marfan syndrome or a history of systemic hypertension. Three members died of ruptured aortic dissecting aneurysm between 14 to 24 years of age while two other members underwent aortic repair for aortic dilatation. Histologic examination of the aortic wall at autopsy or surgery in three patients revealed a pattern of cystic medial necrosis. Milewicz and colleagues described six families with aortic aneurysms and dissections, all of which demonstrated autosomal dominant inheritance associated with decreased penetrance and variable age of onset of the aortic disease.[38] Pathological studies of the aortas in affected individuals demonstrated medial degeneration. Careful clinical follow-up identified individuals as young as 11 years to have aortic root dilatation. Linkage analysis excluded *FBN1* and *COL3A1* as the cause of disease in these families.

Various studies indicated that the aortic disease in the majority of these families is not due to a mutation in the *FBN1* gene or other genes encoding vascular proteins, such as *COL3A1*.[27,39,40] The dramatic familial clustering of aortic disease in these described families suggested a genetic basis to the disease transmitted primarily in an autosomal dominant mode. Numerous families with autosomal dominant inheritance have been identified.[41] The majority of affected individuals in these families present with asymptomatic aortic root dilatation or acute type I dissections. Decreased penetrance is evident by the fact that the aortic disease can skip generations, i.e., an individual carries the defective gene but never presents with aortic disease. Variable age of onset can be seen with the aortic disease presenting in individuals ranging from age 14 to 80 years of age.

Recent studies are beginning to elucidate the genes that predispose individuals without a known syndrome to TAAD. Initially it was determined that *FBN1* mutations can lead to TAAD in individuals who fail to meet the diagnostic criteria for MFS. The *FBN1* mutation G1127S in exon 27 was found in 9 of the 10 members of a family in whom ascending aortic disease had been identified. None of the individuals had MFS although the affected individuals had limited skeletal features of MFS.[40,42] Two mutations, D1155N in *FBN1* exon 28 and P1837S in *FBN1* exon 44, were identified in two unrelated patients with TAAD, who also had minimal skeletal features of MFS.[39] In addition, *FBN1* screening was done on 16 adult patients who had TAAD, either alone (11 patients) or associated with minor skeletal features of MFS[4] or striae,[1] and no *FBN1* mutations were identified.[43] The studies indicate that *FBN1* mutations are a rare cause of TAAD in individuals and families that fail to meet the diagnostic criteria for MFS, and these individuals may have mild skeletal features of MFS.

Through reported cases and familial aggregation studies, large families with dominant inheritance of TAAD were identified, and a parametric positional cloning approach was applied to map the defective genes causing nonsyndromic TAAD. The mapping of first locus, termed the *TAAD1* locus, was achieved through a genome-wide scan using the DNA obtained from members affected with the aortic disease from two large families[41] (Fig. 1). Linkage analysis using closely spaced markers at the chromosomal loci included in the genome-wide scan using DNA from multiple families (n=15) confirmed linkage of some of the families to 5q13-14. Genetic heterogeneity was confirmed with the cosegregation of the disease and the *TAAD1*

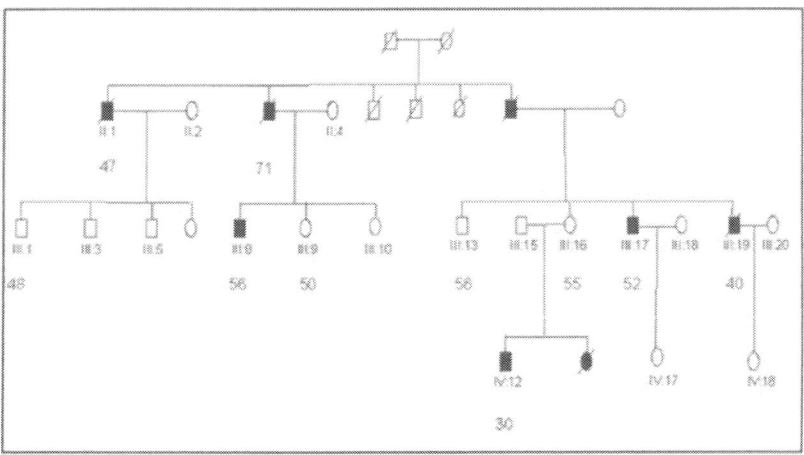

Figure 1. Pedigree of a family with multiple members with thoracic aortic aneurysms and dissections linked to *TAAD1*. Round symbols indicate females; square symbols indicate males. Symbols with a line crossed through it symbolize a deceased person. Blackened symbols indicate affected individuals (affected status was assigned to individuals with aortic aneurysms/dissections, aortic surgical repair, or aortic dissection reported in autopsy report). Open symbols indicate normal ascending aortic size at the sinus of Valsalva. An autosomal dominant pattern of inheritance is observed for the disease in the family. Variable age of onset is exemplified by individuals II:1 and IV:12 (age of onset 61 and 24 years, respectively). Skipping of generations as seen in individuals II:8 (affected), III:16 (unaffected), and IV:12 (affected) indicates the decreased penetrance of the condition in the family.

locus, with alpha, the proportion of families showing evidence of linkage as 0.45. The maximum LOD score of 4.74 was obtained with marker D5S2029. Recently, this locus has been confirmed by an independent study in a Finnish population where approximately one half of the families studied show evidence of linkage to *TAAD1*.[44] The critical interval containing the defective gene maps to a 7.8 cM region. Multiple genes in the critical interval encoding the ECM proteins or the ECM interacting proteins have been sequenced, but the defective gene has not been identified.

Another locus for familial aortic aneurysms and dissections has been mapped to the long arm of chromosome 11(11q23-24) using a single large family.[45] Contrary to the families linked to *TAAD1*, the clinical phenotype of the family linked to *FAA1* locus, indicates a diffuse vascular etiology. Apart from dilatation in the sinuses of Valsalva, involvement of other aortic segments and arteries were also observed, such as dilatation in the abdominal aorta and left subclavian artery. Therefore, contrary to the clinical findings in the families linked to *TAAD1*, the disease linked to *FAA1* is characterized by the involvement of other segments of aorta and smaller arteries along with aneurysms of thoracic aorta. The disease is fully penetrant and is a rare cause of the vascular condition as no other family is linked to this locus. These studies confirm the genetic heterogeneity involved in the etiology of TAAD.

Recently, a second locus for TAAD has been mapped using another large family with multiple members with aneurysms and dissections of the thoracic aorta. The disease in the family was characterized as autosomal dominant with decreased penetrance and variable age of onset. Using a positional cloning approach similar to that used for TAAD1, the locus has been mapped to a 25cM region on 3p24-25, termed *TAAD2* locus (Fig. 2).[46] A maximum LOD score of 4.27 was obtained with D3S2336. Eighteen TAAD families described previously failed to show linkage to 3p24-25, indicating that *TAAD2* is a minor locus for TAAD.[41,45] Interestingly, the *TAAD2* interval encompasses the *MFS2* locus previously mapped to a 9cM region between D3S1293 and D3S1283 using a large French family. The phenotype of the affected members

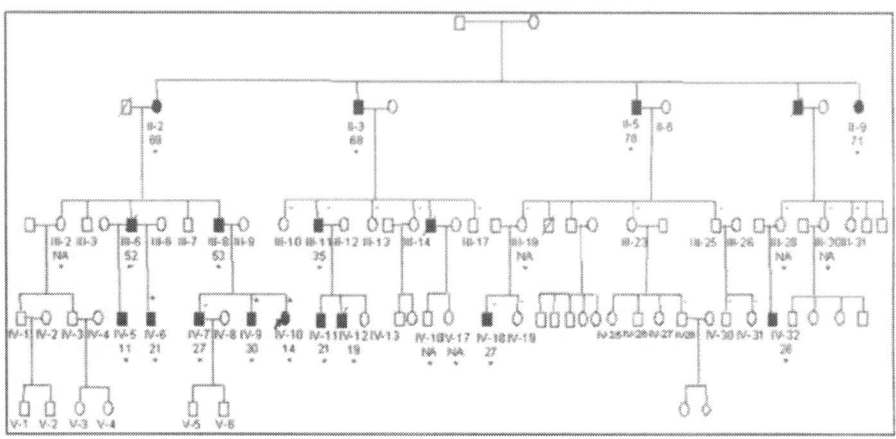

Figure 2. Pedigree of a family with multiple members with thoracic aortic aneurysms and dissections linked to *TAAD2*. Round symbols indicate females; square symbols indicate males. Symbols with a line crossed through it symbolize a deceased person. Blackened symbols indicate affected individuals (affected status was assigned to individuals with aortic aneurysms/dissections, aortic surgical repair, or aortic dissection reported in autopsy report). Open symbols indicate normal ascending aortic size at the sinus of Valsalva. The arrow indicates the proband. An autosomal dominant pattern of inheritance is observed for the disease in the family. Variable age of onset is exemplified by individuals II:2 and IV:11 (age of onset 69 and 21 years, respectively). Skipping of generations as seen in individuals II:5 (affected), III:19 (unaffected), and IV:18 (affected) indicates the decreased penetrance. The asterisk indicates the individuals who carry the affected haplotype.

in the French family included some findings similar to MFS complications in the skeletal and cardiovascular systems but no ocular complications. In the family used to map the *TAAD2* locus, cardiovascular features alone were used for assigning the affected status. Although some of the individuals from this family had minor skeletal features of MFS, none of the individuals met the diagnosis of MFS based on Ghent criteria.[47] The similarities of the cardiovascular phenotype suggest that the segregation of TAAD and MFS-like features in these two families may result from mutations in the same gene. Identification of the gene will decipher if *TAAD2* is allelic to *MFS2*.

Bicuspid aortic valve (BAV) is a common congenital heart defect. Familial aggregation studies of patients with BAV have indicated that 9% have first-degree relatives with BAV and that the disease segregates in an autosomal dominant fashion. Multiple studies have shown that BAV is associated with TAAD[48-50] Roberts and Roberts (1991) studied 186 patients with aortic dissection at autopsy.[51] Pathological examinations have shown severe medial degeneration in the aortas of the patients with BAV, indicating that these two conditions have a common developmental defect. In contrast to MFS where the initial dilatation of the aorta is at the sinuses of Valsalva, the initial aortic dilatation in patients with BAV is typically above the sinuses of Valsalva. Numerous pedigrees have been identified, with individuals either with TAAD, BAV or both indicating that BAV is associated with TAAD and that it segregates in an autosomal dominant fashion with variable penetrance and expression.[52,53]

Patent ductus arteriosus (PDA) is the second most common congenital heart defect, present in approximately 1/1600-5000 live births. A genetic contribution to this condition is suggested by the fact that 5% of the of PDA cases have a relative with PDA.[54] The genes involved in the defect have not been well characterized with the exception of syndromic form of PDA called Char syndrome (169100), an autosomal dominant disorder caused by mutations in the

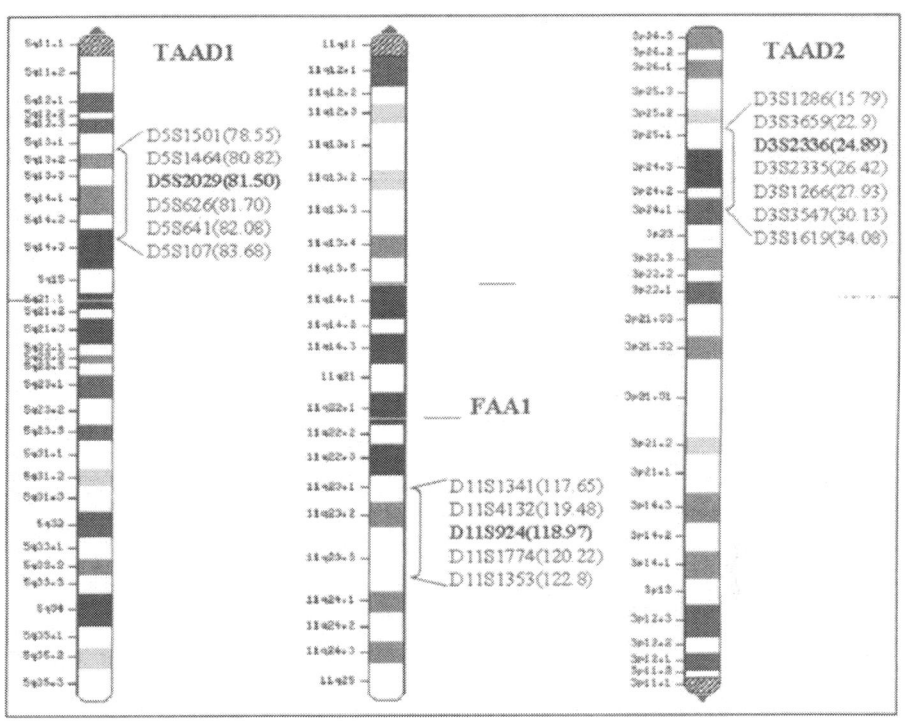

Figure 3. The three different loci mapped for TAAD. The three loci, *TAAD1* (5q13-14), *FAA1* (11q23-24) and *TAAD2* (3p24-25) are indicated along with their location. The chromosomal image of only the arm which harbors the defective gene is represented. The arrow at the tip of the chromosomal arm indicates the centromeric end. The markers used in the mapping studies are shown according to their location in Mb, based on the UCSC Genome browser, July 2003. The marker in bold indicates the marker which gave the maximum LOD score at that specific locus in the published study.

transcription factor TFAP2B (601601). More recently, an autosomal recessive model with decreased penetrance has been shown and a gene for PDA (*PDA1*) has been mapped to 12q24 in Iranian kindreds using homozygosity mapping.[55] There have been multiple reports of aortic aneurysms being associated with PDA[56,57] and an autosomal dominant model of inheritance has been proposed based on the available pedigrees.[58] However, studies are needed to see if this associated condition is caused due to mutations in the same gene as the isolated PDA cases or a TAAD gene, or a novel locus yet to be identified.

In summary, three loci, *TAAD1*, *FAA1*, and *TAAD2* are known to harbor the genes defective in nonsyndromic familial TAAD (Fig. 3). A major locus for TAAD has been mapped to the long arm of chromosome 5q and termed the *TAAD1* locus.[41] One half of the families studied demonstrate evidence of linkage of the aortic disease to this locus. A second locus, *FAA1*, has been identified at 11q with only one family linked to this locus.[45] A third locus, *TAAD2* has been mapped in a single large family to 3p24-25, encompassing the *MFS2* locus previously mapped for MFS. Since approximately one-half of the families identified still fail to map to a known locus, there exists at least one more locus for TAAD to be characterized. Thus the studies till date have established the genetic heterogeneity for the aortic disease, TAAD in the absence of associated genetic disorders. Identification of the TAAD genes will allow for timely identification of individuals at risk for this life threatening condition and also allow for new insights into the proteins which maintain the structural integrity of the aortic wall.

References

1. Roman MJ, Devereux RB, Kramer-Fox R et al. Two-dimensional echocardiographic aortic root dimensions in normal children and adults. Am J Cardiol 1989; 64(8):507-512.
2. Karmy-Jones R, Aldea G, Boyle Jr EM. The continuing evolution in the management of thoracic aortic dissection. Chest 2000; 117(5):1221-1223.
3. Roberts WC. Aortic dissection: Anatomy, consequences, and causes. Am Heart J 1981; 101(2):195-214.
4. Langille BL. Arterial remodeling: Relation to hemodynamics. Can J Physiol Pharmacol 1996; 74(7):834-841.
5. Erdheim J. Medionecrosis aortae idiopathica cystica. Virchows Arch Path Anat 1930; 276:187.
6. Carlson RG, Lillehei CW, Edwards JE. Cystic medial necrosis of the ascending aorta in relation to age and hypertension. Am J Cardiol 1970; 25(4):411-415.
7. Schlatmann TJ, Becker AE. Histologic changes in the normal aging aorta: Implications for dissecting aortic aneurysm. Am J Cardiol 1977; 39(1):13-20.
8. Segura AM, Luna RE, Horiba K et al. Immunohistochemistry of matrix metalloproteinases and their inhibitors in thoracic aortic aneurysms and aortic valves of patients with Marfan's syndrome. Circulation 1998; 98(19 Suppl):II331-7.
9. Lesauskaite V, Tanganelli P, Sassi C et al. Smooth muscle cells of the media in the dilatative pathology of ascending thoracic aorta: Morphology, immunoreactivity for osteopontin, matrix metalloproteinases, and their inhibitors. Hum Pathol 2001; 32(9):1003-1011.
10. Pressler V, McNamara JJ. Aneurysm of the thoracic aorta. Review of 260 cases. J Thorac Cardiovasc Surg 1985; 89(1):50-54.
11. Daily PO, Trueblood HW, Stinson EB et al. Management of acute aortic dissections. Ann Thorac Surg 1970; 10(3):237-247.
12. Sakomura Y, Nagashima H, Aoka Y et al. Expression of peroxisome proliferator-activated receptor-gamma in vascular smooth muscle cells is upregulated in cystic medial degeneration of annuloaortic ectasia in Marfan syndrome. Circulation 2002; 106(12 Suppl 1):I259-I263.
13. Leier CV, Call TD, Fulkerson PK et al. The spectrum of cardiac defects in the Ehlers-Danlos syndrome, types I and III. Ann Intern Med 1980; 92(2 Pt 1):171-178.
14. Gupta PA, Putnam EA, Carmical SG et al. Ten novel FBN2 mutations in congenital contractural arachnodactyly: Delineation of the molecular pathogenesis and clinical phenotype. Hum Mutat 2002; 19(1):39-48.
15. McKusick VA, Logue RB, Bahson HT. Association of aortic valvular disease and cystic medial necrosis of the ascending aorta. Circulation 1957; 16:188-194.
16. Lie JT. Aortic dissection in Turner's syndrome. Am Heart J 1982; 103(6):1077-1080.
17. Shachter N, Perloff JK, Mulder DG. Aortic dissection in Noonan's syndrome (46 XY turner). Am J Cardiol 1984; 54(3):464-465.
18. Collod G, Babron MC, Jondeau G et al. A second locus for Marfan syndrome maps to chromosome 3p24.2-p25 [see comments]. Nat Genet 1994; 8(3):264-268.
19. Boileau C, Jondeau G, Babron MC et al. Autosomal dominant Marfan-like connective-tissue disorder with aortic dilation and skeletal anomalies not linked to the fibrillin genes [see comments]. Am J Hum Genet 1993; 53(1):46-54.
20. Gilchrist DM. Marfan syndrome or Marfan-like connective-tissue disorder. Am J Hum Genet 1994; 54(3):553-554.
21. Dietz H, Francke U, Furthmayr H et al. The question of heterogeneity in Marfan syndrome. Nat Genet 1995; 9(3):228-231.
22. Putnam EA, Zhang H, Ramirez F et al. Fibrillin-2 (FBN2) mutations result in the Marfan-like disorder, congenital contractural arachnodactyly. Nat Genet 1995; 11(4):456-458.
23. Park ES, Putnam EA, Chitayat D et al. Clustering of FBN2 mutations in patients with congenital contractural arachnodactyly indicates an important role of the domains encoded by exons 24 through 34 during human development. Am J Med Genet 1998; 78(4):350-355.
24. Pepin M, Schwarze U, Superti-Furga A et al. Clinical and genetic features of Ehlers-Danlos syndrome type IV, the vascular type [see comments]. N Engl J Med 2000; 342(10):673-680.
25. Kontusaari S, Tromp G, Kuivaniemi H et al. A mutation in the gene for type III procollagen (COL3A1) in a family with aortic aneurysms. J Clin Invest 1990; 86(5):1465-1473.
26. Biddinger A, Rocklin M, Coselli J et al. Familial thoracic aortic dilatations and dissections: A case control study. J Vasc Surg 1997; 25(3):506-511.
27. Tromp G, Wu Y, Prockop DJ et al. Sequencing of cDNA from 50 unrelated patients reveals that mutations in the triple-helical domain of type III procollagen are an infrequent cause of aortic aneurysms. J Clin Invest 1993; 91(6):2539-2545.

28. Beighton P, De Paepe A, Steinmann B et al. Ehlers-Danlos syndromes: Revised nosology, Villefranche, 1997. Ehlers-Danlos National Foundation (USA) and Ehlers-Danlos Support Group (UK). Am J Med Genet 1998; 77(1):31-37.
29. Burrows NP, Nicholls AC, Yates JR et al. The gene encoding collagen alpha1(V)(COL5A1) is linked to mixed Ehlers-Danlos syndrome type I/II. J Invest Dermatol 1996; 106(6):1273-1276.
30. Wenstrup RJ, Langland GT, Willing MC et al. A splice-junction mutation in the region of COL5A1 that codes for the carboxyl propeptide of pro alpha 1(V) chains results in the gravis form of the Ehlers-Danlos syndrome (type I). Hum Mol Genet 1996; 5(11):1733-1736.
31. Narcisi P, Richards AJ, Ferguson SD et al. A family with Ehlers-Danlos syndrome type III/articular hypermobility syndrome has a glycine 637 to serine substitution in type III collagen. Hum Mol Genet 1994; 3(9):1617-1620.
32. Wenstrup RJ, Meyer RA, Lyle JS et al. Prevalence of aortic root dilation in the Ehlers-Danlos syndrome. Genet Med 2002; 4(3):112-117.
33. Tiller GE, Cassidy SB, Wensel C et al. Aortic root dilatation in Ehlers-Danlos syndrome types I, II and III. A report of five cases. Clin Genet 1998; 53(6):460-465.
34. Elsheikh M, Dunger DB, Conway GS et al. Turner's syndrome in adulthood. Endocr Rev 2002; 23(1):120-140.
35. Hagan PG, Nienaber CA, Isselbacher EM et al. The International Registry of Acute Aortic Dissection (IRAD): New insights into an old disease. JAMA 2000; 283(7):897-903.
36. Coady MA, Rizzo JA, Goldstein LJ et al. Natural history, pathogenesis, and etiology of thoracic aortic aneurysms and dissections. Cardiol Clin 1999; 17(4):615-635.
37. Nicod P, Bloor C, Godfrey M et al. Familial aortic dissecting aneurysm. J Am Coll Cardiol 1989; 13(4):811-819.
38. Milewicz DM, Chen H, Park ES et al. Reduced penetrance and variable expressivity of familial thoracic aortic aneurysms/dissections. Am J Cardiol 1998; 82(4):474-479.
39. Milewicz DM, Michael K, Fisher N et al. Fibrillin-1 (FBN1) mutations in patients with thoracic aortic aneurysms. Circulation 1996; 94(11):2708-2711.
40. Francke U, Berg MA, Tynan K et al. A Gly1127Ser mutation in an EGF-like domain of the fibrillin-1 gene is a risk factor for ascending aortic aneurysm and dissection. Am J Hum Genet 1995; 56(6):1287-1296.
41. Guo D, Hasham S, Kuang SQ et al. Familial thoracic aortic aneurysms and dissections: Genetic heterogeneity with a major locus mapping to 5q13-14. Circulation 2001; 103(20):2461-2468.
42. Whiteman P, Downing AK, Smallridge R et al. A Gly —> Ser change causes defective folding in vitro of calcium-binding epidermal growth factor-like domains from factor IX and fibrillin-1. J Biol Chem 1998; 273(14):7807-7813.
43. Loeys B, Nuytinck L, Delvaux I et al. Genotype and phenotype analysis of 171 patients referred for molecular study of the fibrillin-1 gene FBN1 because of suspected Marfan syndrome. Arch Intern Med 2001; 161(20):2447-2454.
44. Kakko S, Raisanen T, Tamminen M et al. Candidate locus analysis of familial ascending aortic aneurysms and dissections confirms the linkage to the chromosome 5q13-14 in Finnish families. J Thorac Cardiovasc Surg 2003; 126(1):106-113.
45. Vaughan CJ, Casey M, He J et al. Identification of a chromosome 11q23.2-q24 locus for familial aortic aneurysm disease, a genetically heterogeneous disorder. Circulation 2001; 103(20):2469-2475.
46. Hasham S, Willing M, Guo D et al. Mapping a locus for familial thoracic aortic aneurysms and dissections to 3p24-25. Circulation 2003; 107(25):3184-90.
47. De Paepe A, Devereux RB, Dietz HC et al. Revised diagnostic criteria for the Marfan syndrome. Am J Med Genet 1996; 62(4):417-426.
48. Burks JM, Illes RW, Keating EC et al. Ascending aortic aneurysm and dissection in young adults with bicuspid aortic valve: Implications for echocardiographic surveillance. Clin Cardiol 1998; 21(6):439-443.
49. Pachulski RT, Weinberg AL, Chan KL. Aortic aneurysm in patients with functionally normal or minimally stenotic bicuspid aortic valve. Am J Cardiol 1991; 67(8):781-782.
50. Nkomo VT, Enriquez-Sarano M, Ammash NM et al. Bicuspid aortic valve associated with aortic dilatation: A community-based study. Arterioscler Thromb Vasc Biol 2003; 23(2):351-356.
51. Roberts CS, Roberts WC. Dissection of the aorta associated with congenital malformation of the aortic valve. J Am Coll Cardiol 1991; 17(3):712-716.
52. McKusick VA. Association of congenital bicuspid aortic valve and erdheim's cystic medial necrosis. Lancet 1972; 1(7758):1026-1027.
53. Schievink WI, Mokri B. Familial aorto-cervicocephalic arterial dissections and congenitally bicuspid aortic valve. Stroke 1995; 26(10):1935-1940.

54. Polini PE, Campell M. Factors in the causation of persistent ductus arteriosus. Ann Hum Genet 1960; 24:343-357.
55. Mani A, Meraji SM, Houshyar R et al. Finding genetic contributions to sporadic disease: A recessive locus at 12q24 commonly contributes to patent ductus arteriosus. Proc Natl Acad Sci USA 2002; 99(23):15054-15059.
56. Lund JT, Jensen MB, Hjelms E. Aneurysm of the ductus arteriosus. A review of the literature and the surgical implications. Eur J Cardiothorac Surg 1991; 5(11):566-570.
57. Cusick DA, Frederiksen JW, Mehlman DJ. Acute aortic dissection: Association with patent ductus arteriosus. Am J Card Imaging 1996; 10(3):200-203.
58. Glancy DL, Wegmann M, Dhurandhar RW. Aortic dissection and patent ductus arteriosus in three generations. Am J Cardiol 2001; 87(6):813-5, A9.

CHAPTER 10

Fibrillin-2 Mutations in Congenital Contractural Arachnodactyly

Maurice Godfrey

Introduction and Clinical Phenotype

Congential contractural arachnodactyly (CCA) or Beals syndrome is characterized by a marfanoid habitus. In addition to the tall, slender, asthenic appearance, most individuals with CCA have "crumpled" ears, flexion contractures, severe kyphoscoliosis, and muscular hypoplasia (Figs. 1 and 2).[1-4] The ear abnormalities are characterized as a folded upper helix of the external ear. Although significantly less frequent, additional craniofacial abnormalities include mild micrognathia, high arched palate, scaphocephaly, brachycephaly, dolichocephaly, and frontal bossing. In most patients, contractions of major joints (knees, elbows, ankles) are present at birth. The proximal interphalangeal joints display flexion contractures (i.e., camptodactyly). The toes are similar. Contractures of the hip, adducted thumbs, and clubfoot may also occur. Bowed long bones and muscular hypoplasia are additional musculoskeletal findings in CCA. Contractures usually resolve with time. Arachnodactyly (long slender fingers and toes) is present in most individuals with CCA. The greatest morbidity in CCA is caused by progressive kypho/scoliosis that can begin in early infancy. It is present in about half of all affected individuals. The spinal abnormalities are progressive. Severe thoracic cage abnormalities with associated scoliosis may cause restrictive pulmonary disease.[5]

While CCA shares some clinical characteristics with the Marfan syndrome (Table 1), it does not share the usually shortened life expectancy.

Diagnosis and Genetic Counseling

Diagnosis of congenital contractural arachnodactyly is based on a constellation of clinical signs noted above and in Table 1. Linkage and mutation analysis for the *FBN2* gene encoding fibrillin-2 are available in specialized centers. CCA is inherited as an autosomal dominant condition. Affected individuals often have an affected parent. However, individuals with CCA may appear as the result of a new dominant mutation. Like most autosomal dominant conditions, risk to siblings is 50% if a parent is affected. Germline mosaicism has been observed in CCA.[6;7] An affected individual has a 50% risk of passing the abnormal *FBN2* allele to each offspring. CCA appears to be fully penetrant.

Like the Marfan syndrome, CCA does not have gender or ethnic predilection. The prevalence of CCA has not formally been assessed, but it is clearly several fold less common than that of the Marfan syndrome. Prenatal testing has been successfully performed with both linkage analysis[8] and fetal ultrasound.[9]

Management

Management involves physical therapy for joint contractures to increase joint mobility and ameliorate the effects of muscle hypoplasia (usually calf muscles). This type of therapy is best

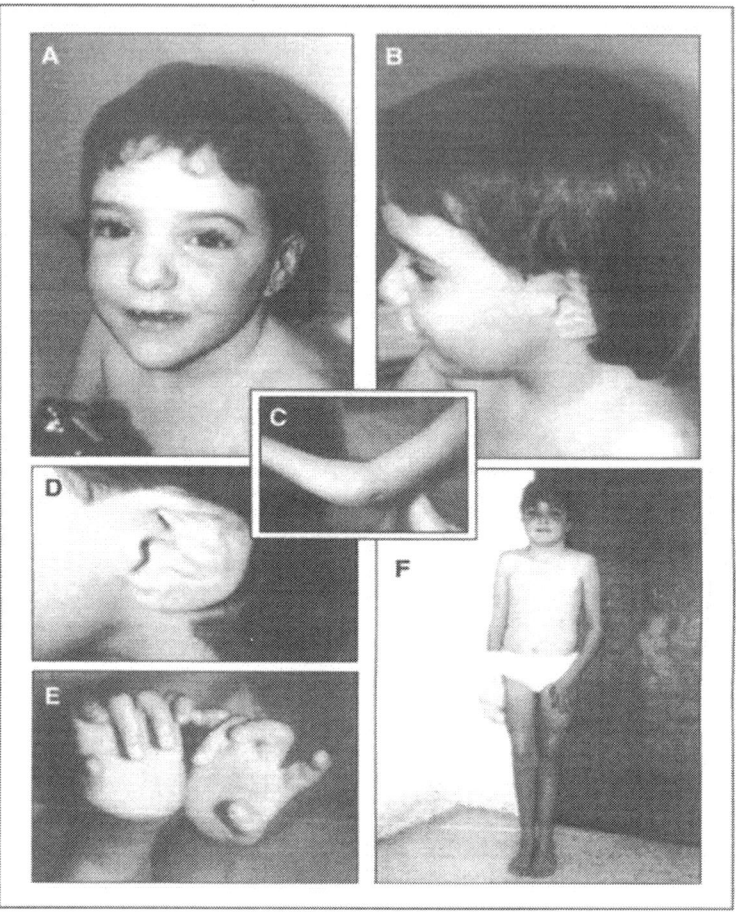

Figure 1. Patient with congenital contractural arachnodactyly. A) Face at three years. B) Profile at three years. C) Left elbow at three years (note contracture). D) Left ear at three years (note crumpled upper pinna). E) Hands at three years (note arachnodactyly and camptodactyly). F) Total body (note bandaged hand after surgery for camptodactyly and also the thin legs below the knees).(Reprinted with permission from: Belleh et al. Am J Med Genet 2000.)

instituted in childhood. As affected individuals age, spontaneous improvement in camptodactyly is frequently observed. If necessary, surgical release of contractures may be performed. Because aortic root dilatation has been observed in CCA and because of the difficulty distinguishing CCA from the Marfan syndrome, it is prudent to recommend that an echocardiogram be performed on patients being evaluated for CCA. The kyphoscoliosis tends to be progressive requiring bracing and/or surgical correction. Consultation with an orthopedist is encouraged. Since the severe/lethal CCA form of CCA is so rare (see below), no general recommendations exist and problems need to be managed as they arise.

Severe/Lethal CCA

Most individuals with CCA have normal life expectancies. However, a rare —only about half dozen cases are reported in the literature— variant termed severe/lethal CCA does have a

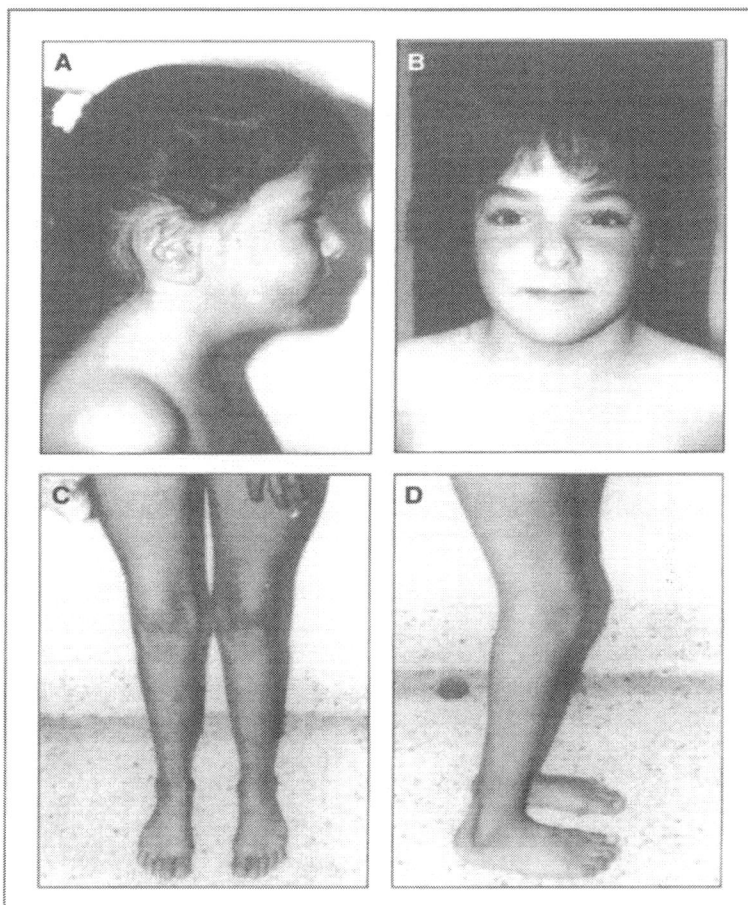

Figure 2. Patient with congenital contractural arachnodactyly at eight years. A) Profile (note the crumpled upper pinna). B) Face. C) Anterior aspect of lower extremities. (Note thin legs below the knees). D) Lateral aspect of lower extremities. (Reprinted with permission from: Belleh et al. Am J Med Genet 2000.)

significantly reduced life span. In fact, this group can be compared to individuals at the most severe end of the Marfan syndrome spectrum, i.e., neonatal Marfan syndrome (Table 2).

Individuals with severe/lethal CCA have the same skeletal and ear findings as those with the classical form of CCA.[7,10-12] In addition, they have cardiovascular and gastrointestinal abnor-

---

Table 1. Clinical features of congenital contractural arachnodactyly

Marfanoid habitus
Flexion contractures of multiple joints including elbows, knees, hips, fingers
Kyphoscoliosis (sometimes severe)
Muscular hypoplasia
Abnormal pinnae (presenting as crumpled outer helices)

Table 2. Comparison of clinical manifestations of severe/lethal CCA and neonatal MFS

Manifestation	Neonatal Marfan Syndrome[a]	Severe Lethal CCA[b]
Cardiovascular		
Mitral valve prolapse	9/9	0/6
Tricuspid valve prolapse	9/9	0/6
Mitral valve insufficiency	9/9	0/6
Aortic root dilatation	9/9	1/6
Atrial septal defect	0/9	4/6
Ventricular septal defect	0/9	4/6
Interrupted aortic arch	0/9	4/6
Single umbilical artery	0/9	3/6
Skeletal		
Arachnodactyly	9/9	6/6
Flexion contractures	8/9	6/6
Scoliosis	0/6	3/6
Vertebral anomalies	0/6	2/5[c]
Gastrointestinal		
Duodenal atresia	0/9	3/5[c]
Esophageal atresia	0/9	1/5[c]
Intestinal malrotation	0/9	2/5[c]

[a] Data are from Godfrey et al (1995); [b] Data are from Lipson et al (1974), Currarino and Friedman (1986), Macnab et al (1991), Godfrey et al (1995), and Wang et al (1996); [c] Data from one patient are not known.

malities that are related to abnormal development. The cardiovascular abnormalities include atrial and ventricular septal defects, interrupted aortic arch, and a single umbilical artery. The gastrointestinal anomalies are duodenal or esophageal atresia and intestinal malrotation. In fact, it was these differences that led Lipson and colleagues[11] to presciently suggest some three decades ago that "the cardiovascular defects in CCA reflect incomplete organ development" when compared to the phenotype in the Marfan syndrome. We know now that they were only referring to a small number of CCA cases, yet their observation, decades before the discovery of the fibrillins, is valid and insightful.

Molecular Genetics of CCA

The discovery of a second fibrillin gene led to the genetic association of fibrillin-2 encoded by *FBN2* with CCA. Studies showed significant genetic linkage in several CCA families.[13,14] However, the ultimate proof had to await the identification of specific *FBN2* mutations in people with CCA. Putnam, et al[15] were the first to identify mutations in two unrelated individuals with CCA. Both of the mutations they identified were cysteine substitutions. It is important to note that in contrast to the distribution of *FBN1* mutations causing the Marfan syndrome throughout the coding region, the *FBN2* mutations so far identified in CCA, appear to cluster. The cluster of the *FBN2* mutations is between exons 23-34 (Table 3). The homologous region of *FBN1*, the so called "neonatal region", contains the greatest percentage of mutations from Marfan syndrome patients at the most severe end of that disorder's clinical spectrum.[16]

The known *FBN2* mutations are listed in Table 3. Virtually all of the known *FBN2* mutations are of the calcium binding epidermal growth factor-like (cbEGF) domains.[17-21] (For discussion of the structure of the fibrillins, see Chapters 14 and 15).

Table 3. Mutations in the gene encoding fibrillin 2 (FBN2) that are known to cause CCA

Amino Acid Change	Exon	Reference
ins343bp	23-24	Gupta et al, 2002
G1056D	24	Park et al, 1998
I1092T	25	Park et al, 1998
A1114H	25	Babcock et al, 1998
C1141F	26	Belleh et al, 2000
del exon 26[a]	26	Gupta et al, 2002
G1177C	27	Gupta et al, 2002
C1197Y	27	Gupta et al, 2002
C1239R	28	Gupta et al, 2002
C1252Y	29	Putnam et al, 1995
C1252W	29	Belleh et al, 2000
C1256W	29	Gupta et al, 2002
C1267R	29	Gupta et al, 2002
del exon 29	29	Putnam et al, 1997
del exon 31	31	Maslen et al, 1997; Park et al, 1998
Y1421X	33	Gupta et al, 2002
C1433S	33	Putnam et al, 1995
del exon 34[b]	34[b]	Wang et al, 1996

[a] This mutation is described in three unrelated probands; [b] The deletion of exon 34 was found in an individual with severe/lethal CCA and her mosaic mother (Wang et al 1996).

Molecular studies of only one individual with severe/lethal CCA have been performed.[7] This individual had an exon splicing mutation that caused the skipping of exon 34, a cbEGF-like domain. Significantly, this individual's mother was a somatic mosaic with one-third of her fibroblasts also harboring the same exon 34 mis-splicing mutation. Therefore, one can speculate that there is a threshold for certain mutations causing skeletal perturbations versus severe developmental abnormalities in the cardiovascular and gastrointestinal systems.

Fibrillin-2 in Development and Animal Models

The temporal and spatial expression of fibrillin-2 has been examined in several species. In human fetal aorta, antibodies to fibrillin-2 were found to stain most intensely in the media, where elastic fibers are most abundant. In human elastic cartilage, fibrillin-2 localized to the cartilaginous core while fibrillin-1 localized primarily to the surrounding connective tissue.[22] Quondamatteo and coworkers compared the spatial and temporal distribution of fibrillin-1 and fibrillin-2 in early human embryonic development. Their data show that in most tissues both fibrillins are similarly expressed. Exceptions included the kidney, liver, rib anlage, and notochord.[23] Similar studies in the developing mouse showed that in most tissues fibrillin-2 was expressed earlier than fibrillin-1.[24]

Studies in the chick have shown that fibrillin-2 (called JB-3 in the early literature) is expressed very early in development and is found in the regions of heart development.[25] The early expression of the fibrillins has led to the speculation that they may mediate the tensile forces that shape the early embryo.[26] A possible role for fibrillin-2 in lung development has been shown in a rat model. Studies of fetal lung explants demonstrated abnormal branch morphogenesis when the explants were incubated with antisense oligonucleotides to fibrillin-2.[27]

Browning et al[28] described a mouse with syndactyly (*sne*) that was derived from chemical mutagenesis of murine embryonic stem cells. They showed that *sne* (now renamed *syfp-3J*) was an allele of the *sy* locus. *sy* is the shaker-with-syndactylism mouse, a radiation mutant with a chromosome 18 (syntenic to human chromosome 5) contiguous gene deletion syndrome.[29] The deleted region contains the gene encoding fibrillin-2. Some spontaneously occurring mouse models with syndactyly also mapped to the *sy* locus (*syfp* and *syfp-2J*). All three *syfp* mutations are *FBN2* mutations.[28,30] Additional evidence that absence of *FBN2* leads to syndactyly came from gene targeting studies.[31] The *FBN2⁻/⁻* knockout mouse displayed the same type of syndactyly observed in the *sy* mice. (For detailed description see Chapter 16). Interestingly, two of the fibrillin-2 mutations in the *syfp* mice were outside the "neonatal region", i.e., the area in which all of the human CCA mutations have been found.[30] These findings have lead to the obvious speculation that fibrillin-2 mutations outside the neonatal region, for example, may lead to other human phenotypes.[30]

References

1. Hecht F, Beals RK. "New" syndrome of congenital contractural arachnodactyly originally described by Marfan in 1896. Pediatrics 1972; 49:574-579.
2. Epstein CJ, Graham CB, Hodgkin WE et al. Hereditary dysplasia of bone with kyphoscliosis, contractures, and abnormally shaped ears. J Pediatr 1968; 73:379-386.
3. Ramos Arroyo MA, Weaver DD, Beals RK. Congenital contractural arachnodactyly. Report of four additional families and review of literature. Clin Genet 1985; 27:570-581.
4. Viljoen D. Congenital contractural arachnodactyly (Beals syndrome). J Med Genet 1994; 31:640-643.
5. Jones JL, Lane JE, Logan JJ et al. Beals-Hecht syndrome. South Med J 2002; 95(7):753-755.
6. Putnam EA, Park E-S, Aalfs CM et al. Parental somatic and germ-line mosaicism for a FBN2 mutation and analysis of FBN2 transcript levels in dermal fibroblasts. Am J Hum Genet 1997; 60:818-827.
7. Wang M, Clericuzio Cl, Godfrey M. Familial occurrence of typical and severe lethal congenital contractural arachnodactyly caused by mis-splicing of exon 34 of fibrillin-2 (FBN2). Am J Hum Genet 1996; 59:1027-1034.
8. Belleh S, Spooner L, Allanson J et al. Prenatal diagnosis in congenital contractural arachnodactyly. Genet Test 1997; 1(4):293-296.
9. Kolble N, Wisser J, Babcock D et al. Prenatal ultrasound findings in a fetus with congenital contractural arachnodactyly. Ultrasound Obstet Gynecol 2002; 20(4):395-399.
10. Currarino G, Friedman JM. A severe form of congenital contractural arachnodactyly in two new-born infants. Am J Med Genet 1986; 25:763-773.
11. Lipson EH, Viseskul C, Herrmann J. The clinical spectrum of congenital contractural arachnodactyly: A case with congenital heart disease. Z Kinderheilk 1974; 118:1-8.
12. Macnab AJ, D'Orsogna L, Cole DE et al. Cardiac anomalies complicating congenital contractural arachnodactyly. Arch Dis Child 1991; 66(10 Spec No):1143-1146.
13. Lee B, Godfrey M, Vitale E et al. Linkage of Marfan syndrome and a phenotypically related disorder to two different fibrillin genes. Nature 1991; 352:330-334.
14. Tsipouras P, Del Mastro R, Sarfarazi M et al. Linkage analysis demonstrates that Marfan syndrome, dominant ectopia lentis, and congenital contractural arachnodactyly are linked to the fibrillin genes on chromosmes 15 and 5. N Engl J Med 1992; 326:905-909.
15. Putnam EA, Zhang H, Ramirez F et al. Fibrillin-2 (FBN2) mutations result in the Marfan-like disorder, congenital contractural arachnodactyly. Nat Genet 1995; 11:456-458.
16. Kainulainen K, Karttunen L, Puhakka L et al. Mutations in the fibrillin gene responsible for dominant ectopia lentis and neonatal Marfan syndrome. Nat Genet 1994; 6:64-69.
17. Gupta PA, Putnam EA, Carmical SG et al. Ten novel FBN2 mutations in congenital contractural arachnodactyly: Delineation of the molecular pathogenesis and clinical phenotype. Hum Mutat 2002; 19(1):39-48.
18. Park ES, Putnam EA, Chitayat D et al. Clustering of FBN2 mutations in patients with congenital contractural arachnodactyly indicates an important role of the domains encoded by exons 24 through 34 during human development. Am J Med Genet 1998; 78(4):350-355.
19. Belleh S, Zhou G, Wang M et al. Two novel fibrillin-2 mutations in congenital contractural arachnodactyly. Am J Med Genet 2000; 92:7-12.
20. Maslen C, Babcock D, Raghunath M et al. A rare branch-point mutation is associated with misspicing of fibrillin-2 in a large family with congenital contractural arachnodactyly. Am J Hum Genet 1997; 60:1389-1398.

21. Babcock D, Gasner C, Francke U et al. A single mutation that results in an Asp to His substitution and partial exon skipping in a family with congenital contractural arachnodactyly. Hum Genet 1998; 103(1):22-28.
22. Zhang H, Apfelroth SD, Hu W et al. Structure and expression of fibrillin-2 a novel microfibrillar component preferentially located in elastic matrices. J Cell Biol 1994; 124:855-863.
23. Quondamatteo F, Reinhardt DP, Charbonneau NL et al. Fibrillin-1 and fibrillin-2 in human embryonic and early fetal development. Matrix Biol 2002; 21(8):637-646.
24. Mariencheck MC, Davis EC, Zhang H et al. Fibrillin-1 and fibrillin-2 show temporal and tissue-specific expression in developing elastic tissues. Connect Tiss Res 1995; 31:87-97.
25. Wunsch AM, Little CD, Markwald RR. Cardiac endothelial heterogeneity defines valvular development as demonstrated by the diverse expression of JB3, an antigen of the endocardial cushion tissue. Dev Biol 1994; 165(2):585-601.
26. Rongish BJ, Drake CJ, Argraves WS et al. Identification of the developmental marker, JB3-antigen, as fibrillin-2 and its de novo organization into embryonic microfibrous arrays. Dev Dyn 1998; 212(3):461-471.
27. Yang Q, Ota K, Tian Y et al. Cloning of rat fibrillin-2 cDNA and its role in branching morphogenesis of embryonic lung. Dev Biol 1999; 212(1):229-242.
28. Browning VL, Chaudhry SS, Planchart A et al. Mutations of the mouse Twist and sy (fibrillin 2) genes induced by chemical mutagenesis of ES cells. Genomics 2001; 73(3):291-298.
29. Johnson KR, Cook SA, Zheng QY. The original shaker-with-syndactylism mutation (sy) is a contiguous gene deletion syndrome. Mamm. Genome 2003; 9:889-892.
30. Chaudhry S-S, Gazzard J, Baldock C et al. Mutation of the gene encoding fibrillin-2 results in syndactyly in mice. Hum Mol Genet 2001; 10(8):835-843.
31. Arteaga Solis E, Gayraud B, Lee S-Y et al. Regulation of limb patterning by extracellular microfibrils. J Cell Biol 2001; 154(2):275-281.

CHAPTER 11

Assembly of Microfibrils

Kerstin Tiedemann, Boris Bätge and Dieter P. Reinhardt

Introduction

Fibrillins are physiologically secreted as extended thread-like monomers into the extracellular matrix by many cell types. The mature functional entity, however, is constituted by higher order aggregates which are called microfibrils.[1] The assembly process of forming these supramolecular structures is a continuous process that probably requires a number of molecules and molecular events on its way from monomeric fibrillin to mature supramolecular microfibrils. It is likely that the molecular assembly of microfibrils is disturbed by at least some mutations in the gene for fibrillin-1 (*FBN1*) leading to Marfan syndrome (MFS) and other microfibrillopathies. This chapter summarizes the current knowledge of fibrillin assembly into microfibrils and discusses this understanding in the light of *FBN1* mutations.

Methodological Approaches to Study Microfibril Assembly

To facilitate the discussion of the molecular assembly mechanisms of monomeric fibrillin and other molecules into supramolecular microfibrils, we introduce in this chapter arbitrary groups of events based on methodological accessibility of the involved processes. Early events take place within minutes to several hours, intermediate events within several days, and late events within weeks to months (Fig. 1). In the following, an overview is given about the methods used to monitor assembly in each of these stages.

The early events in microfibril assembly are typically analyzed by metabolic labeling techniques and biochemical approaches. Fibrillins are rich in cysteine residues (~13%), a property that facilitates pulse chase metabolic labeling of cultured cells or organ cultures with radiolabeled [^{35}S]cysteine. Analysis of radiolabeled culture medium and extracted extracellular matrices by autoradiography after gel electrophoresis allows conclusions on molecular events based on changes of the molecular size or concentrations of fibrillin monomers or multimers. With this approach, it is possible to analyze the role of profibrillin processing, and of fibrillin multimerization mediated by formation of reducible and nonreducible cross-links. Biochemical analyses of assembly intermediates and of molecular properties important for fibrillin assembly have been hampered by the propensity of fibrillin to form reducible and nonreducible cross-links in tissues. Therefore, extraction procedures for fibrillin from tissues usually include reducing agents resulting in denaturation of the protein.[2] Although it is in principal possible to purify fibrillin monomers from cell culture sources using nondenaturing buffer systems, the required methods are time consuming and costly.[3] To solve this dilemma, various recombinant expression techniques have been developed over the past few years for production of fibrillin fragments, such as expression in eukaryotic cells, in bacteria, as well as expression with in vitro translation systems. Recombinant fragments of fibrillin-1 and fibrillin-2 produced by these systems have been extensively used in various biochemical assembly assays.

Within a few days fibrillin assembles into a loosely structured network in cell culture assays (Fig. 1).[3] Primary cells of mesenchymal origin such as dermal fibroblasts, smooth muscle cells,

Marfan Syndrome: A Primer for Clinicians and Scientists, edited by Peter N. Robinson and and Maurice Godfrey. ©2004 Eurekah.com and Kluwer Academic / Plenum Publishers.

chondrocytes, ligament cells and established cell lines such as MG-63 (human osteosarcoma) and others secrete fibrillin into the culture medium and assemble microfibrillar networks (Fig. 1). These fibrillin networks can be easily visualized by indirect immunofluorescence with specific mono- or polyclonal antibodies against the specific fibrillin isoform. This type of assembly assay is often used to demonstrate quantitative and qualitative differences of the assembled fibrillin network in various situations. Additionally, potential modifiers and inhibitors can be analyzed presuming that these reagents are not toxic for the cells.

After a few weeks of cell culture, "beads-on-a-string" microfibrils can be extracted from the extracellular layer of cells in culture and visualized by transmission electron microscopy after rotary shadowing or negative staining of the samples (Fig. 1).[4,5] Microfibrils extracted from cell culture systems are very similar, if not identical, to microfibrils extracted from tissue sources such as ocular tissues,[6,7] placenta extracts,[8] or bovine nuchal ligaments.[9] Identification of antibody epitopes for fibrillin has led to a relatively precise localization and arrangement of fibrillin monomers within the beaded microfibrils as described in detail in the chapter "Organization and Biomechanical Properties of Fibrillin Microfibrils". While extracted microfibrils can be used for example to test the structural integrity in various situations, they are of limited use to analyze dynamic assembly mechanisms. Beaded microfibrils do not represent the final stage of microfibrils in tissues. Tissue microfibrils often do not appear as beaded structures but simply as bundles of extended thread-like aggregates (Fig. 1). It is likely that components that are peripherally associated with tissue microfibrils are released by the extraction procedures, and thus converting the thread-like structure to the beaded form. Analysis of tissue microfibrils is helpful to uncover mechanisms relevant to assembly in mouse models with modified fibrillin genes.[10-12]

Microfibril Assembly Is Often Disturbed in Individuals with MFS

Pulse chase experiments with dermal fibroblasts from individuals with MFS identified defects in fibrillin synthesis, secretion and aggregation into the extracellular matrix, indicating that at least for some fibrillin mutations early assembly events are compromised.[13,14] In a detailed study, five different groups of MFS fibroblasts have been identified based on the parameters of fibrillin synthesis and deposition into the extracellular matrix.[15] Despite differences in the amount of secreted fibrillin, fibroblasts in three of the five groups (43 of 55 analyzed) showed impaired incorporation of the mutant fibrillin into extracellular aggregates at various levels, suggesting functional problems in early assembly steps.[15] Consequently, the intermediate assembly stages of the relevant groups, assessed by indirect immunofluorescence of fibrillin, showed considerably reduced or thinner microfibrillar networks as compared to normal controls.[16] Originally, immunofluorescence studies with skin biopsies from individuals with MFS or with dermal fibroblasts cultivated from such tissues samples revealed that the intermediate microfibrillar networks in cell culture or the mature microfibrils in skin were reduced in amount as compared to age-matched controls.[17-19] One possible interpretation of these data is that fibrillin assembly is impaired at a stage upstream of the intermediate stage. However, reduced or qualitatively altered fibrillin staining patterns in skin or dermal fibroblasts from individuals clinically diagnosed with MFS are not always observed. In one study, about 70% (16 of 23) of affected skin samples and 89% (16 of 18) of the respective fibroblast cultures showed altered fluorescence patterns.[19] In contrast, in another study, none (0 of 7) of the analyzed fibroblasts cultures established from individuals with MFS showed alterations in the immunostainable fibrillin patterns.[20] Therefore, in terms of fibrillin assembly, it is likely that only a subset of fibrillin mutations resulting in MFS disturbs the assembly mechanism. In the severe neonatal form of MFS, which is caused by mutations in the central region of fibrillin-1 coded by exons 24-32, reduced immunostainable fibrillin-1 depositions have also been reported.[21-23] Additionally, in these samples, fibrillin fibers appeared short, fragmented and frayed, indicating that the central part of the fibrillin-1 molecule has a role in elongation and/or alignment of fibrillin within microfibrils (see also below, Fig. 2).

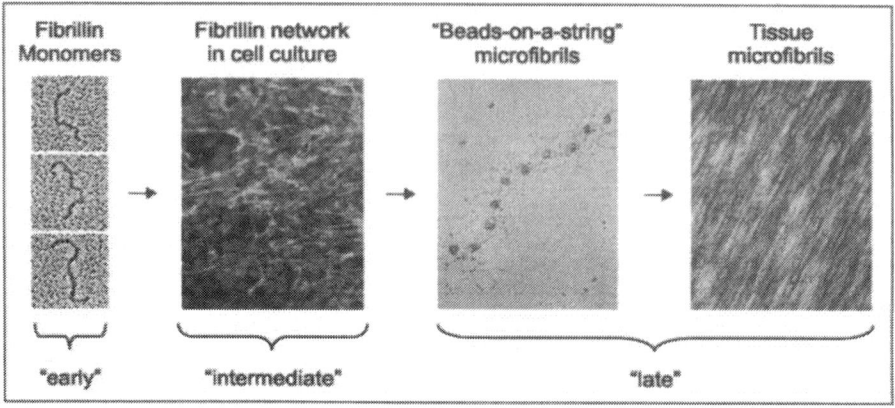

Figure 1. Assembly of Microfibrils Involves Many Steps. After secretion of monomeric fibrillin, early assembly events takes place within minutes to hours and include for example disulfide-bond formation. After a few days (intermediate stages) a typical fibrillin network can be monitored by indirect immunofluorescence (green fibrillin network versus red cell nuclei). It takes a few weeks (late stages) until "beads-on-a-string" microfibrils can be extracted from cell cultures. Mature microfibrils in tissue probably take even longer (months) to form and typically appear as bundles. The beaded structure is often not visible in tissue microfibrils. Images are shown in different magnifications.

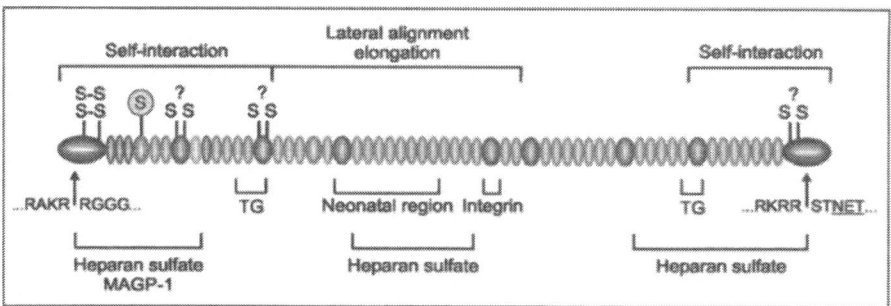

Figure 2. Regions in Fibrillin-1 Potentially Important for Assembly. Shown is a schematic model of fibrillin-1 composed of various modules. "S" represent thiol groups in cysteine residues potentially involved in intermolecular disulfide bond formation. In the first hybrid module (blue) a free thiol has been identified. Four cysteine residues in the N-terminal domain have been determined to form intramolecular disulfide bonds. Currently it is not clear whether cysteine residues in the first and second 8-Cys/TB module (red) can undergo disulfide reshuffling, and whether the two cysteine residues in the C-terminal region participate in intermolecular disulfide bridges. Cross-linking regions for transglutaminase cross-links are indicated (TG). The fibrillin self-interaction regions are currently not exactly defined, the most likely regions are indicated. The profibrillin-1 recognition motifs for furin/PACE type proteases are shown in the unique N- and C-terminal domains (gray). An N-linked glycosylation site potentially important in regulation of C-terminal processing is underlined. Mutations in the neonatal region can result in a very severe form of MFS. The binding epitopes for heparan sulfate, MAGP-1, and integrin $\alpha_v\beta_3$ are marked. Identification of the central region important for lateral alignment and elongation results from genetic studies in mice.

A number of studies have addressed the ultrastructural appearance of beaded microfibrils obtained from long term cell cultures of MFS fibroblasts. Four types of abnormalities have been characterized when compared with normal controls: The interbead domains were diffuse and poorly defined,[5,24,25] and cell cultures produced nearly no beaded structures at all,[24,26] the interbead domains appeared frayed,[24,25] and variable interbead periodicities were observed.[24,25] All of these consequences on the morphology of beaded microfibrils could potentially arise from various defects in fibrillin assembly.

In summary, it is clear for early assembly steps, analyzed by pulse chase experiments, that a subset of the MFS mutations directly affect assembly events. Based on the correlation of this subset with intermediate consequences, it is likely that other abnormalities observed in the intermediate and the late stages are also results of problems in the assembly process. All of the observed consequences on the early, intermediate and late assembly stages are not conclusive as to which molecular event may be compromised. However, not all MFS mutations result in defects in the assembly mechanism, but this observation is not addressed in this chapter.

Role of Propeptide Processing in Fibrillin Assembly

Originally, pulse chase experiments with human skin fibroblasts suggested that fibrillin-1 is secreted from the cells as a precursor protein of ~350 kDa, which is processed to a mature ~320 kDa form.[13,27] Conversion of profibrillin-1 to fibrillin-1 usually is completed within several hours. Only the processed form of fibrillin-1 is deposited into the extracellular matrix, suggesting that profibrillin-1 conversion to mature fibrillin-1 plays a regulatory role in assembly into higher order aggregates.[13] The reduction of about 30 kDa upon processing implied that a relatively small fragment of fibrillin-1 was removed from either one or both terminal ends, indicating that the cleavage sites are located within the unique terminal domains (Fig. 2). The conversion of profibrillin-1 to fibrillin-1 was experimentally inhibited by calcium depletion of the culture medium, indicating that calcium-dependent enzymes are responsible for the conversion.[26,28,29] A mutation in fibrillin-1 (R2726W) associated with isolated skeletal features of MFS was shown to interfere with normal processing at the C-terminal end and thus disturbed the incorporation of the mutated protein into the extracellular matrix.[27] Since the amino acid residue altered by this mutation occurs immediately adjacent to a sequence motif (RKRR), that matches the tribasic recognition site for proprotein proteases of the furin/PACE type (RX(K/R)R), it has been suggested that processing is mediated by this type of endoproteases.[27] Furin/PACE type proteases are responsible for cleavage of propeptides of a variety of secreted proteins (for review see ref. 30). The predicted propeptide removed from the C-terminal end of fibrillin-1 by processing spans 140 amino acid residues. Essentially similar conclusions were drawn based on the analysis of a mutation harboring a premature stop codon (W2756ter), which also prevents profibrillin-1 processing in the C-terminal domain.[28] It is interesting that the W2756ter mutation caused over-glycosylation of the mutated fibrillin-1 due to intracellular retention. On the other hand, elimination of the N-glycosylation site (NET) directly C-terminal of the furin/PACE recognition motif in the C-terminal domain increased processing by 15-20 % (Fig. 2).[31] These data indicate a regulatory role of this N-glycosylation site for processing of profibrillin-1.

Direct evidence for utilization of this recognition sequence was obtained by analysis of a recombinant fibrillin-1 fragment comprising the last calcium binding epidermal growth factor (cbEGF) like module and the unique C-terminal domain.[32] This recombinant fragment was cleaved during or after secretion directly after the tribasic recognition sequence (RKRR↓STNET, Fig. 2). In another recombinant system, processing was demonstrated by specific antibodies against the C-propeptide.[29] In addition, site directed mutagenesis of the furin/PACE recognition sequence at various positions inhibited processing and thus further substantiated that processing is mediated by furin/PACE-type proteases.[29,32,33] Furthermore, it was shown that various members of the furin/PACE family, namely furin/PACE, PACE4, PC1/3, and PC2, were able to mediate processing and that an inhibitor for furin/PACE-type proteases prevented processing as well as matrix deposition.[33] The importance of profibrillin-1 processing as a crucial assembly requirement is further highlighted by the fact that the tribasic furin/PACE recognition sequence in the C-terminal domain is conserved in human fibrillin-2 (RQKR) and fibrillin-3 (RPRR) as well as in fibrillins from other species, indicating that those fibrillin isoforms are probably processed in an identical or similar fashion.[34-38]

The discussion at what topological site the C-terminal processing takes place is controversial. Pulse chase methods have led to the conclusion that the site of profibrillin-1 processing is extracellular.[13,27] In contrast, experiments with secretion blocking agents and recombinant constructs indicated intracellular cleavage of the profibrillin-1 carboxy-terminal region in an early secretory pathway compartment.[29] Although furin/PACE-type proteases are typically localized to the trans-Golgi network, there is evidence that these proteins cycle between the cell surface, where they would be able to process extracellular proteins, and the Golgi apparatus (for review see ref. 30). For understanding of the assembly mechanism, it is important to clarify these controversial issues in the future. Dependent on the local environment for the initial assembly steps (secretory pathway versus extracellular matrix), the consequences of mutations in fibrillin-1 in the region of the processing site may lead to a different outcome in terms of how they interfere with assembly.

Recognition sequence motifs for furin/PACE-type proteases do not only occur in the unique C-termini of fibrillins but also in the unique N-termini of all fibrillins.[34,35,39-41] For fibrillin-1, N-terminal sequencing of the full length authentic protein, isolated from cell culture medium of dermal fibroblasts, produced a sequence starting directly after the tribasic recognition sequence motif RAKR↓R45GGG (Fig. 2).[42] In addition, it has been shown that recombinant N-terminal fragments expressed in mammalian cells are cleaved at the same position.[43,44] On the other hand, the cleavage site for the signal peptide, responsible for secretion of fibrillin-1 into the endoplasmic reticulum, was experimentally determined using recombinant systems. The preferred site of cleavage is located between glycine at position 24 and the adjacent alanine (G24↓A25) and a secondary minor site between alanine at position 27 and the adjacent asparagine (A27↓N28).[29,42] Therefore, removal of the propeptide by furin/PACE-type proteases only removes 20 or 17 amino acid residues, respectively. Due to this very small difference, experimental analyses to address whether processing of the N-terminal propeptide is also a prerequisite for the fibrillin-1 assembly have been hampered. No mutation leading to MFS has been detected within the cleaved propeptide.[45] It is not clear whether other mutations, which are located relatively close C-terminal to the processing site, have any adverse effect on profibrillin-1 processing at the N-terminus. Interestingly, fibrillin deposition into the extracellular matrix was markedly diminished by over-expression of a construct in which exon 2 of the coding sequence for fibrillin-1 was deleted.[46] This deletion predicts a mutant polypeptide consisting of 100 amino acid residues composed of the authentic fibrillin-1 sequence 1-55 followed by 45 out of frame nonsense residues. If this mutant polypeptide is processed like the wild-type fibrillin-1 after position 44, then only 11 functional residues would be able to disrupt normal fibrillin assembly. However, a potential "toxic" effect of the nonsense sequence (resulting from the frameshift), that is normally not part of the fibrillin-1 molecule, also needs to be considered.

Role of Intermolecular Cross-Link Formation in Fibrillin Assembly

Fibrillin can be immunoprecipitated from the medium and the cell layer of fibroblast and smooth muscle cell cultures in the form of monomers and higher molecular weight disulfide-bonded aggregates.[4] These results are in agreement with metabolic labeling assays of organ cultures, demonstrating rapid formation of disulfide-bonded aggregates as one of the initial steps in fibrillin assembly.[42] These concepts are supported by the observation that reducing agents are necessary to extract fibrillin from tissues.[2] Based on the structures of individual modules contained in fibrillins, it is believed that almost all cysteine residues contribute to intramolecular disulfide-bonds.[47,48] Only a few candidate cysteine residues are potentially available for intermolecular disulfide-bonds (Fig. 2): 4 cysteine residues in the unique N-terminal domain, 2 cysteine residues within the unique C-terminal domain, and, additionally, the first hybrid (Fib-like) module contains one unpaired cysteine residue. All of these cysteine residues are conserved between the fibrillin isoforms and between species.[34-38,41,49-51] It has been shown using a recombinant approach that the 4 cysteine residues in the unique N-terminal domain are involved in intramolecular disulfide-bonds, while cysteine 204 in the first hybrid domain occurs as a free thiol on or close to the surface of the molecule (Fig. 2).[42] Currently, it is still unclear whether the 2 cysteine residues in the C-terminal domain can contribute to intermolecular disulfide bridges. It is very likely that more than one cysteine residue participates in intermolecular disulfide bonding, since molecular aggregates with more than two fibrillin molecules require more than one connecting residue per molecule. It is tempting to speculate that the two cysteine residues in the C-terminal domain, in addition to cysteine 204, can fulfill such a bridging function. The two candidate cysteine residues in the C-terminal domain are located very close to the furin/PACE processing site (Fig. 2). Thus, they potentially could be involved in regulatory mechanisms connecting profibrillin processing with intermolecular disulfide bond formation in early assembly stages. At present, however, experimental evidence is lacking for this hypothesis. No mutations leading to MFS have been reported for either cysteine 204 or for the cysteine residues in the unique N- and C-terminal domains.[45] It is possible that loss of these cysteine residues is not compatible with life.

Other experimental evidence, however, suggests that other cysteine residues may be involved in intermolecular disulfide bond formation. Reducible homodimer formation was observed for recombinant fragments of fibrillin-1 and fibrillin-2 spanning from the proline and glycine-rich modules, respectively, to the second 8-Cys/TB module.[52] The authors hypothesized that cysteine residues in the second 8-Cys/TB module are involved in formation of intermolecular disulfide bridges (Fig. 2). Interestingly, in this study, the homodimer formation occurred early during biosynthesis in the endoplasmic reticulum.[52] Another study has utilized recombinant fragments of the proline-rich region of fibrillin-1 and the glycine-rich region of fibrillin-2 including flanking domains.[53] These fragments also had the tendency to dimerize, although dimer formation was assigned to the extracellular compartment. The authors suggested that molecular recognition might be mediated through the proline and glycine domains, respectively, and that the dimers become covalently stabilized by a hypothetic disulfide exchange mechanism involving cysteine residues in the first 8-Cys/TB module (Fig. 2).[53] In general, disulfide reshuffling mechanisms for extracellular matrix proteins have not been extensively studied. However, some protein disulfide isomerases, which mediate disulfide bond exchanges, can be found on the cell surface and in the extracellular matrix (for review see ref. 54). It may be even possible that disulfide isomerases exist, that specifically mediate disulfide exchanges in fibrillins and microfibrils.

Further support for dimer formation comes from ultrastructural studies of microfibrils. In some cases interbead regions appear as two defined strands[43,55] rather than the 6-8 strands, which are believed to form the mature microfibrils.[6,7,56] However, it is not clear at present whether the two-stranded interbead regions represent early assembly intermediates or rather degradation products.

Typically, extraction of microfibrillar components from mature tissues using reducing conditions results in relatively small amounts, indicating the presence of additional nonreducible cross-links in microfibrillar aggregates. Nonreducible ε(γ-glutamyl)lysine cross-links have in fact been identified in microfibrils extracted from various tissues.[57-59] A family of Ca^{2+}-dependent enzymes, the transglutaminases, catalyze the formation of such cross-links between a γ-carboxyamide group of a glutamine and a lysine residue (for review see ref. 60). In ocular tissue, the zonular fibers have been shown to be a target for a member of this family, transglutaminase 2.[61] The analysis of extracted microfibrils from human amnion indicated that 10-15% of the lysine residues in microfibrils are involved in cross-links.[58] One cross-link region between a N-terminal region of the fibrillin-1 molecule starting at position 580 and a C-terminal region starting at position 2312 has been identified (Fig. 2).[58] Based on biomechanical approaches, it has been suggested that transglutaminase cross-links play a major role in strengthening microfibrillar networks.[62] In terms of assembly mechanisms, it is conceivable that transglutaminase cross-links may facilitate correct molecular alignments at fixed positions during early microfibrillar assembly stages as a prerequisite for further assembly. At present however, the time course for the formation of transglutaminase cross-links is not known and needs to be clarified. In addition to fibrillin-1, the microfibril associated glycoprotein (MAGP)-1 (see below) was also characterized as a substrate for transglutaminase.[63] Thus, it is possible that besides homotypic fibrillin-1 transglutaminase cross-links, heterotypic fibrillin-1-MAGP-1 cross-links may be present in microfibrils. However, the transglutaminase cross-links in microfibrils have been localized to the interbead domain of beaded microfibrils, while MAGP-1 was primarily localized to the beads.[58,64] Mutations that disrupt transglutaminase cross-link sites are predicted to result in serious consequences for microfibril assembly and stability, which would likely contribute to the pathogenesis of MFS. To advance the understanding of the role of transglutaminase cross-links in MFS, it is important to exactly identify the amino acid residues involved.

Interaction Epitopes Important for Fibrillin Assembly

A relatively large number of protein ligands have been immunolocalized to microfibrils (for review see ref. 65). However, on the molecular level, only a few fibrillin-protein interactions have been identified. Fibulin-2 interacts with high affinity with fibrillin-1 and are found on some but not all microfibrils.[44] Similar observations have been made with versican,[66] and LTBP-1 and -4.[67] While the precise functional roles of these microfibril-associated proteins is not clear at present, it appears unlikely that they fulfill an essential role in the assembly mechanism, since (i) they are not present in all microfibrils, and (ii) they are not regularly spaced in microfibrils. If a microfibril-associated protein serves an important role in the assembly process, a regular arrangement of this ligand within microfibrils would be expected. A periodic labeling pattern similar to the pattern of fibrillins have been demonstrated for MAGP-1 and -2.[64,68] While MAGP-1 codistributes with microfibrils in most if not all tissues where they occur,[69,70] MAGP-2 appears to be expressed in a more tissue-specific fashion and is thus not a good candidate for an assembly mediator.[68] MAGP-1 can interact with fibrillin-1 either directly,[71] or mediated by the proteoglycan decorin.[72] The interaction epitope of MAGP-1 with fibrillin-1 is localized at the N-terminal region of the fibrillin-1 molecule (Fig. 2).[71]

During elastic fiber synthesis, the microfibrils appear before the amorphous elastin core and are believed to act as a scaffold for the deposition of tropoelastin.[73] Therefore it is unlikely that direct[74] or indirect interactions, possibly mediated by MAGP-1,[63,71,75] of fibrillins with tropoelastin play a critical role in the assembly of microfibrillar structures. Moreover tropoelastin is not present in nonelastic tissues such as the ciliary zonules in the eye where abundant bead-on-a-string structures can be found.

In summary, although various proteins ligands have been identified to interact with fibrillins, at present there is no conclusive evidence that one or more of these fibrillin binding ligands play an important role in the fibrillin assembly process. These ligands rather may have modulating functional roles in the biology of microfibrils.

In another scenario, the molecules that assist in the assembly process are not permanently associated with microfibrils but rather transiently "catalyze" the assembly process. Such molecules would bind to fibrillins or other structural microfibrillar proteins probably at an early stage of the assembly process and release as soon as the specific assembly event has been completed. In this light, it is important to mention that three regions of fibrillin-1 have been found to interact with heparin/heparan sulfate with high affinity (Fig. 2).[76] In a cell culture assembly assay with dermal fibroblasts, these glycosaminoglycans can very efficiently inhibit microfibrillar assembly leading to the hypothesis that binding of fibrillin to highly sulfated heparan side chains of proteoglycans is important to nucleate the assembly process (Fig. 3).[76] This hypothesis is further supported by the fact that inhibition of sulfation or formation of heparan sulfate chains compromises microfibrillar assembly.[72,76] Since assembly of microfibrils is believed to take place close to the cell surface, it is possible that heparan sulfate containing proteoglycans located on the plasma membrane, such as syndecans or glypicans, may fulfill such a catalyzing function. Binding of fibrillins to cell surface glycosaminoglycans may be necessary to generate

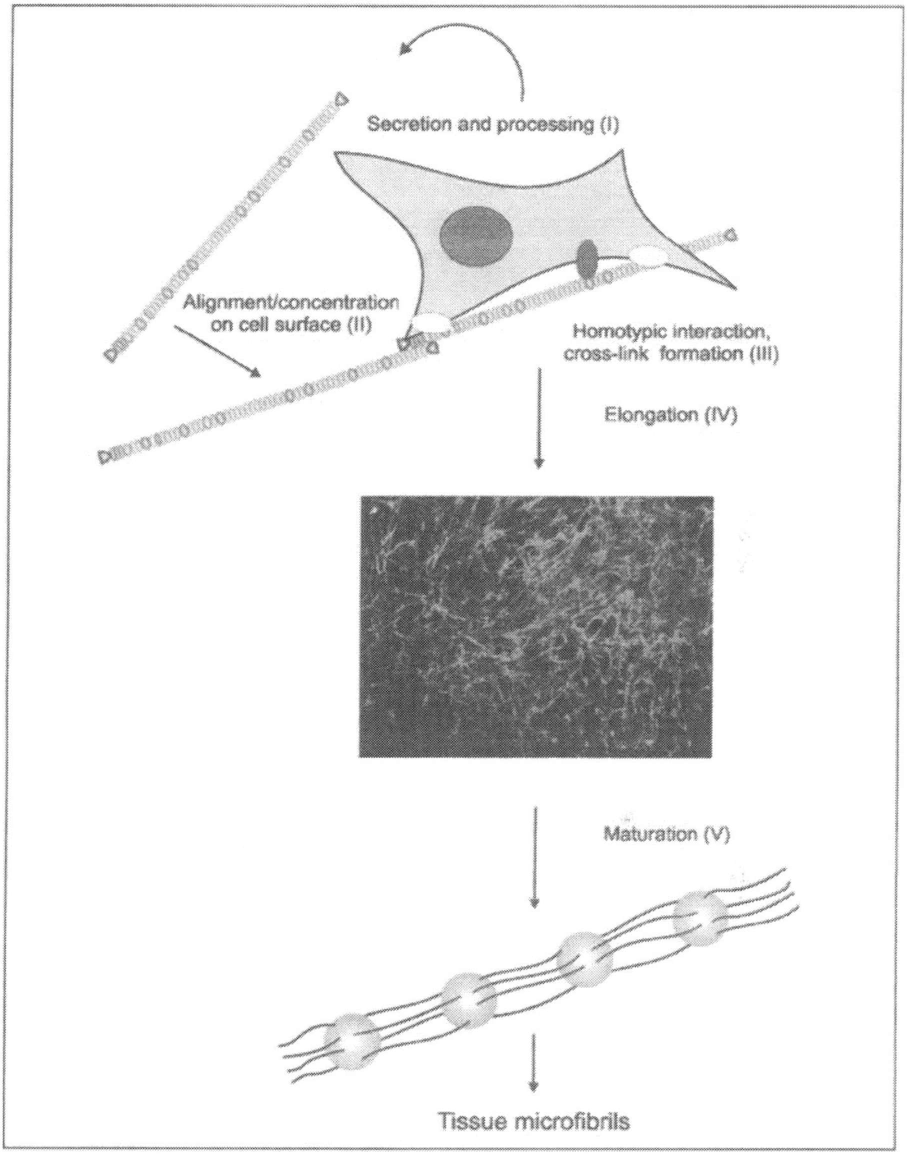

Figure 3. Hypothetical Sequence of the Fibrillin Assembly Steps into Microfibrils. Various cell types such as fibroblasts or smooth muscle cells synthesize and secrete fibrillin. Processing of the terminal ends by furin/ PACE type proteases occur either intra- or extracellularly (I). Processed molecules interact with cell surface proteoglycans (light gray ellipses) or integrins (dark grey ellipses) for alignment and concentration (II). Homotypic fibrillin-1 and/or heterotypic fibrillin-1/fibrillin-2 interaction occur at the terminal ends followed by reducible (disulfide bonds) and nonreducible transglutaminase mediated cross-link formation (III). Elongation is mediated through the central parts of the fibrillin molecules (IV) leading to an observable fibrillin network by indirect immunofluorescence. Further maturation events lead to the "beads-on-a-string" microfibrils and finally to tissue microfibrils (V).

high local concentrations required for assembly or for the alignment of fibrillin molecules in the correct spatial orientation, which would allow formation of intermolecular disulfide bonds. Alternatively, binding of fibrillin to heparan sulfate chains potentially confer conformational changes necessary to expose assembly epitopes.

Binding of fibrillins to cell surface integrin receptors has been clearly defined. Fibrillin-1 and -2 interact via the fourth 8-Cys/TB domain, which contains a RGD sequence, with $\alpha_v\beta_3$ integrin as shown with recombinant polypeptides[77] and with purified fibrillin from tissues using denaturing agents (Fig. 2).[78] In addition, members of the $\beta1$ integrin dimers and possibly $\alpha_5\beta_1$ integrin may also play a role in the cellular interactions of fibrillins.[78,79] Potentially, fibrillin-integrin interactions may play a distinct role in the assembly process similar to the role of $\alpha_5\beta_1$ in fibronectin assembly (for review see ref. 80). Interactions of fibronectin with the $\alpha_5\beta_1$ integrin introduce conformational changes in the fibronectin molecules resulting in exposed assembly epitopes. Much of the discussion about the role of cell surface receptors in fibrillin assembly is of speculative nature. More research efforts are necessary to answer the underlying questions and to clarify potential genotype-biochemical phenotype relationships in terms of cellular interaction of mutated fibrillin-1 in individuals with MFS. At present no point mutation, other than the generation of a stop signal, has been found within the RGD sequence of fibrillin-1 that would allow conclusions on these important questions.[45]

The static organization of fibrillin-1 in microfibrils has been examined by several groups and various techniques (see the chapter "Organization and Biomechanical Properties of Fibrillin Microfibrils" in this volume). Labeling of microfibrils with specific antibodies, high resolution structure of cbEGF modules, and analysis of transglutaminase cross-links have lead to various models of fibrillin alignment in microfibrils.[43,47,56,58,81] Despite the controversy whether fibrillin molecules are arranged in a nonstaggered or in a staggered fashion, common to all models is a head-to-tail arrangement of fibrillin molecules originally proposed by Sakai and coworkers in 1991.[81] Mapping of monoclonal antibody epitopes in fibrillin-1 molecules and correlation with the epitopes in microfibrils revealed that the N- and the C-terminal ends of the fibrillin-1 molecules are located closely together in or close to the bead structures.[43,56,81] To address the question whether fibrillin molecules directly interact with each other or whether they need adapter molecules such as MAGP-1 for example, a recombinant approach has been utilized.[82] These studies demonstrated homotypic self-interaction properties of fibrillin-1 and heterotypic interaction properties between fibrillin-1 and -2 in a N- to C-terminal fashion.[82] The interactions were of high affinity in the low nanomolar range. Aggregation of full length recombinant fibrillin-1 via its terminal regions further supported these data, emphasizing that at least early assembly events do not require adapter molecules.[82] Heterotypic interactions between fibrillin-1 and fibrillin-2 in a similar fashion suggested that both isoforms can be colocalized within the same microfibril. Immunogold localization of fibrillin-1 and -2 within the same microfibril further supported this hypothesis.[83] Interestingly, fibrillin-2 was not able to homotypically interact with itself, suggesting that fibrillin-2 alone might use a different mechanism to assemble into microfibrils.[82] For these studies, relatively large recombinant polypeptides have been utilized. Interestingly, homo- and heterotypic interactions could not be observed using smaller subfragments, suggesting that the assembly epitopes are stabilized by long range structural effects.[82] In the light of mutations leading to MFS, this observation may indicate that in addition to the mutations located in close vicinity to the assembly epitopes, mutations located farther away may result in conformational changes of the assembly epitopes. Further work with mutation constructs will be necessary to clarify this hypothesis.

Mouse Models and Fibrillin Assembly

The contribution of the central part of the fibrillin-1 molecule to microfibrillar assembly was analyzed in different mouse models. In one model a sequence coding for residues 770-1042 (coded by exons 19-24) of the fibrillin-1 molecule was genetically deleted, and due to transcriptional interference by the neo cassette, the mutated fibrillin-1 was expressed at low levels.[10] Immunohistochemical analysis of fibrillin-1 assembly by homozygous mutant dermal fibroblasts revealed fewer and more primitive appearing microfibrils. However, immunoelectron microscopy of skin sections of homozygous mutant animals demonstrated the presence of beaded microfibrils.[10] Despite the uncertainty whether these microfibrils represented fully functional aggregates, these data clearly demonstrated that deletion of the region between the cbEGF module 8 and 8-Cys/TB module 3 does not preclude microfibrillar assembly.

The Tsk mutation is a genomic in-frame duplication of a relatively large central part of the mouse fibrillin-1 gene resulting in the "tight skin" phenotype.[84] This mutation predicts a larger fibrillin-1 molecule with 984 additional amino acid residues and a molecular mass of ~418 kDa instead of the normal ~350 kDa fibrillin-1 product. The assembly properties of this mutant fibrillin-1 has been analyzed by two groups: Kielty and coworkers analyzed beaded microfibrils isolated from skin of the heterozygous Tsk/+ mouse and found the occurrence of two mutually exclusive populations of beaded microfibrils.[11] One population appeared normal and was suggested to be composed of normal fibrillin-1, while the other population displayed longer than normal bead-to-bead periodicities, as well as an altered morphology and organization, and thus was suggested to contain the abnormally long fibrillin-1. The authors concluded a molecular selection based on the length of the molecules and suggested that lateral alignment of fibrillin molecules is a crucial assembly step which precedes or facilitates linear polymerization. Based on analysis of Tsk/Tsk fibroblasts, Gayraud and coworkers also found that the abnormally long fibrillin-1 can form homopolymeric beaded microfibrils with abnormal morphologies such as irregularities in the size and shape of the globular beads, as well as in the distances between the beads.[12] However, these authors demonstrated that the longer Tsk fibrillin-1 is also able to copolymerize together with normal wild-type fibrillin-1 into abnormal beaded microfibrils. It was concluded that the central long stretch of cbEGF-like domains, which is duplicated in the abnormal Tsk fibrillin-1, has an important role in the alignment of fibrillin-1 molecules within microfibrils and in the elongation process.[12] Collectively, these data further strengthen the idea that primary assembly epitopes closer to the terminal ends of the fibrillin-1 molecules, which are not affected by the mutated Tsk fibrillin-1, are important for the initial assembly processes while regions in the center of the molecule may be important for lateral alignment and elongation.

Future Directions

Despite the advances over the last years in understanding the assembly mechanism of fibrillins into microfibrils, only a fragmented picture has evolved. This is in part due to the lack of feasible and modifiable assays to monitor fibrillin assembly. Therefore, the development of such assays will be crucial for more in depth understanding of the principle mechanisms. The major goal of such assay systems will be to precisely define (i) the assembly epitopes in fibrillins, (ii) the structural components important for assembly, and (iii) catalysts, which only transiently participate in microfibrillar assembly. As our understanding of the assembly mechanism increases, so will the potential to correlate the genotype with the biochemical phenotype in MFS. The hope is that from a detailed understanding of this correlation, more mechanism-based approaches for the development of therapeutic strategies will evolve.

Acknowledgements

This work was supported by the Deutsche Forschungsgemeinschaft (Grants SFB367-A1, Re1021/3-2, and Re1021/4-2). We are grateful to Lynn Y. Sakai and Douglas R. Keene for generously providing parts of Figure 1.

References

1. Low FN. The extracellular portion of the human blood-air barrier and its relation to tissue space. Anat Rec 1961; 139:105-124.
2. Gibson MA, Kumaratilake JS, Cleary EG. The protein components of the 12- nanometer microfibrils of elastic and nonelastic tissues. J Biol Chem 1989; 264:4590-4598.
3. Sakai LY, Keene DR. Fibrillin: Monomers and microfibrils. Methods Enzymol 1994; 245:29-52.

4. Kielty CM, Shuttleworth CA. Synthesis and assembly of fibrillin by fibroblasts and smooth muscle cells. J Cell Sci 1993; 106:167-173.
5. Kielty CM, Shuttleworth CA. Abnormal fibrillin assembly by dermal fibroblasts from two patients with Marfan syndrome. J Cell Biol 1994; 124:997-1004.
6. Wright DW, Mayne R. Vitreous humor of chicken contains two fibrillar systems: An analysis of their structure. J Ultrastruct Mol Struct Res 1988; 100:224-234.
7. Wallace RN, Streeten BW, Hanna RB. Rotary shadowing of elastic system microfibrils in the ocular zonule, vitreous, and ligament nuchae. Curr Eye Res 1991; 10:99-109.
8. Keene DR, Maddox BK, Kuo HJ et al. Extraction of extendable beaded structures and their identification as fibrillin-containing extracellular matrix microfibrils. J Histochem Cytochem 1991; 39(4):441-449.
9. Kielty CM, Cummings C, Whittaker SP et al. Isolation and ultrastructural analysis of microfibrillar structures from foetal bovine elastic tissues. J Cell Sci 1991; 99:797-807.
10. Pereira L, Andrikopoulos K, Tian J et al. Targeting of fibrillin-1 recapitulates the vascular phenotype of Marfan syndrome in the mouse. Nat Genet 1997; 17(2):218-222.
11. Kielty CM, Raghunath M, Siracusa LD et al. The tight skin mouse: Demonstration of mutant fibrillin-1 production and assembly into abnormal microfibrils. J Cell Biol 1998; 140(5):1159-1166.
12. Gayraud B, Keene DR, Sakai LY et al. New insights into the assembly of extracellular microfibrils from the analysis of the fibrillin 1 mutation in the tight skin mouse. J Cell Biol 2000; 150(3):667-680.
13. Milewicz D, Pyeritz RE, Crawford ES et al. Marfan syndrome: Defective synthesis, secretion, and extracellular matrix formation of fibrillin by cultured dermal fibroblasts. J Clin Invest 1992; 89:79-86.
14. Aoyama T, Tynan K, Dietz HC et al. Missense mutations impair intracellular processing of fibrillin and microfibril assembly in Marfan syndrome. Hum Mol Genet 1993; 2:2135-2140.
15. Aoyama T, Franke U, Dietz HC et al. Quantitative differences in biosynthesis and extracellular deposition of fibrillin in cultured fibroblasts distinguish five groups of Marfan syndrome patients and suggest distinct pathogenetic mechanisms. J Clin Invest 1994; 94:130-137.
16. Brenn T, Aoyama T, Francke U et al. Dermal fibroblast culture as a model system for studies of fibrillin assembly and pathogenetic mechanisms: Defects in distinct groups of individuals with Marfan's syndrome. Lab Invest 1996; 75(3):389-402.
17. Godfrey M, Olson S, Burgio RG et al. Unilateral microfibrillar abnormalities in a case of asymetric Marfan syndrome. Am J Hum Genet 1990; 46:661-671.
18. Godfrey M, Menashe V, Weleber RG et al. Cosegregation of elastin-associated microfibrillar abnormalities with the Marfan phenotype in families. Am J Hum Genet 1990; 46:652-660.
19. Hollister DW, Godfrey M, Sakai LY et al. Immunohistologic abnormalities of the microfibrillar-fiber system in the Marfan syndrome. N Engl J Med 1990; 323:152-159.
20. Halliday D, Hutchinson S, Kettle S et al. Molecular analysis of eight mutations in FBN1. Hum Genet 1999; 105(6):587-597.
21. Raghunath M, Superti-Furga A, Godfrey M et al. Decreased extracellular deposition of fibrillin and decorin in neonatal Marfan syndrome fibroblasts. Hum Genet 1993; 90:511-515.
22. Godfrey M, Raghunath M, Cisler J et al. Abnormal morphology of fibrillin microfibrils in fibroblast cultures from patients with neonatal Marfan syndrome. Am J Pathol 1995; 146(6):1414-1421.
23. Wang M, Price C, Han J et al. Recurrent mis-splicing of fibrillin exon 32 in two patients with neonatal Marfan syndrome. Hum Mol Genet 1995; 4:607-613.
24. Kielty CM, Phillips JE, Child AH et al. Fibrillin secretion and microfibril assembly by Marfan dermal fibroblasts. Matrix Biol 1994; 14(2):191-199.
25. Kielty CM, Davies SJ, Phillips JE et al. Marfan syndrome: Fibrillin expression and microfibrillar abnormalities in a family with predominant ocular defects. J Med Genet 1995; 32(1):1-6.
26. Raghunath M, Kielty CM, Kainulainen K et al. Analyses of truncated fibrillin caused by a 366 bp deletion in the FBN1 gene resulting in Marfan syndrome. Biochem J 1994; 302:889-896.
27. Milewicz DM, Grossfield J, Cao SN et al. A mutation in FBN1 disrupts profibrillin processing and results in isolated skeletal features of the Marfan syndrome. J Clin Invest 1995; 95:2373-2378.
28. Raghunath M, Kielty CM, Steinmann B. Truncated profibrillin of a Marfan patient is of apparent similar size as fibrillin: Intracellular retention leads to over-N-glycosylation. J Mol Biol 1995; 248:901-909.
29. Ritty TM, Broekelmann T, Tisdale C et al. Processing of the Fibrillin-1 Carboxyl-terminal Domain. J Biol Chem 1999; 274(13):8933-8940.
30. Nakayama K. Furin: A mammalian subtilisin/Kex2p-like endoprotease involved in processing of a wide variety of precursor proteins. Biochem J 1997; 327 (Pt 3):625-635.
31. Ashworth JL, Kelly V, Rock MJ et al. Regulation of fibrillin carboxy-terminal furin processing by N-glycosylation, and association of amino- and carboxy-terminal sequences. J Cell Sci 1999; 112:4163-4171.

32. Lönnqvist L, Reinhardt DP, Sakai LY et al. Evidence for furin-type activity-mediated C-terminal processing of profibrillin-1 and interference in the processing by certain mutations. Hum Mol Genet 1998; 7:2039-2044.
33. Raghunath M, Putnam EA, Ritty T et al. Carboxy-terminal conversion of profibrillin to fibrillin at a basic site by PACE/furin-like activity required for incorporation in the matrix. J Cell Sci 1999; 112:1093-1100.
34. Zhang H, Apfelroth SD, Hu W et al. Structure and expression of fibrillin-2, a novel microfibrillar component preferentially located in elastic matrices. J Cell Biol 1994; 124:855-863.
35. Nagase T, Nakayama M, Nakajima D et al. Prediction of the coding sequences of unidentified human genes. XX. The complete sequences of 100 new cDNA clones from brain which code for large proteins in vitro. DNA Res 2001; 8(2):85-95.
36. Tilstra DJ, Li L, Potter KA et al. Sequence of the coding region of the bovine fibrillin cDNA and localization to bovine chromosome 10. Genomics 1994; 23:480-485.
37. Yin W, Smiley E, Germiller J et al. Primary structure and developmental expression of *Fbn-1*, the mouse fibrillin gene. J Biol Chem 1995; 270:1798-1806.
38. Zhang H, Hu W, Ramirez F. Developmental expression of fibrillin genes suggests heterogeneity of extracellular microfibrils. J Cell Biol 1995; 129:1165-1176.
39. Maslen CL, Corson GM, Maddox BK et al. Partial sequence of a candidate gene for the Marfan syndrome. Nature 1991; 352:334-337.
40. Corson GM, Chalberg SC, Dietz HC et al. Fibrillin binds calcium and is coded by cDNAs that reveal a multidomain structure and alternatively spliced exons at the 5' end. Genomics 1993; 17:476-484.
41. Pereira L, D'Alessio M, Ramirez F et al. Genomic organization of the sequence coding for fibrillin, the defective gene product in Marfan syndrome. Hum Mol Genet 1993; 2:961-968.
42. Reinhardt DP, Gambee JE, Ono RN et al. Initial steps in assembly of microfibrils. Formation of disulfide-cross-linked multimers containing fibrillin-1. J Biol Chem 2000; 275(3):2205-2210.
43. Reinhardt DP, Keene DR, Corson GM et al. Fibrillin 1: Organization in microfibrils and structural properties. J Mol Biol 1996; 258:104-116.
44. Reinhardt DP, Sasaki T, Dzamba BJ et al. Fibrillin-1 and fibulin-2 interact and are colocalized in some tissues. J Biol Chem 1996; 271:19489-19496.
45. Robinson PN, Booms P, Katzke S et al. Mutations of FBN1 and genotype-phenotype correlations in Marfan syndrome and related fibrillinopathies. Hum Mutat 2002; 20(3):153-161.
46. Eldadah ZA, Brenn T, Furthmayr H et al. Expression of a mutant human fibrillin allele upon a normal human or murine genetic background recapitulates a Marfan cellular phenotype. J Clin Invest 1995; 95:874-880.
47. Downing AK, Knott V, Werner JM et al. Solution structure of a pair of calcium-binding epidermal growth factor-like domains: implications for the Marfan syndrome and other genetic disorders. Cell 1996; 85:597-605.
48. Yuan X, Downing AK, Knott V et al. Solution structure of the transforming growth factor β-binding protein-like module, a domain associated with matrix fibrils. EMBO J 1997; 16(22):6659-6666.
49. Kanwar YS, Ota K, Yang Q et al. Isolation of rat fibrillin-1 cDNA and its relevance in metanephric development. Am J Physiol 1998; 275(5):F710-F723.
50. Biery NJ, Eldadah ZA, Moore CS et al. Revised genomic organization of FBN1 and significance for regulated gene expression. Genomics 1999; 56(1):70-77.
51. Yang Q, Ota K, Tian Y et al. Cloning of rat fibrillin-2 cDNA and its role in branching morphogenesis of embryonic lung. Dev Biol 1999; 212(1):229-242.
52. Trask TM, Ritty TM, Broekelmann T et al. N-terminal domains of fibrillin 1 and fibrillin 2 direct the formation of homodimers: A possible first step in microfibril assembly. Biochem J 1999; 340:693-701.
53. Ashworth JL, Kelly V, Wilson R et al. Fibrillin assembly: Dimer formation mediated by amino-terminal sequences. J Cell Sci 1999; 112:3549-3558.
54. Turano C, Coppari S, Altieri F et al. Proteins of the PDI family: Unpredicted nonER locations and functions. J Cell Physiol 2002; 193(2):154-163.
55. Hanssen E, Franc S, Garrone R. Atomic force microscopy and modeling of natural elastic fibrillin polymers. Biol Cell 1998; 90(3):223-228.
56. Baldock C, Koster AJ, Ziese U et al. The supramolecular organization of fibrillin-rich microfibrils. J Cell Biol 2001; 152(5):1045-1056.
57. Bowness JM, Tarr AH. Epsilon(gamma-Glutamyl)lysine crosslinks are concentrated in a noncollagenous microfibrillar fraction of cartilage. Biochem Cell Biol 1997; 75(1):89-91.
58. Qian RQ, Glanville RW. Alignment of fibrillin molecules in elastic microfibrils is defined by transglutaminase-derived cross-links. Biochemistry 1997; 36:15841-15847.

59. Thurmond FA, Koob TJ, Bowness JM et al. Partial biochemical and immunologic characterization of fibrillin microfibrils from sea cucumber dermis. Connect Tissue Res 1997; 36(3):211-222.
60. Lorand L, Graham RM. Transglutaminases: Crosslinking enzymes with pleiotropic functions. Nat Rev Mol Cell Biol 2003; 4(2):140-156.
61. Raghunath M, Cankay R, Kubitscheck U et al. Transglutaminase activity in the eye: Cross-linking in epithelia and connective tissue structures. Invest Ophthalmol Vis Sci 1999; 40(12):2780-2787.
62. Thurmond FA, Trotter JA. Morphology and biomechanics of the microfibrillar network of sea cucumber dermis. J Exp Biol 1996; 199:1817-1828.
63. Brown-Augsburger P, Broekelmann T, Mecham L et al. Microfibril-associated glycoprotein binds to the carboxyl-terminal domain of tropoelastin and is a substrate for transglutaminase. J Biol Chem 1994; 269:28443-28449.
64. Henderson M, Polewski R, Fanning JC et al. Microfibril-associated glycoprotein-1 (MAGP-1) is specifically located on the beads of the beaded-filament structure for fibrillin-containing microfibrils as visualized by the rotary shadowing technique. J Histochem Cytochem 1996; 44(12):1389-1397.
65. Kielty CM, Sherratt MJ, Shuttleworth CA. Elastic fibres. J Cell Sci 2002; 115(Pt 14):2817-2828.
66. Isogai Z, Aspberg A, Keene DR et al. Versican interacts with fibrillin-1 and links extracellular microfibrils to other connective tissue networks. J Biol Chem 2002; 277(6):4565-4572.
67. Isogai Z, Ono RN, Ushiro S et al. Latent transforming growth factor β-binding protein 1 interacts with fibrillin and is a microfibril-associated protein. J Biol Chem 2003; 278(4):2750-2757.
68. Gibson MA, Finnis ML, Kumaratilake JS et al. Microfibril-associated glycoprotein-2 (MAGP-2) is specifically associated with fibrillin-containing microfibrils but exhibits more restricted patterns of tissue localization and developmental expression than its structural relative MAGP-1. J Histochem Cytochem 1998; 46(8):871-886.
69. Gibson MA, Cleary EG. The immunohistochemical localization of microfibril- associated glycoprotein (MAGP) in elastic and nonelastic tissues. Immunol Cell Biol 1987; 65:345-356.
70. Kumaratilake JS, Gibson MA, Fanning JC et al. The tissue distribution of microfibrils reacting with a monospecific antibody to MAGP, the major glycoprotein antigen of elastin-associated microfibrils. Eur J Cell Biol 1989; 50(1):117-127.
71. Jensen SA, Reinhardt DP, Gibson MA et al. MAGP-1, Protein interaction studies with tropoelastin and fibrillin-1. J Biol Chem 2001; 276(43):39661-39666.
72. Trask BC, Trask TM, Broekelmann T et al. The microfibrillar proteins MAGP-1 and fibrillin-1 form a ternary complex with the chondroitin sulfate proteoglycan decorin. Mol Biol Cell 2000; 11(5):1499-1507.
73. Mecham RP, Davis E. Elastic fiber structure and assembly. In: Yurchenco PD, ed. Extracellular Matrix Assembly and Structure. New York:Academic Press, 1994: 281-314.
74. Trask TM, Crippes Trask B, Ritty TM et al. Interaction of tropoelastin with the amino-terminal domains of fibrillin-1 and fibrillin-2 suggests a role for the fibrillins in elastic fiber assembly. J Biol Chem 2000; 275(32):24400-24406.
75. Brown-Augsburger P, Broekelmann T, Rosenbloom J et al. Functional domains on elastin and microfibril-associated glycoprotein involved in elastic fibre assembly. Biochem J 1996; 318:149-155.
76. Tiedemann K, Bätge B, Müller PK et al. Interactions of fibrillin-1 with heparin/heparan sulfate: Implications for microfibrillar assembly. J Biol Chem 2001; 276(38):36035-36042.
77. Pfaff M, Reinhardt DP, Sakai LY et al. Cell adhesion and integrin binding to recombinant human fibrillin-1. FEBS Lett 1996; 384:247-250.
78. Sakamoto H, Broekelmann T, Cheresh DA et al. Cell-type specific recognition of RGD- and nonRGD-containing cell binding domains in fibrillin-1. J Biol Chem 1996; 271:4916-4922.
79. D'Arrigo C, Burl S, Withers AP et al. TGF-beta1 binding protein-like modules of fibrillin-1 and -2 mediate integrin-dependent cell adhesion. Connect Tissue Res 1998; 37(1-2):29-51.
80. Schwarzbauer JE, Sechler JL. Fibronectin fibrillogenesis: A paradigm for extracellular matrix assembly. Curr Opin Cell Biol 1999; 11(5):622-627.
81. Sakai LY, Keene DR, Glanville RW et al. Purification and partial characterization of fibrillin, a cysteine-rich structural component of connective tissue microfibrils. J Biol Chem 1991; 266:14763-14770.
82. Lin G, Tiedemann K, Vollbrandt T et al. Homo- and heterotypic fibrillin-1 and -2 interactions constitute the basis for the assembly of microfibrils. J Biol Chem 2002; 277(52):50795-50804.
83. Charbonneau NL, Dzamba BJ, Ono RN et al. Fibrillins can coassemble in fibrils, but fibrillin fibril composition displays cell-specific differences. J Biol Chem 2003; 278(4):2740-2749.
84. Siracusa LD, McGrath R, Ma Q et al. A tandem duplication within the fibrillin 1 gene is associated with the mouse tight skin mutation. Genome Res 1996; 6:300-313.

CHAPTER 12

Organization and Biomechanical Properties of Fibrillin Microfibrils

Cay M. Kielty, Tim J. Wess, J. Louise Haston, Michael J. Sherratt, Clair Baldock and C. Adrian Shuttleworth

Introduction

The evolution and function of multicellular organisms has required tissue flexibility and the necessity to withstand stretch. These properties underpin movement and allow responses to changing environments. The elastic fibre is a bio-composite material composed of two insoluble extracellular matrix assemblies; fibrillin-rich microfibrils and elastin. Elastic fibres permit long-range deformability and recoil within dynamic tissues such as blood vessels, lungs, ligaments and skin.

Fibrillin-rich microfibrils are among the oldest elastomers in the animal kingdom. They have conferred long-range elasticity on connective tissues for at least 550 million years. Fibrillin molecules and microfibrils are remarkably conserved throughout multicellular evolution from medusa jellyfish (*Podocoryne carnea M. Sars*), sea cucumber (*Cucumaria frondosa*) and other invertebrates, to man, which confirms their critical biomechanical importance.[1-5] In all these organisms, the tissue and developmental distributions of fibrillins correlate closely with the biomechanical needs of the organism. In the low pressure circulatory system of the lobster, aortic elasticity is due almost entirely to microfibril arrays which reorientate and align at lower pressures then deform and elongate at higher pressures.[6] In man, microfibrils are abundant in nonelastic tissues such as ciliary zonules that hold the lens in dynamic suspension. They are also, together with elastin, components of elastic fibres.[7] Elastin, a resilient rubber-like protein, evolved more recently to reinforce high pressure closed circulatory systems in which the interplay between highly resilient elastin and stiff fibrillar collagen is a critical feature.[8] Elastin appeared first in cartilaginous fish, and is abundantly expressed in other elastic tissues such as blood vessels, lung, ligaments and skin. The distinct mechanical properties of aorta in the human high pressure closed circulation are due to the additional presence of elastic fibres that can keep their elastic properties up to 140% extension. Interactions with elastin undoubtedly modulate the biomechanical properties of microfibrils.

Elastic fibres have evolved to maintain elastic function for a lifetime. Loss of connective tissue elasticity in ageing and disease occurs as a consequence of degeneration of the microfibrillar and elastic fibre network. Fibrillin molecules and microfibrils are susceptible to degradation by various matrix metalloproteinases (MMPs) and serine proteinases, and accumulative proteolytic damage is a major mechanism underlying degenerative changes in ageing.[9-11] Some of the fibrillin-1 proteolytic cleavage sites have been identified, and MMP treatment of isolated microfibrils has been shown to compromise the ~56 nm periodic repeat structure of microfibrils.[10,12,13] Degradation of these extracellular polymers is a major contributing factor in blood vessel ageing and development of aortic aneurysms, in lung emphysema, and in degenerative changes in sun-damaged skin. In ageing eyes, degeneration of microfibrils compromises

Marfan Syndrome: A Primer for Clinicians and Scientists, edited by Peter N. Robinson and Maurice Godfrey. ©2004 Eurekah.com and Kluwer Academic / Plenum Publishers.

the biomechanical properties of the ciliary zonules.[14] Accelerated damage to the microfibrillar apparatus is thought to contribute to the pathology in some Marfan syndrome patients as a consequence of specific fibrillin-1 mutations that may predispose to increased fibrillin-1 proteolytic susceptibility (see also "Stretched Microfibrils from Marfan Tissues").[15] One such example is the ectopia lentis-causing mutation E2447K which introduces or exposes a matrix metalloproteinase cleavage site.[10] Recombinant fibrillin-1 peptides containing certain disease-causing Marfan mutations increase their susceptibility to trypsin due to conformational changes, although the physiological significance of these observations in terms of tissue degradation is unclear.[16-18]

Microfibrils are multimolecular assemblies based on a linear fibrillin polymer.[13,19] Early attempts to isolate microfibrillar proteins utilised a series of steps which included collagenase digestion and guanidine extraction of tissues in the presence of an agent to reduce disulphide bonds.[20] In 1986, a monoclonal antibody raised to a pepsin extract of human amnion was shown to immunolocalise to microfibrils and to recognise a large glycoprotein (M_r 350,000) in the medium of human fibroblast cultures.[21] This molecule was designated fibrillin, and subsequent biochemical and electron microscopy studies showed that it is the principal structural element of this class of connective tissue microfibrils. That this was the case was further reinforced by linkage of mutations in the *FBN1* gene on chromosome 15 to Marfan syndrome which is associated with pleiotrophic connective tissue manifestations.

Microfibrils may have tissue-specific compositions that influence their structure and function.[22] Fibrillins, the principal structural molecules of microfibrils, are encoded by three genes.[3-5] They are large glycoproteins (~350 kDa) with calcium-binding epidermal growth factor-like (cbEGF) domain arrays that, in the presence of Ca^{2+}, adopt a rod-like arrangement.[23,24] These domains are interspersed with 8-cysteine motifs (also called 'TB' modules because of their homology to TGF-β binding modules in latent TGF-β binding proteins). Towards the fibrillin-1 N-terminus is a proline (pro)-rich sequence that may act as a 'hinge'; fibrillin-2 has a glycine (gly)-rich sequence, and fibrillin-3 a pro- and gly-rich sequence. Transglutaminase cross-links between fibrillin peptides have been identified that probably influence microfibril stability and elasticity.[25] Fibrillin -1 and fibrillin-2 have distinct, but overlapping expression patterns, with fibrillin-2 generally expressed earlier in development than fibrillin-1.[26-28] In some tissues, microfibrils may be assembled from different fibrillin isoforms. Immuno-EM and tissue extraction studies have identified several molecules that colocalise with microfibrils; they include microfibril-associated glycoproteins (MAGPs) -1 and -2, microfibril-associated proteins (MFAPs) -1, -3 and -4, LTBPs -1, -2 and -4, and proteoglycans/ glycosaminoglycans (decorin, biglycan, versican, heparan sulphate).[22,29] Other molecules localise at the interface between the outer microfibril mantle and the inner elastin core; they include emilin-1, collagen VIII, proteoglycans, fibulins -1 and-2, and nidogen-2.

Current intense interest in the biology of fibrillin-rich microfibrils is driven by the need to understand their contribution to tissue elasticity, genotype-to-phenotype correlations in Marfan syndrome, and the possibility of exploiting their biomechanical properties in tissue engineering and wound repair strategies. In this chapter, we outline current understanding of microfibril organisation and biomechanical properties, and discuss structural implications of loss of microfibril integrity in Marfan syndrome.

Tissue Organisation of Fibrillin-Rich Microfibrils

Tissue Microfibril Bundles

Microfibrillar elements of connective tissues, which have a specific association with amorphous elastic tissues, have been recognised histologically for many years (Fig. 1).[30,31] Analysis by transmission electron microscopy revealed these microfibrils to have a diameter of 8-12 nm, a tubular appearance and beaded periodicity, and to contain glycoproteins. They are commonly found in tissue locations that are subject to repeated mechanical stresses, and in the proximity of basement membranes. This distribution gave rise to speculation that these mi-

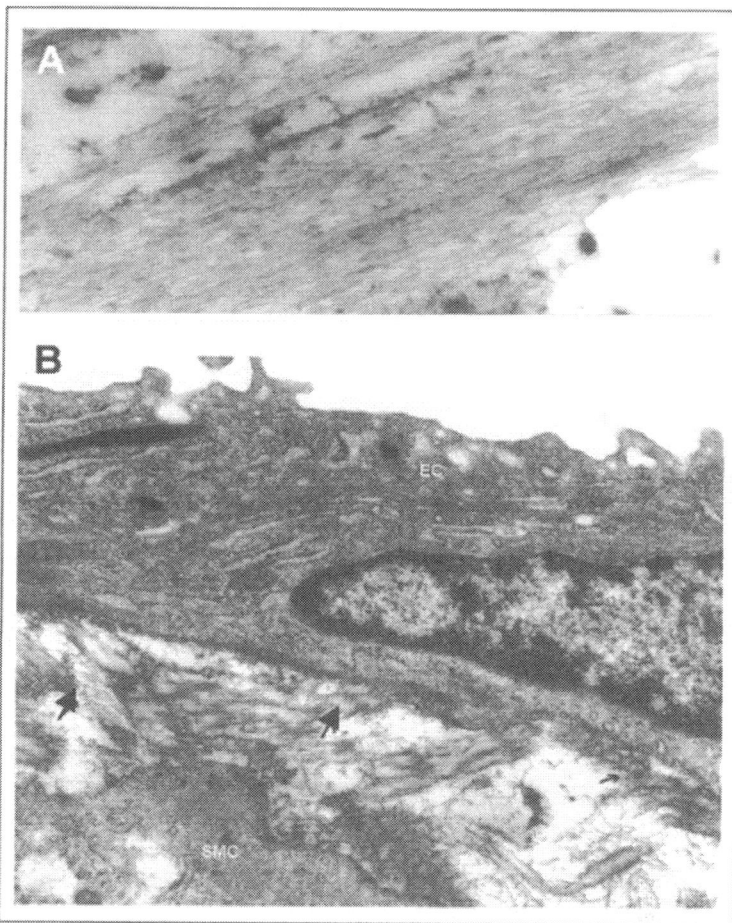

Figure 1. Transmission electron microscopy of tissue microfibrils. A) Parallel arrays of fibrillin-rich mi-crofibrils deposited by bovine nuchal ligament fibroblasts in culture. B) Elastic fibers assembling in the subendothelial space between vascular endothelial cell (EC) and smooth muscle cell (SMC). Elastin is deposited upon a microfibril framework.

crofibrils might serve a critical biomechanical anchoring role in dynamic connective tissues. They were also found at the periphery of elastic fibres, but could be clearly distinguished from the elastin core on the basis of major differences in their staining properties, due to their strongly anionic nature.

The concept of arrays or 'bundles' of loosely packed microfibrils in roughly parallel align-ments was established from transmission electron microscopy of fixed dehydrated tissues (Fig. 1A).[7,31] The fine structure of microfibrils has been difficult to discern using conventional elec-tron microscopy, although they often reveal a periodic nature using this technique. More re-cently, other approaches have revealed further details of microfibril packing within hydrated tissues. They include high pressure freezing followed by freeze substitution, environmental scanning electron microscopy (ESEM), and quick freeze-deep etch (QFDE) microscopy.[13,32,33] ESEM has visualised the intercalation of microfibril-based filaments into the lens capsule base-ment membrane, and shown that branches occur within whole zonular filaments.[13] QFDE has provided 3D surface views of microfibrils that have not been fixed, dehydrated or stained with heavy metals.[33] By this approach, microfibrils appeared as tightly packed rows of small bead-like subunits that did not display the interbead filamentous links seen in isolated microfibrils (see 'Organisation of Isolated Microfibrils' below). At regular 50 nm intervals, a larger bead was often seen which tended to be aligned with those from adjacent microfibrils within bundles.[33] This QFDE study also identified small filaments of unknown molecular composition spanning between adjacent zonular microfibrils at irregular intervals.

Tissue-specific fibrillin-rich microfibril organisations occur that are probably dictated by cells and by the strength and direction of forces put upon the tissue, and reflect the particular mechanical functions of different tissues. For example, the zonular filaments of the eye, which comprise higher-order packed arrays of microfibril bundles (Fig. 2), function to distort the lens mechanically during accomodation of the eye.[34] Oxytalan fibres comprise small microfibril bundles cascading from the dermal-epidermal junction and merging within the papillary dermis, and their role is to provide an elastic anchor linking the epidermis and dermis. Subendothelial microfibril bundles are seen to run parallel with the direction of blood flow, in developing aorta which provide elastic anchorage for endothelial cells in flow conditions.[35] Bundles of microfibrils in developing aorta are seen to intercalate with the smooth muscle cells within the medial layer, prior to deposition of tropoelastin and formation of medial elastic fibres that provide vessel elasticity and resilience (Fig. 3A).[36] In Marfan syndrome, such microfibril and elastic fibre arrangements in aorta and other elastic tissues are often disordered due to mutational effects or remodelled such that they become dysfunctional (Fig. 3B-E).

Small angle X-ray fibre diffraction has provided important information on the structure and behaviour of zonular microfibrils in the hydrated state. Fibre diffraction is the analysis of

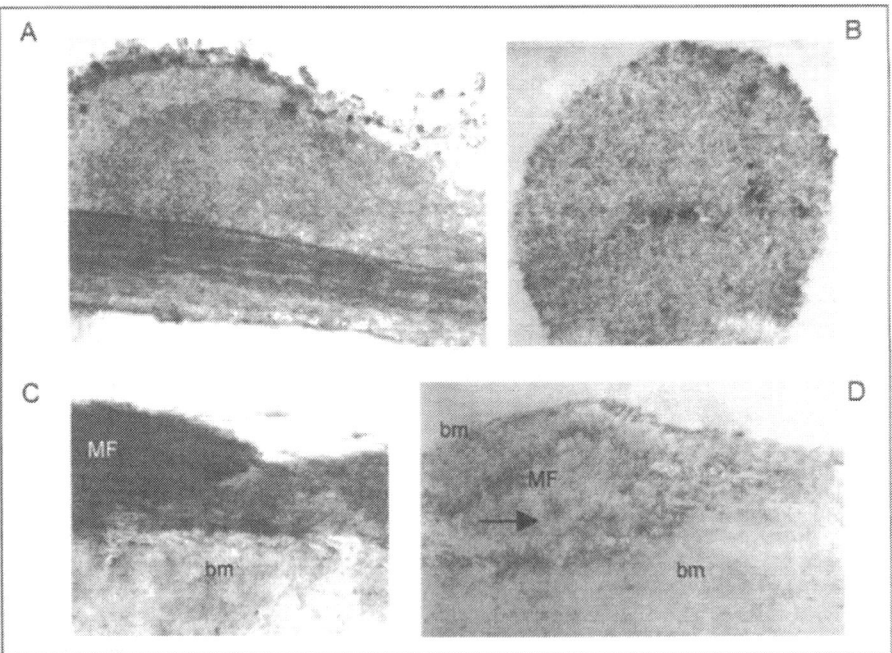

Figure 2. Transmission electron micrographs of bovine ciliary zonules. A, B) Distinct microfibril bundles can be seen within the zonular filaments. C) Microfibril bundles aligned on the lens capsule basement membrane. D) A microfibril bundle intercalated within the lens capsule basement membrane.

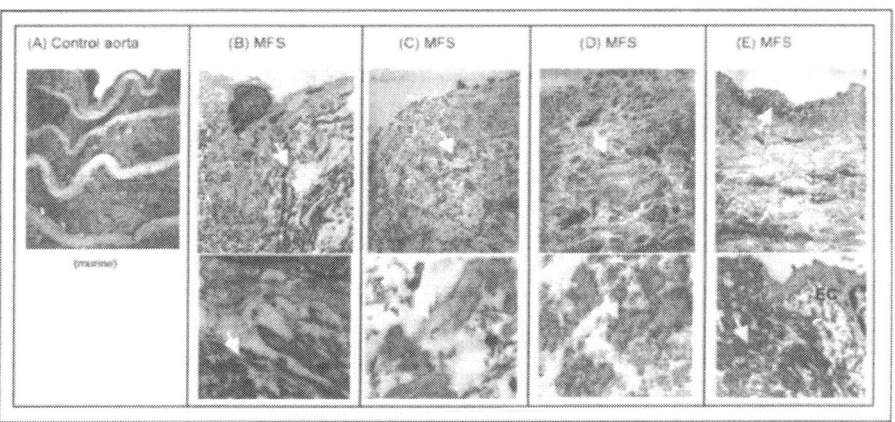

Figure 3. Transmission electron micrographs of cross-sections through control and Marfan aortae. A) Control aorta (murine), showing the ordered lamellar arrangement of elastic fibers and SMC in the medial layer. B-E) Four Marfan aortae, showing gross structural changes especially in the organisation of microfibrils and elastic fibres. In (B) and (E), abundant microfibrils and small forming elastic fibres were observed especially in the subendothelium and inner medial layer. In (C) and (D), medial elastic fibre organisation was grossly disrupted. In (C), (D) and (E), weakly attached endothelial cells were lost during preparation. Arrows indicate abundant disordered microfibrils and associated elastin.

X-ray diffraction patterns obtained from fibrous macromolecules. This technique provides information on the supramolecular architecture of fibrillar structures. Since the molecular organisation is examined using an X-ray beam that covers a macroscopic area of specimen, the results are statistically reliable. Information about axial packing is derived from the meridional series (which runs parallel to the fibre orientation), whilst detail about lateral packing can be obtained from the equatorial series (running at right angles to the fibre). In the case of mammalian zonular filaments, a series of meridional diffraction peaks (orders) that index on a fundamental axial periodicity of 56 nm have been characterised. Furthermore, in the native state this series has been found to exhibit dominant 3rd and 6th orders, which correspond to a spacing of 56/3 and 56/6 nm respectively.[37,38,60] These findings have provided evidence for one-third staggered junctional regions within zonular bundles that may be important in modulating force transmission through the bundles, and for changes in diffraction patterns upon chelation of calcium.[37] This characteristic 3rd order staggered array organisation is recoverable even after 150% tissue extension, which is probably beyond the physiological range of microfibril elasticity.[29,38,63]

Microfibrils in Elastic Fibres

Elastic fibrillogenesis is a highly developmentally regulated process in which tropoelastin (the soluble precursor of mature elastin) is deposited on a preformed template of fibrillin-rich microfibrils.[7] Mature elastic fibres are thus a composite biomaterial comprising an outer microfibrillar mantle and an inner core of amorphous cross-linked elastin with some embedded microfibrils (Fig. 1B).

The first signs of elastic fibres in fetal development are parallel arrays of fibrillin-rich microfibrils close to the cell membrane, often occupying infoldings of the cell membrane.[39,40] Elastin is then deposited as small clumps of amorphous materials within these microfibril arrays, which subsequently coalesce to form mature elastic fibres. The observable proportion of microfibrils to elastin appears to decline with age, with adult elastic fibres having only a sparse peripheral mantle of microfibrillar material. Microfibrils probably serve to align tropoelastin molecules, the soluble secreted form of elastin, in precise register, so that cross-linking regions are juxtaposed prior to oxidation by lysyl oxidase. The interaction of elastin with microfibrils may be directly with fibrillin-1 or with microfibril-associated molecules such as MAGP-1 or biglycan.[41-43] Since the microfibrillar arrays take the form and orientation of presumptive elastic fibres, they are thought to direct the morphogenesis and architecture of elastic fibres by acting as a scaffold on which elastin is deposited.[40]

Just as microfibrillar organisations reflect tissue mechanics, distinct tissues can have profoundly different elastic fibre organisations that reflect specific functions. In aorta, elastic fibres form concentric fenestrated lamellae separated by smooth muscle cells in the medial layer, and they provide essential elasticity and resilience. Skin elasticity is endowed by a continuous elastic network from the reticular dermis that contains thick horizontally arranged elastic fibres, to the papillary dermis that contains thinner perpendicular elastic fibres (elaunin fibres), to oxytalan fibres (microfibrillar arrays) that intercalate into the dermal-epidermal junction. In auricular cartilage, a thin network of elastic fibres interspersed with collagen fibrils in the interterritorial zone contributes to tissue deformability. In lung alveoli, thin highly branched elastic fibres throughout the respiratory tree support alveolar expansion and recoil during breathing. In ligaments, elastic fibres are abundant, small, rope-like and variable in length.

Organisation of Isolated Microfibrils

Isolation Protocols
Microfibrils can be liberated from tissues such as ciliary zonules by a simple mechanical homogenisation step.[44] A more commonly used protocol is based on tissue extraction with bacterial collagenase followed by elution in the void volume of a Sepharose CL-2B size fractionation column, and provides a highly microfibril-enriched preparation.[45] This methodology is particularly effective with crude bacterial collagenase, possibly due to additional protease activity. It has been shown that the large chondroitin sulphate proteoglycan, versican, is removed from microfibrils during this preparation, and that LTBP-1 is not an integral component of microfibrils.[46-48] Isolated microfibrils can be readily purified from these preparations using a CsCl density gradient centrifugation protocol.[47]

Beaded Structure
Rotary shadowing of microfibrils isolated from mammalian tissues such as amnion, skin and ligament by various different protocols all show a prominent repeating 'beads-on-a-string' appearance with an average, but variable, periodicity of 56 nm[13] (Fig. 4A) (see Chapter 13 by M. Gibson). In some preparations, six to eight 'arms' are observed to emanate from the beaded structure. Periodic peaks and troughs of mass corresponding to beads and interbeads were identified by scanning transmission electron microscopy (STEM) mass mapping (Fig. 4A).[13] The atomic force microscope (AFM) employs an extremely fine tip mounted at the end of a flexible cantilever to detect variations in surface topography with a vertical resolution better than 0.1 nm. Fibrillin-rich microfibrils imaged by intermittent contact mode and non-contact mode AFM also also exhibit the characteristic 56 nm beaded periodicity, where beads correspond to height maxima and interbeads to height minima (Fig. 4A).[29,49] Our cryoEM and AFM studies of isolated hydrated microfibrils have confirmed that the beaded structure is apparent in hydrated isolated microfibrils state (Fig. 4A).[22] Interestingly, QFDE of hydrated microfibrils also revealed the periodic bead structure, but only after sonication, suggesting that removal of other matrix molecules is required to expose the beaded framework.[33]

Fibrillin molecules and microfibrils bind calcium, as predicted by the numerous cbEGF domains.[3] The presence of calcium within contiguous cbEGF domain arrays leads to adoption of a rod-like conformation.[23,24] Velocity sedimentation and rotary shadowing have revealed a 20-25% decrease in the length of the rod-shaped fibrillin molecule in the presence of 5mM EDTA.[50] X-ray diffraction, dark field STEM and rotary shadowing of isolated and tissue microfibrils all showed that chelation of calcium reduces beaded periodicity by 15-30% and pro-

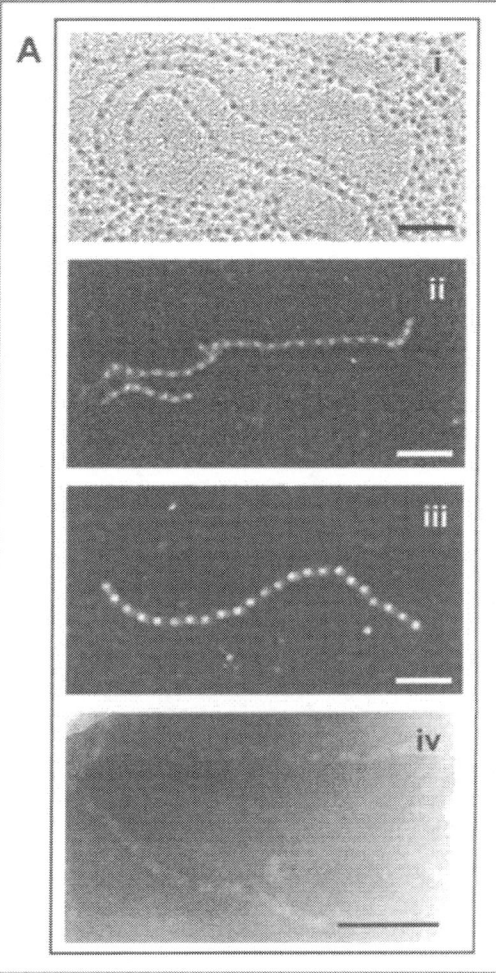

Figure 4A. Isolated microfibrils imaged by four different microscopy approaches, all of which demonstrate the ~56 nm untensioned beaded periodicity. Panel A) Ultrastructural appearance of isolated bovine ciliary zonule fibrillin microfibrils visualised by four microscopical techniques. i-iii) Dehydrated microfibrils. iv) Hydrated microfibrils. Bars = 200 nm. i) TEM micrograph of rotary shadowed microfibrils adsorbed onto a thin carbon film. ii) Scanning transmission electron microscopy (STEM) mass map of an unstained and unshadowed microfibril adsorbed onto a thin carbon film. iii) Intermittent contact mode atomic force microscopy (AFM) of unstained and unshadowed microfibrils adsorbed onto a poly-L-lysine modified mica substrate. iv) Cryo-EM of unstained and unshadowed microfibrils, imaged within vitreous ice, showing that the beaded structure and periodicity is also apparent in fully hydrated isolated microfibrils.

duces a diffuse appearance with evidence for increased molecular flexibility.[37,51,52] The observed decrease in fibrillin molecule and microfibril rest length is partly accounted for by an increase in the degrees of freedom of the linkages between consecutive calcium-binding epidermal growth factor-like (cbEGF) domains. NMR studies showed that, in the presence of bound calcium, a fibrillin-1 cbEGF domain pair is in a rigid, rod-like arrangement, stabilised by interdomain calcium binding and hydrophobic interactions.[23,24]

Molecular Packing in Microfibrils

Details of the alignment of fibrillin molecules within microfibrils have emerged from ultra-structural studies of isolated microfibrils, and several models have been described. One model, based on antibody epitope mapping and measured molecular dimensions, suggested a parallel head-to-tail alignment of unstaggered fibrillin molecules with amino- and carboxy-termini at, or close to the beads.[53] STEM and AFM have shown that isolated microfibrils have an asymmetrical repeating bead-interbead organisation, supporting the concept that microfibrils are directional and that fibrillin molecules are parallel.[13,29,49] Staggered arrangements, based on extrapolation of molecular length from cbEGF-like domain dimensions and untensioned microfibril periodicity (approximately one-half stagger), or on alignment of the fibrillin-1 homo-typic transglutaminase cross-links in an approximate one-third stagger, have also been proposed.[23,25] None of these models are compatible with the axial mass distribution of untensioned microfibrils revealed by STEM.

Automated electron tomography (AET) generated 3D reconstructions provided further insights into the complex organisation of untensioned fibrillin-rich microfibrils, and the alignment of fibrillin molecules in the untensioned state (Fig. 4B).[54,55] Twisting was seen to occur within untensioned microfibrils and detailed dimensions and volumes of heart-shaped beads and interbead organisation were determined. Localisation of fibrillin-1 antibody and gold-binding epitopes, and mapping of bead and interbead mass changes in untensioned and extended microfibrils provided compelling evidence that fibrillin molecules undergo significant intramolecular folding within each untensioned beaded repeat. STEM analysis defined microfibril mass and its axial distribution, and predicted up to eight fibrillin molecules in cross-section. A model based on these data predicted that microfibril assembly progresses from initial parallel head-to-tail alignment to an approximate one-third stagger that can be stabilised by transglutaminase cross-links and further folds to generate the ~56 nm periodic repeat.[54-56] STEM analysis of unstretched and extended microfibrils revealed that there was often a sharp periodicity transition from under 95 nm to over 110 nm and a progressive loss of interbead mass at periodicities between 56-90 nm and of bead mass above 100 nm.[54] Using intermittent contact mode AFM height measurements, we have also shown that, within the periodic range 44-68 nm, interbead height is reduced at increasing periodicities but bead height remains constant along a 35nm region centred on the bead.[29] These observations suggest that interbead conformational changes account for extension between ~56-90 nm. If individual microfibrils are elastic, interbead changes would account for extension and recoil between ~56-90 nm, whilst irreversible bead changes probably occur at periodicities above 100 nm (Fig. 5A).

Mechanical Properties of Microfibrils

Early studies of isolated untensioned microfibrils revealed that, while the vast majority of beaded repeats had a regular periodicity of approximately 56 nm, there were also seen some highly stretched beaded repeat regions with periodicities of up to ~160 nm (Fig. 5B).[13,54,57] These highly stretched regions were presumed to have arisen due to tangling in debris during preparation, and led to predictions that microfibrils are elastic. Since then, studies of microfibril bundles and of hydrated microfibril-based tissues such as ocular ciliary zonules, have delineated the elastic properties of tissue microfibrils.

Invertebrate Studies

Important insights into microfibril elasticity have emerged from invertebrate studies. Microfibril networks extracted from sea cucumber dermis using guanidine and bacterial collagenase were shown, by tensile testing, to be reversibly extensible up to approximately 300% of their initial length.[2] These networks behaved like viscoelastic solids, with a long-range elastic component as well as a time-dependent viscous component. The strength of the network appeared to be due to nonreducible cross-links, but its elasticity was dependent on disulphide

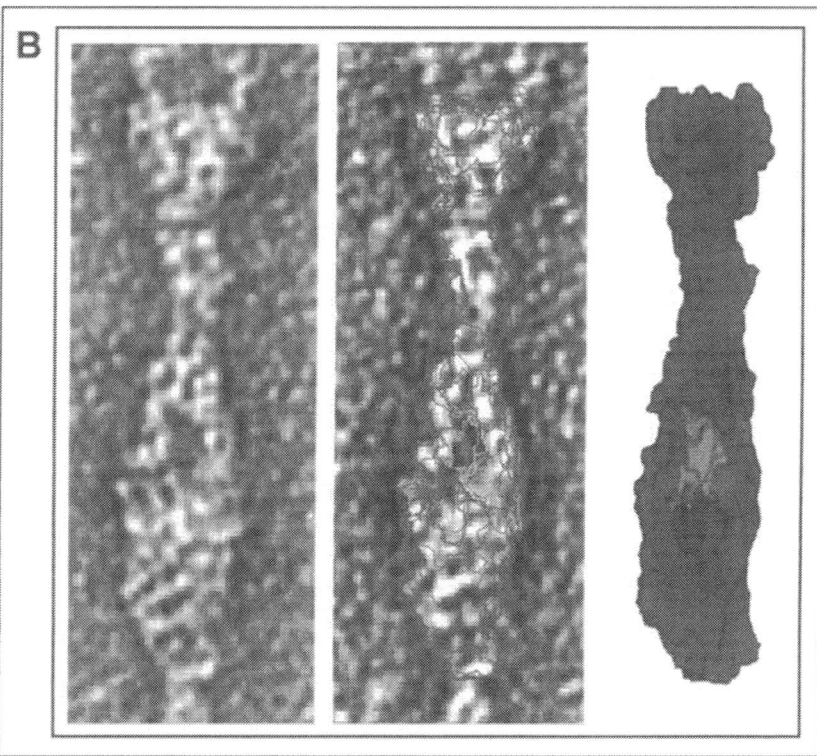

Figure 4B. Isolated microfibrils imaged by four different microscopy approaches, all of which demonstrate the ~56 nm untensioned beaded periodicity. Panel B) three-dimensional reconstruction of a microfibril by automated electron tomography, shown with volume rendering. Arrows indicate the beads; bead-to-bead distance is 57 nm. A color version of this figure can be seen online at http://www.eurekah.com/ eurekahlogin.php?chapid=1750&bookid=119&catid=28.

bonds. The modulus of elasticity of lobster (*Homarus americanus*) aorta (1.0 MPa), a viscoelastic tissue based on microfibril bundles and smooth muscle cells, was found to be similar to that of the vertebrate rubber-like protein elastin (1.1 MPa).[6] Lobster aorta elastic properties are nonlinear; low stiffness allows arterial volume to increase, but then stiffer vessels correspond with arterial volume plateauing. Further assessment of this system revealed that microfibrils alone characterise the stress-strain behaviour of the vessel, when initial reorientation and subsequent deformation are accounted for.[58]

Mammalian Studies

Mammalian microfibril studies have focussed on the zonular filaments of the eye. X-ray fibre diffraction has provided valuable insights into the organisation and elasticity of hydrated whole ciliary zonules, whilst biomechanical testing has shown that microfibril bundles teased out from zonules are reversibly extensible.[37,59]

Tensile Tests

Tensile tests and stress-relaxation tests were used on whole ciliary zonular filaments to examine the relationship between their mechanical behaviour and the molecular organisation of the fibrillin-rich microfibrils upon which the tissue is based. By this approach, zonular fila-

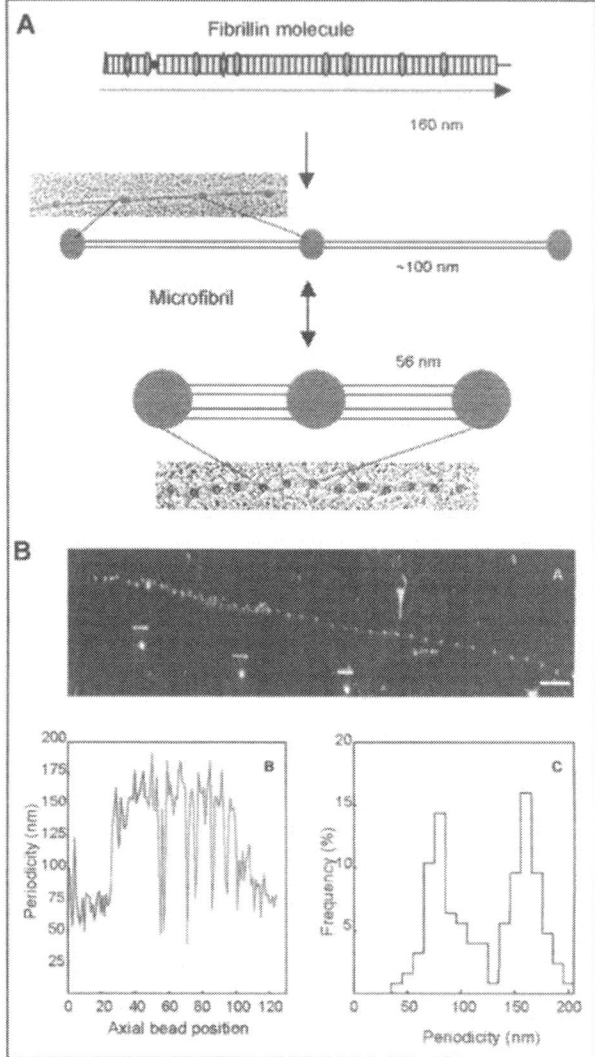

Figure 5. Stretched microfibrils. A) Schematic diagram indicating how microfibrils may assemble initially in a head-to-tail arrangement to give microfibrils with a beaded periodicity of ~160 nm. Subsequently, molecular packing may lead to a reduction of periodicity to 56 nm. The elastic range has been predicted to be between 56-90 nm.[54] B) STEM analysis of a stretched microfibril isolated from the zonular filaments of a dog with ectopia lentis. Numerous examples of stretched microfibrils were apparent in this tissue. Periodicities were either in the range 56-100 nm, or 140-180 nm. A color version of Figure 5B, distinguishing unstretched and extended periodic repeat regions within this microfibril, can be seen online at http://www.eurekah.com/eurekahlogin.php?chapid=1750&bookid=119&catid=28.

ments also revealed a nonlinear (J-shaped) stress-strain curve and appreciable stress-relaxation, and it was proposed that this behaviour may reflect an initial strain-induced lateral alignment of untensioned microfibrils prior to elastic extension and recoil.[59] A similar nonlinear relationship between force and strain was observed for microfibril bundles of less than 1μm in diameter.[62]

X-Ray Diffraction

Macroscopic tissue extensions of up to approximately 50% produce only minor changes in the axial unit cell periodicity of fibrillin-rich microfibrils, whereas the application of much larger strains to the zonules induces major periodicity changes. Mechanical testing of zonular filaments from bovine eye had shown the stress-strain curve to be J-shaped (see above), with small strains corresponding to a nonlinear region of the curve and larger strains inducing linearity.[59] An interpretation of the underlying molecular behaviour in this initial nonlinear region is required to fully understand the viscoelastic properties of fibrillin-rich tissues.

Recent studies performed at the European Synchrotron Radiation Facility (ESRF, Grenoble, France) included X-ray diffraction analysis of zonular filaments, particularly in the toe region of the stress-strain curve (Fig. 6). Investigation over this region of the stress-strain curve allows detailed measurements to be made of any changes which occur in the supramolecular architecture of fibrillin-rich microfibrils prior to any change in periodicity. The meridional X-ray diffraction data peaks (orders) arise from the axial periodic electron density of the microfibril. Their position, intensity, breadth and angular misorientation give information about the periodicity, electron density coherence (and distortion) and angular distribution of the microfibrillar repeating structure respectively. Here, we describe an analysis of changes in the meridional peak angular distribution and breadth that reflect early changes in the stress strain regime.

Figure 6A-C shows the diffraction images obtained following the application of increasing strains to zonular filaments obtained from *Cervus elaphus hip.* (red deer) eye. These images were obtained using a 5m camera and a 2D detector on beamline ID02 at ESRF. Peak breadth analysis was carried out using the CCP13 program Xfit.[38] Figure 6A was obtained at rest, Figure 6B following extension by 33% and Figure 6C after 66% extension from the rest length. Figure 6F shows a linear integration of the images in Figures 6A-C). The application of relatively small strains has been shown previously to generate only minor differences in overall fibril periodicity.[29,38,60] This is also true in this system, when the observed 3rd order shifts from approximately 55nm (0% tissue extension) to 56nm (33% tissue extension) and finally 58nm (67% extension). This corresponds to an overall shift of 5% in periodicity for a 67% change in macroscopic tissue length.

The more noticeable changes which occur following the application of relatively low strains are alterations in fibrillar alignment and lateral coherence. As can be seen from Figures 6A-C, the intensity of each meridional peak is distributed over an arc, corresponding to a misalignment of microfibrillar bundles. The correlation between macroscopic tissue extension with no periodicity change at low levels of strain, with periodicity changes at higher levels of strain implies that elasticity in the lower regions of the stress-strain curve arises from microfibrillar bundle rearrangements and realignments as opposed to alteration in microfibrillar axial periodicity. However, measurement of the degree of realignment is complicated by a broadening of the lateral dimension of the meridional peaks.

Analysis of peak breadth in X-ray diffraction images has been made to give an indication of diffracting 'particle size'. In this case this parameter would correspond to the number of unit cells acting as coherent diffracting units along a fibrillin rich microfibril. Highly ordered fibres such as collagen fibrillar structures from rat tendon display a negligible peak broadening and can be used here as a guide to the broadening that occurs from the profile of the X-ray beam itself. An X-ray diffraction image obtained from rat tail tendon is shown for comparison purposes (Fig. 6D).

Particle size (D) can be determined using the formula:

$$D = \frac{0.9 * \lambda}{\beta \cos \theta}$$

λ is the wavelength, θ is obtained from the diffraction angle (2θ) and is given in radians, β corresponds to the peak breadth (FWHM, full-width half-maximum) in radians (following correction for beam broadening).[61]

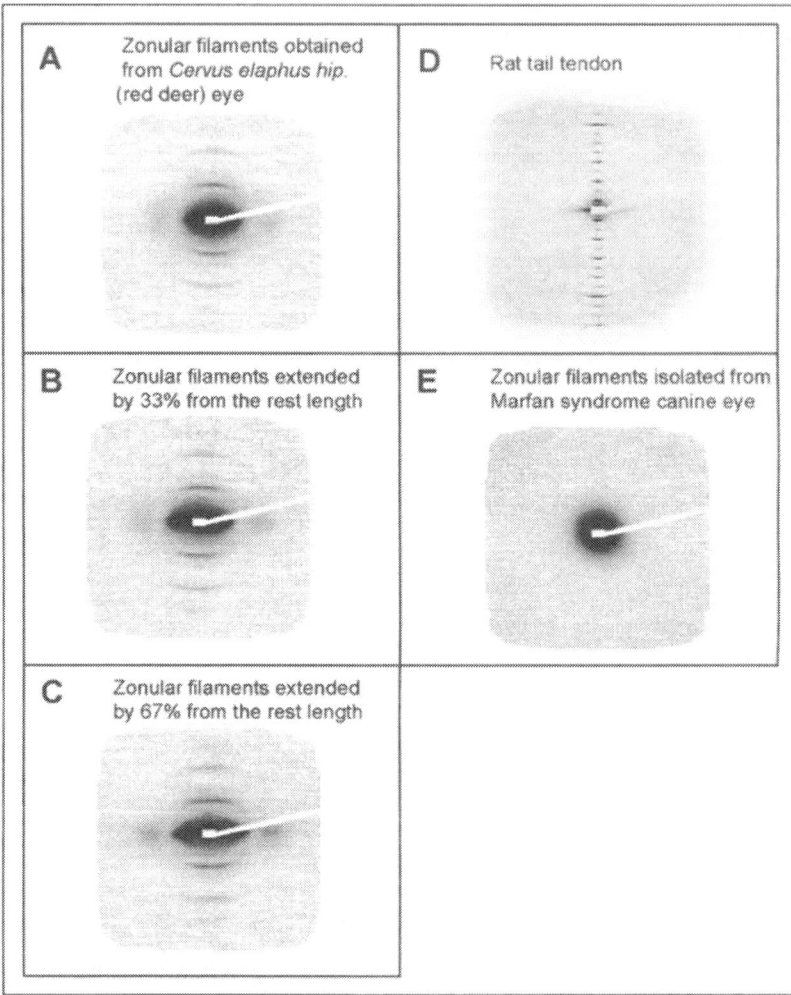

Figure 6. X-ray fibre diffraction analysis of the 'toe' of the J-shaped curve of elastic tissue microfibrils. X-ray diffraction analyses of zonular filaments performed at the European Synchrotron Radiation Facility (ESRF, Grenoble, France) focussed on resolving structural features relating to the toe region of the stress-strain curve and changes in the supramolecular architecture of microfibrils prior to any change in their periodicity. The meridional X-ray diffraction data peaks arise from the axial periodic electron density of the microfibril. Their position, intensity, breadth and angular misorientation give information about the periodicity, electron density coherence (and distortion) and angular distribution of the microfibrillar repeating structure respectively. A-C) X-ray diffraction patterns of zonular filaments (red deer) at rest length (A), and extended 33% (B) and 67% (C) from rest length. D) X-ray diffraction pattern of rat tail collagen. E) X-ray diffraction pattern of human zonules from a Marfan syndrome patient. F) Linear trace of Figs. 6A-C in radial direction.

In the examples illustrated here, the axial peak breadth of the 3rd order of fibrillin demonstrated broadening following tissue extension. Use of the above equation estimates the 'effective crystalline' length of the fibrillin microfibril to range from approximately 346 nm at rest, 338 nm after 33% extension to 255 nm after 67% extension. This corresponds to a diffracting unit of about 6 times the size of the established 56 nm repeating unit at rest and a unit of approximately 4.5 times following extension to 67%. A decrease in the size of the 'effective crystalline' length of fibrillin-rich microfibrils following tissue extension in the toe region of the stress-strain curve implies that a corresponding decrease arises in the axial coherence of the microfibrils. It should be noted, however, that evidence exists for an increased regularity of structure with a corresponding peak narrowing at higher extension levels (data not shown).

This relatively small persistence length for a fibre suggests that other factors may be contributing to the breadth of the peaks seen with fibrillin-rich microfibrils. Other reasons for peak broadening include distortions within the lattice and differences in the unit cell periodicity. An indication of differences in unit cell periodicity is the increased broadening of peaks at higher orders of diffraction. In contrast, lattice distortions would produce equivalent broadening in all diffraction peaks. The meridional series of fibrillin-rich microfibrils is known to truncate after the 8th order, making detailed analysis of any increasing peak broadening difficult. From analysis of this data, however, the 6th order does not appear to be any broader than the 3rd, suggesting that variability in unit cell periodicity is not as dominant as distortions within the unit cells themselves.

Zonular filaments obtained from a Marfan syndrome eye were analysed in an identical manner (see Fig. 6E). This tissue produced X-ray scattering, which is characteristic of aggregated protein structures. In direct contrast to zonular filaments from healthy mammals however, no discernible X-ray diffraction peaks were observed, implying the lack of a more ordered hierarchical structure. It is known that ciliary zonules in this disease become much more fragile than in healthy zonules. It appears, therefore, that defects in fibrillin deposition produce changes in microfibrillar structure, with a loss of structural coherence.

Stretching Isolated Microfibrils

Stretching isolated microfibrils in solution under controlled conditions has proved a challenge, and reproducible approaches have yet to be established. Methods such as centrifugation and aspiring through fine-gauge needles have not reproducibly extended microfibrils.[13] When microfibrils were adsorbed onto carbon-coated electron microscopy grids that had not been glow-discharged, surface forces were shown to extend beaded periodicity up to ~90 nm and to exert a profound effect on their organisation.[54] Incubation of isolated microfibrils with a monoclonal anti-fibrillin-1 antibody generated extensive microfibril arrays with a double striated pattern corresponding to the beads and to the monoclonal antibody interbead epitope which, in untensioned microfibrils, is ~41.1% of the bead-to-bead distance.[54] However, after stretching these antibody-generated arrays with surface forces, the striation corresponding to the bound antibody was only detected in microfibrils that were stretched up to ~70 nm periodicity. These data imply that the antibody epitope does not move significantly relative to the bead until periodicity exceeds 70 nm. At higher periodicities, the alignment of the epitope is lost, presumably due to a major structural rearrangement. Thus, microfibrils can stretch from 56-70 nm before this epitope has to move.

Recently, we used molecular combing techniques to determine Young's modulus for individual native microfibrils (Fig. 7).[63] The surface tension force generated by a receding meniscus aligns and stretches partially adsorbed molecules.[64] Young's modulus (E) can be determined using the formula:

$$E = \frac{\gamma \pi D}{A\left(\dfrac{l}{l_0} - 1\right)}$$

Where γ is the surface tension between a liquid / air interface, D is the hydrated diameter, A is the cross-sectional area derived from D, l_0 and l are the unstretched and stretched periodicities respectively measured from intermittent contact mode AFM micrographs of combed and noncombed microfibril suspensions. This study revealed that microfibril periodicity is not altered at physiological zonular tissue extensions, and Young's modulus is between 78 to 96 MPa, which is two orders of magnitude stiffer than the value for elastin and invertebrate microfibril-

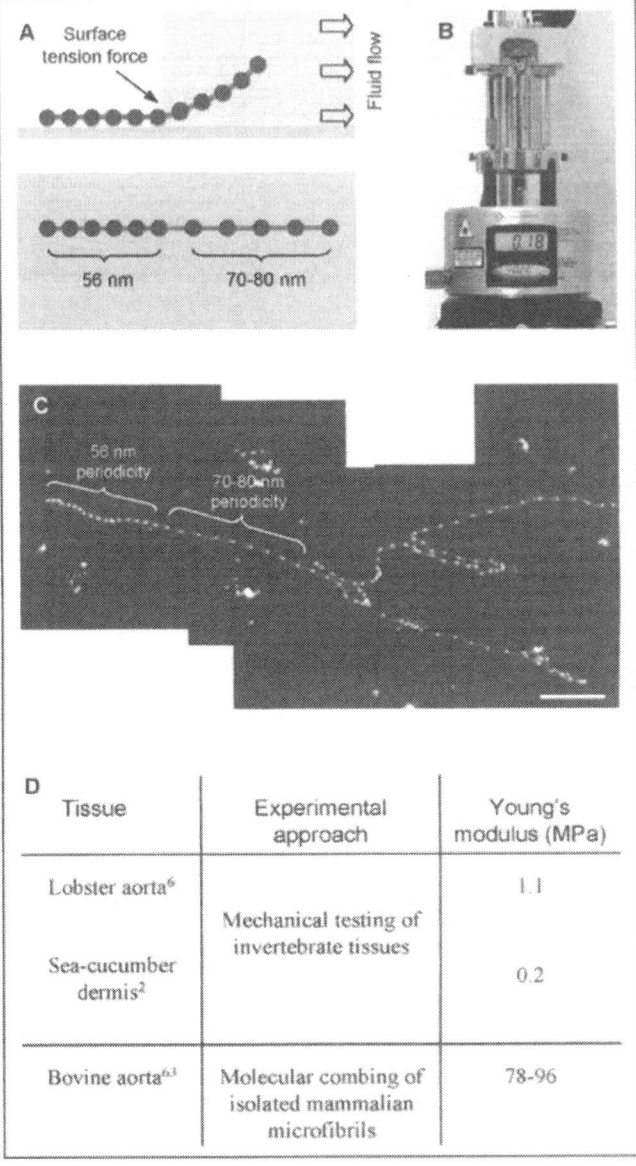

Figure 7. Young's modulus of fibrillin microfibrils determined by molecular combing. A) Partially adsorbed molecules or supra-molecular assemblies on a substrate are subject to a surface tension force, which acts locally at the position of the meniscus. The nonadsorbed region of the molecule or assembly may be extended by the surface tension force, adsorbed regions retain the original length/periodicity. B) Atomic force microscope. Extended microfibrils on a mica substrate are imaged by intermittent contact mode AFM. C) Intermittent contact mode height image composite of a combed fibrillin-rich microfibril. Adsorbed regions retain the untensioned 56 nm periodicity. Nonadsorbed regions are extended to periodicities within the 70-80 nm range. Scale bar = 400 nm. D) Young's modulus has been determined by mechanical testing of invertebrate tissues containing fibrillin-rich microfibrils and by molecular combing of isolated mammalian microfibrils.

lar tissues. Therefore fibrillin-rich microfibrils behave as relatively stiff fibres which can perform anchoring roles in the ciliary zonules and oxytalan fibres at the dermal epidermal junction. Futhermore elastic fibres are likely to act as microfibril reinforced fibrous composites. These data imply that elasticity in microfibril-containing tissues arises primarily from reversible alterations in supra-microfibrillar arrangements rather than from intrinsic elastic properties of individual microfibrils.

Effects of Calcium on Periodicity and Microfibril Elasticity

As predicted from the primary structure of fibrillin-1 with its preponderance of cbEGF domains, bound calcium exerts a profound effect on the elastic properties of tissues which comprise fibrillin-rich microfibrils. X-ray diffraction zonular stretching experiments have shown that bound calcium influences microfibril load deformation but is not necessary for high extensibility and elasticity.[37,60] The influence of calcium has also been examined in mechanical tests of microfibril bundles using the micro-needle technique.[62] Calcium depletion resulted in a 50% decrease in rest length and a reduction in microfibril stiffness; this effect was reversible upon addition of calcium. However, at high strain, irreversible damage occurred irrespective of the presence or absence of calcium. Thus, microfibril elasticity is modified by, but not dependent on calcium-induced beaded periodic changes. These studies also confirmed that reversible extension occurs only within a specific periodic range and that irreversible transitions in the quaternary structure occur at high stretch.

Stretched Microfibrils from Pathological Tissues

STEM analysis of microfibrils isolated from a canine ectopia lentis model that contained numerous examples of both untensioned and highly stretched regions that were stable in physiological buffers and did not exhibit elastic recoil. Similar highly stretched microfibrils have also been seen following treatment of microfibrils from control tissues with matrix metalloproteinases (MMPs 9 and 12).[13] Thus, highly stretched microfibrils from disease tissues may reflect proteolytic damage, or possibly alterations in the molecular mechanism of microfibril elasticity.

Conclusion

Studies of invertebrate and mammalian microfibril-rich tissues have identified an essential role for tissue microfibrils in the provision of long-range elasticity. Recent data suggest that tissue elasticity is not a consequence of inherent elasticity within individual microfibrils but due to inter or intra-microfibril bundle rearrangements. It remains to be determined whether microfibril elasticity varies in different tissues due to influences of tissue-specific microfibril-associated molecules. In some Marfan individuals, microfibrils have increased susceptibility to proteolysis which alters their tissue organisation and damages their elastic properties. Studies are now addressing the molecular basis of the elastic mechanism of microfibrils in order to clarify genotype-phenotype relationships in this disease.

Acknowledgements

We thank Dr Carolyn Jones for her expert assistance with the tissue transmission electron microscopy studies. CMK and MJS are supported by the MRC (UK). The support of ESRF in providing funding and access is gratefully acknowledged. JLH acknowledges the support of BBSRC (UK) (grant reference number 98/S15326).

References

1. Reber-Muller S, Spissinger T, Schubert P et al. An extracellular matrix protein of jellyfish homologous to mammalian fibrillins forms different fibrils depending on the life stage of the animal. Dev Biol 1995; 169:662-672.
2. Thurmond FA, Trotter JA. Morphology and biomechanics of the microfibrillar network of sea cucumber dermis. J Exptl Biol 1996; 199:1817-1828.
3. Pereira L, D'Alessio M, Ramirez F et al. Genomic organization of the sequence coding for fibrillin, the defective gene product in Marfan syndrome. Hum Mol Genet 1993; 2:961-968.

4. Zhang H, Apfelroth SD, Hu W et al. Structure and expression of fibrillin-2, a novel microfibrillar component preferentially located in elastic matrices. J Cell Biol 1994; 124:855-863.

5. Nagase T, Nakayama M, Nakajima D et al. Prediction of the coding sequences of unidentified human genes. XX. The complete sequences of 100 new cDNA clones from brain which code for large proteins in vitro. DNA Res 2001; 8:85-95.

6. McConnell CJ, Wright GM, DeMont ME. The modulus of elasticity of lobster aorta microfibrils. Experientia 1996; 52:918-921.

7. Mecham RP, Davis EC. Elastic fiber structure and assembly. In: Yurchenco PD, Birk DE, Mecham RP, eds. Extracellular Matrix Assembly and Structure. New York USA: Academic Press, 1994; 281-314.

8. Faury G. Function-structure relationship of elastic arteries in evolution: From microfibrils to elastin and elastic fibres. Pathol Biol (Paris) 2001; 49:310-25.

9. Kielty CM, Woolley DE, Whittaker SP et al. Catabolism of intact fibrillin microfibrils by neutrophil elastase, chymotrypsin and trypsin. FEBS Lett 1994; 351:85-89.

10. Ashworth JL, Sherratt MJ, Rock MJ et al. Fibrillin turnover by metalloproteinases: Implications for connective tissue remodelling. Biochem J 1999; 340:171-181.

11. Pereira L, Lee SY, Gayraud B et al. Pathogenetic sequence for aneurysm revealed in mice underexpressing fibrillin-1. Proc Natl Acad Sci USA 1999; 96:3819-23.

12. Hindson VJ, Ashworth JL, Rock MJ et al. Fibrillin catabolism by matrix metalloproteinases: Characterisation of amino- and carboxy-terminal cleavage sites. FEBS Lett 1999; 452:195-198.

13. Sherratt MJ, Wess TJ, Baldock C et al. Fibrillin-rich microfibrils of the extracellular matrix: Ultrastructure and assembly. Micron 2001; 32:185-200.

14. Hanssen E, Franc S, Garrone R. Synthesis and structural organization of zonular fibers during development and aging. Matrix Biol 2001; 20:77-85.

15. Sachdev NH, Di Girolamo N, McCluskey PJ et al. Lens dislocation in Marfan syndrome: Potential role of matrix metalloproteinases in fibrillin degradation. Arch Ophthalmol 2002; 120:833-5.

16. McGettrick AJ, Knott V, Willis A et al. Molecular effects of calcium binding mutations in Marfan syndrome depend on domain context. Hum Mol Genet 2000; 9:1987-94.

17. Reinhardt DP, Ono RN, Sakai LY. Calcium stabilizes fibrillin-1 against proteolytic degradation. J Biol Chem 1997; 272:1231-6.

18. Booms P, Tiecke F, Rosenberg T et al. Differential effect of FBN1 mutations on in vitro proteolysis of recombinant fibrillin-1 fragments. Hum Genet 2000; 107:216-24.

19. Kielty CM, Baldock C, Lee D et al. Fibrillin: From microfibril assembly to biomechanical function. Philosoph Trans Royal Soc B 2000; 357:207-217.

20. Gibson MA, Kumaratilake JS, Cleary EG. The protein components of the 12-nanometer microfibrils of elastic and nonelastic tissues. J Biol Chem 1989; 264:4590-4598.

21. Sakai LY, Keene DR, Engvall E. Fibrillin, a new 350-kD glycoprotein, is a component of extracellular microfibrils. J Cell Biol 1986; 103:2499-509.

22. Kielty CM, Sherratt MJ, Shuttleworth CA. Elastic fibres. J Cell Sci 2002; 115;2817-2828.

23. Downing AK, Knott V, Werner JM et al. Solution structure of a pair of calcium-binding epidermal growth factor-like domains: Implications for the Marfan syndrome and other genetic disorders. Cell 1996; 85:597-605.

24. Smallridge RS, Whiteman P, Werner JM et al. Solution structure and dynamics of a calcium binding epidermal growth factor-like domain pair from the "neonatal" region of human fibrillin-1. J Biol Chem 2003; 278:12199-12206.

25. Qian R, Glanville RW. Alignment of fibrillin molecules in elastic microfibrils is defined by transglutaminase-derived cross-links. Biochemistry 1997; 36:15841-15847.

26. Zhang H, Hu W, Ramirez F. Developmental expression of fibrillin genes suggests heterogeneity of extracellular microfibrils. J Cell Biol 1995; 129:1165-1176.

27. Charbonneau NL, Dzamba BJ, Ono RN et al. Fibrillins can coassemble in Fibrils, but fibrillin fibril composition displays cell-specific differences. J Biol Chem 2003; 278:2740-9.

28. Quondamatteo F, Reinhardt DP, Charbonneau NL et al. Fibrillin-1 and fibrillin-2 in human embryonic and early fetal development. Matrix Biol 2002; 21:637-46.

29. Kielty CM, Sherratt MJ, Haston JL et al. Fibrillin-rich microfibrils: Elastic biopolymers of the extracellular matrix. J Muscle Res Cell Motil 2003; 23:581-596.

30. Kielty CM, Shuttleworth CA. Microfibrillar elements of the dermal matrix. Microscopy Res Tech 1997; 38:407-427.

31. Cleary EG, Gibson MA. Elastin-associated microfibrils and microfibrillar proteins. Int Rev Connect Tissue Res 1983; 10:97-209.

32. Keene DR, Marinkovich MP, Sakai LY. Immunodissection of the connective tissue matrix in human skin. Microscopy Res Tech 1997; 38:394-406.

33. Davis EC, Roth RA, Heuser JE et al. Ultrastructural properties of ciliary zonule microfibrils. J Struct Biol 2002; 139:65-75.
34. Ashworth JL, Kielty CM, McLeod D. Fibrillin and the eye. Br J Ophthalmol 2000; 84:1312-7.
35. Davis EC. Endothelial cell connecting filaments anchor endothelial cells to the subjacent elastic lamina in the developing aortic intima of the mouse. Cell Tissue Res 1993; 272:211-9.
36. Dingemans KP, Teeling P, Lagendijk JH et al. Extracellular matrix of the human aortic media: an ultrastructural histochemical and immunohistochemical study of the adult aortic media. Anat Rec 2000; 258:1-14.
37. Wess TJ, Purslow PP, Sherratt MJ et al. Calcium determines the supramolecular organization of fibrillin- rich microfibrils. J Cell Biol 1998; 141:829-837.
38. Haston JL, Engelsen SB, Roessle M et al. Raman microscopy and small angle x-ray scattering: A combined study into the elasticity of fibrillin-rich microfibrils. J Biol Chem 2003; 278:581-596.
39. Jones CJ, Sear CH, Grant ME. An ultrastructural study of fibroblasts derived from bovine ligamentum nuchae and their capacity for elastogenesis in culture. J Pathol 1980; 131:35-53.
40. Mecham RP, Heuser J. Three-dimensional organization of extracellular matrix in elastic cartilage as viewed by quick freeze, deep etch electron microscopy. Connect Tissue Res 1990; 24:83-93.
41. Trask TM, Trask BC, Ritty TM et al. Interaction of tropoelastin with the amino-terminal domains of fibrillin-1 and fibrillin-2 suggests a role for the fibrillins in elastic fiber formation. J Biol Chem 2000; 275:24400-24406.
42. Trask BC, Trask TM, Broekelmann T. The microfibrillar proteins MAGP-1 and fibrillin-1 form a ternary complex with the chondroitin sulfate proteoglycan decorin. Molec Biol Cell 2000; 11:1499-1507.
43. Reinboth B, Hanssen E, Cleary EG et al. Molecular interactions of biglycan and decorin with elastic fiber components. Biglycan forms a ternary complex with tropoelastin and MAGP-1. J Biol Chem 2001; 277:3950-3957.
44. Wallace RN, Streeten BW, Hanna RB. Rotary shadowing of elastic system microfibrils in the ocular zonule, vitreous, and ligamentum nuchae. Curr Eye Res 1991; 10:99-109.
45. Kielty CM, Cummings C, Whittaker SP et al. Isolation and ultrastructural analysis of intact high-M_r assemblies of type VI collagen and fibrillin. J Cell Sci 1991; 99:797-807.
46. Isogai Z, Aspberg A, Keene DR et al. Versican interacts with fibrillin-1 and links extracellular microfibrils to other connective tissue networks. J Biol Chem 2002; 277:4565-4572.
47. Kielty CM, Hanssen E, Shuttleworth CA. Connective tissue microfibrils: Separation and purification of fibrillin and collagen VI by density gradient centrifugation. Anal Biochem 1998; 255:108-112.
48. Isogai Z, Ono RN, Ushiro S et al. Latent transforming growth factor beta-binding protein 1 interacts with fibrillin and is a microfibril-associated protein. J Biol Chem 2003; 278:2750-2757.
49. Hanssen E, Franc S, Garrone R. Atomic force microscopy and modelling of natural elastic fibrillin polymers. Biol Cell 1998; 90:223-228.
50. Reinhardt DP, Mechling DE, Boswell BA et al. Calcium determines the shape of fibrillin. J Biol Chem 1997; 272:7368-7373.
51. Kielty CM, Shuttleworth CA. The role of calcium in the organisation of fibrillin microfibrils. FEBS Lett 1993; 336:323-326.
52. Cardy CM, Handford PA. Metal ion dependency of microfibrils supports a rod-like conformation for fibrillin-1 calcium-binding epidermal growth factor- like domains. J Mol Biol 1998; 276:855-860.
53. Reinhardt DP, Keene DR, Corson GM et al. Fibrillin-1: Organization in microfibrils and structural properties. J Mol Biol 1996; 258:104-114.
54. Baldock C, Koster AJ, Ziese U et al. The supramolecular organisation of fibrillin-rich microfibrils. J Cell Biol 2001; 152:1045-1056.
55. Baldock C, Gilpin C, Koster A et al. Three-dimensional reconstructions of extracellular matrix polymers using automated electron tomography. J Struct Biol 2002; 138:130-135.
56. Alper J. Stretching the limits. Science 2002; 297:329-331.
57. Keene DR, Maddox BK, Kuo H-J et al. (1991) Extraction of extensible beaded structures and their identification as extracellular matrix fibrillin-containing microfibrils. J Histochem Cytochem 1991; 39:441-449.
58. McConnell CJ, DeMont ME, Wright GM. Microfibrils provide nonlinear behaviour in the abdominal artery of the lobster Homarus americanus. J Physiol 1997; 499:513-526.
59. Wright DM, Duance VC, Wess TJ et al. The supramolecular organisation of fibrillin-rich microfibrils determines the mechanical properties of bovine zonular filaments. J Exp Biol 1999; 202:3011-3020.
60. Wess TJ, Purslow PP, Kielty CM. X-ray diffraction studies of fibrillin-rich microfibrils: Effects of tissue extension on axial and lateral packing. J Struct Biol 1998; 122:123-127.
61. Lonsdale K. Crystals and X-rays G. London: Bell and Sons Ltd., 1948.

62. Eriksen TA, Wright DM, Purslow PP et al. Role of Ca^{2+} for the mechanical properties of fibrillin. Proteins: Str Funct Genet 2001; 45:90-95.
63. Sherratt MJ, Baldock C, Haston JL et al. Fibrillin microfibrils are stiff reinforcing fibres in compliant tissues. J Mol Biol 2003; 332:183-193.
64. Bensimon A, Simon A, Chiffaudel A et al. Alignment and sensitive detection of DNA by a moving interface. Science 1994; 265(5181):2096-8.

CHAPTER 13

Microfibril-Associated Glycoprotein-1 (MAGP-1) and Other Non-Fibrillin Macromolecules Which May Possess a Functional Association with the 10 nm Microfibrils

Mark A. Gibson

Introduction

There is growing evidence that fibrillin-containing microfibrils are not just fibrillin polymers but that a variety of additional macromolecules may be associated with these structures. The functions of these molecules may be envisioned to include (a) structural support to stabilize the interaction of fibrillin molecules within the microfibril; (b) mediation of the interaction of adjacent microfibrils within bundles; (c) assembly of elastin on the surface of the microfibrils; (d) interfacing between the microfibrils and other structural elements of different matrices; (e) modulation of the interaction of the microfibrils with cells to influence the deposition, orientation and organization of microfibrils and elastic fibers in different tissue environments; (f) provision and modulation of nonstructural functions of the microfibrils e.g., TGF-beta storage; (g) enzymatic activity e.g., lysyl oxidase; and (h) specific interactions with fibrillin-2-containing microfibrils. In recent years many candidate microfibril-associated macromolecules have been identified. Some of these molecules appear to be found in extracellular matrices associated exclusively with the fibrillin-containing microfibrils indicating that their function(s) are likely to be specific to some aspect of microfibril biology. Other candidates clearly have association with additional structural matrix components and thus they must possess some microfibril-independent functions. The best-characterized matrix macromolecule which is exclusively associated with fibrillin-containing microfibrils is Microfibril-Associated Glycoprotein-1, or MAGP-1, a small glycoprotein with an apparent molecular weight of 31 kDa.[1,2] MAGP-1 was first identified as a component of the elastic-fiber rich tissue, developing bovine nuchal ligament. MAGP-1 was shown to be resistant to chaotropic extraction from tissue homogenates, unless a reducing agent was included in the extraction buffer. Such reductive treatment had been shown to selectively solubilize the microfibrillar component of developing elastic fibers, later to become known as fibrillin-containing microfibrils.[1] MAGP-1 is the strongest candidate for a ubiquitous component of fibrillin-1-containing microfibrils since it extensively and specifically codistributes with these structures in tissues,[3-6] and is localized periodically along these microfibrils in association with the bead regions.[7] A structurally related molecule, MAGP-2 (apparent molecular weight 25 kDa) was later identified in fibrillin-1

Marfan Syndrome: A Primer for Clinicians and Scientists, edited by Peter N. Robinson and Maurice Godfrey. ©2004 Eurekah.com and Kluwer Academic / Plenum Publishers.

enriched extracts of the same tissue along with MAGP-1 and two polypeptides of 78 and 70 kDa.[8,9] MAGP-2 was also found to be exclusively associated with the microfibrils but it exhibited a more restricted distribution than MAGP-1.[5] The 78 kDa and 70 kDa polypeptides, later referred to as MP78 and MP70, were identified as forms of βig-h3,[9] a protein which appears to be associated with type VI collagen microfibrils rather than the fibrillin-containing structures,[10] and thus these two polypeptides are not discussed further in this chapter.

The ultrastructural distribution, tissue expression, structure and function of MAGPs are discussed in more detail below. A number of other microfibril-associated components have been identified including MFAPs,[11-14] several proteoglycans,[15-19] fibulins,[20,21] LTBPs[22-28] and EMILINs.[29,30] Our knowledge of these macromolecules in relation to the fibrillin-containing microfibrils and to genetic disorders is also outlined below.

MAGPs

Ultrastructural Localization and Developmental Tissue Expression Patterns of MAGPs and Their Relationships to Fibrillin-Containing Microfibrils

MAGP-1 has been shown, using immunofluorescence and immunoelectron microscopic techniques, with polyclonal and later monoclonal antibodies, to be specifically localized on fibrillin-containing microfibrils in a wide range of developing and mature elastic and nonelastic tissues. In all tissues examined the localization of MAGP-1 matched that of fibrillin-1.[3-5] Interestingly, only one elastic tissue was identified which did not stain extensively with anti-MAGP-1 antibodies, the elastic ear cartilage, which is considered to contain predominantly fibrillin-2 containing microfibrils.[5,31] The evidence suggests that MAGP-1 is specifically associated with fibrillin-1-containing microfibrils in most if not all situations where they occur.

Using pre-embedding labeling techniques, monoclonal anti-MAGP-1 antibodies were found to localize to fibrillin-containing microfibrils in ocular zonule and vitreous tissues in a cross-banding pattern at intervals of about 50 nm, the periodicity of the beaded filament structure of the microfibrils. Rotary shadowing of isolated microfibrils which had been immunogold-labeled with the anti-MAGP-1 antibody showed specific, apparently symmetrical, localization to the beads rather than the interbead regions (Fig. 1). Occasionally two gold particles were observed to be attached to opposite sides of the same bead, suggesting that multiple MAGP-1 molecules were present in the structure. The study concluded that MAGP-1 is intimately and regularly associated with the bead regions of fibrillin-containing microfibrils and that the findings are consistent with a major role for MAGP-1 in microfibril biology.[7]

MAGP-2 also shows specific ultrastructural localization to fibrillin-containing microfibrils in a variety of developing tissues.[5] However, MAGP-2 has a more limited tissue pattern of expression than MAGP-1 and fibrillin-1 in late-fetal and adult bovine tissues. For instance, in late-trimester nuchal ligament, skeletal muscle and spleen the distribution of MAGP-2 appears to be indistinguishable from that of MAGP-1, but MAGP-2 is not associated with microfibrils of the medial layer of thoracic aorta or peritubular matrix of the kidney. MAGP-2 also appears not to be associated with the microfibrils of the ocular zonule. These observations are consistent with analysis of steady-state mRNA levels in developing bovine tissues. MAGP-2 expression appears to be higher in nuchal ligament, heart and skeletal muscle, but lower in aorta and kidney than that of MAGP-1.[5] In nuchal ligament, MAGP-2 expression appears to peak at around 180 days of fetal development, which correlates with the period of onset of elastinogenesis and maximum expression of fibrillins in this tissue.[5,32] In contrast, MAGP-1 expression remains consistently high throughout nuchal ligament development.

MAGP-2 was also shown to have a periodic association with microfibrils in pre-embedding labeling experiments similar to those described above for MAGP-1[7] but using nuchal ligament in place of zonule which lacks MAGP-2. MAGP-2 was found to copurify on CsCl density gradients with the fibrillin-containing microfibrils even in the presence of the strong chaotrope,

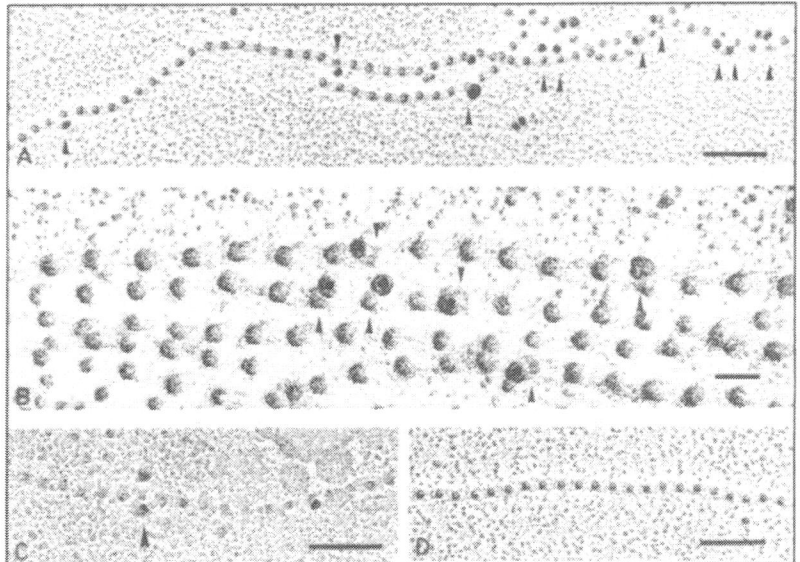

Figure 1. Rotary shadowing and immunolabeling of isolated zonular microfibrils for MAGP-1. A-C) Isolated microfibrils were reacted with anti-MAGP-1 antibodies, followed by 10-nm immunogold particles (visualized as circles with black centers). Small arrows denote attachment of individual immunogold particles to the bead-like structures of the "beads on a string" morphology of the microfibrils. The large arrow in C shows the binding of two immunogold particles to the same bead. D) Control; microfibrils incubated with antibodies to MAGP-2 (which is not detected in zonule microfibrils) show no binding of immunogold particles. Original magnifications A,D x 72000; B x 183,000; C x 81,000. Bars: A,C,D = 0.2 μm; B = 0.5 μm. (Reproduced with permission from Henderson M, Polewski R, Fanning JC et al. Microfibril-associated glycoprotein-1 (MAGP-1) is specifically located on the beads of the beaded-filament structure for fibrillin-containing microfibrils as visualized by the rotary shadowing technique. J Histochem Cytochem 1996; 44:1394.)

6 M guanidinium chloride indicating that the MAGP-2 is strongly associated with these structures. Dual immunolabeling of the isolated microfibrils indicated that individual microfibrils contain both MAGPs. In addition MAGP-2 appeared to be attached at two locations within each period, on the bead and at a specific point between the beads (Hanssen E, Gibson MA, unpublished observations). These findings suggest that in MAGP-2- rich tissues, the glycoprotein is an integral structural component of microfibrils. The ultrastructural and tissue expression data are consistent with a specialized role for MAGP-2 in the biology of the microfibrils, modulating their function in certain tissue-specific matrices during development and differentiation.

The Structure of MAGPs

MAGP-1 and MAGP-2 share a characteristic region of structural similarity and are the only members of this MAGP family to be identified so far. The structural features of MAGP-1 and MAGP-2 are shown in Figure 2. MAGP-1 is synthesized as a 21 kDa polypeptide of 183 amino acids from a 1.1 kb mRNA. The first 17 amino acids comprise the secretion signal peptide which is cleaved from the molecule during intracellular processing.[2,33] Sequence analysis of the MAGP-1 cDNAs has shown that MAGP-1 consists of two structurally dissimilar regions, an amino-terminal segment containing high levels of glutamine, proline and acidic amino acids and a carboxyl-terminal segment containing all 13 of the cysteine residues involved in intra- or inter- molecular disulfide bonding.[2] The amino acid sequence is highly

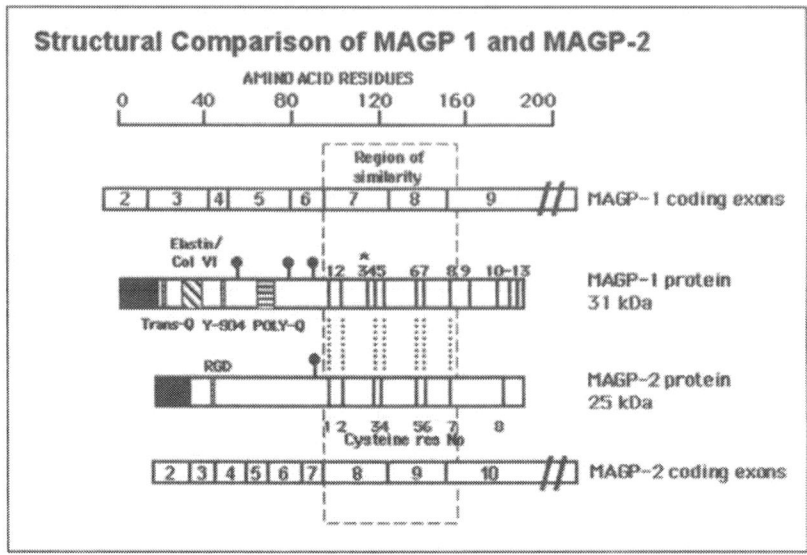

Figure 2. Structural comparison of MAGP-1 and MAGP-2. The structures of MAGP-1 and MAGP-2 are aligned schematically to highlight the region of sequence similarity between the proteins, and to align structural features with the encoding exons in MAGP-1 and MAGP-2 mRNAs respectively. The location of cysteine residues (13 in MAGP-1 and 8 in MAGP-2) are indicated. In the central region of structural similarity between the proteins, seven of the cysteines can be precisely aligned (vertical dotted lines). Note the unmatched cysteine (number 3) in MAGP-1 (star). Other structural features of MAGP-1 and MAGP-2 are also shown. In MAGP-1, sites for transglutamination (thin horizontal stripes), tyrosine sulfation (thin cross-hatches) and binding to elastin and Type VI collagen (thick cross-hatches) are highlighted, as well as a glutamine-rich sequence (thick horizontal stripes). In MAGP-2 the RGD integrin-binding sequence is shown (horizontal stripes). On both proteins the signal peptide (solid black box) and potential glycosylation sites (black circles) are also indicated.

conserved between the human, bovine and murine forms of the protein.[2,34,35] Computer analysis has suggested that MAGP-1 consists of three domains, a small folded amino-terminal domain connected by an extended proline- and glutamine-rich domain to a globular, cysteine-rich, carboxyl-terminal domain.[33] The extended domain appears to be responsible for the anomalous migration of MAGP-1 on polyacrylamide electrophoresis where it exhibits an apparent molecular weight of 31 kDa.[1,2] Post-translational modifications exhibited by MAGP-1 include O-linked glycosylation and tyrosine sulfation, and the molecule also contains a potential amine acceptor site for transglutaminase cross-linking.[1,2,36] These modifications occur within the amino-terminal region of the molecule which also contains a polyglutamine motif possibly involved in self-aggregation of MAGP-1[37] and a putative binding region for tropoelastin and type VI collagen.[38] Since reduction of disulfide bonding is necessary to solubilize MAGP-1 from microfibrils it is likely that the microfibril-binding sequence is present in the cysteine-containing carboxyl-terminal half of the molecule.[2] Recent data suggests that the region containing the first 7 cysteine residues of MAGP-1 is important for its incorporation into extracellular matrix.[39]

MAGP-2 is slightly smaller than MAGP-1, consisting of 173, 170 and 164 amino acids in the human, bovine and murine forms, respectively, including a signal peptide of 18 amino acids. Homology between each of the three species is around 75% and in each case the mature MAGP-2 polypeptide is predicted to be around 17 kDa in size.[9,40] MAGP-2 has little similarity with other known proteins with the exception of MAGP-1. Like MAGP-1, MAGP-2 is a highly hydrophilic protein containing two distinct regions, an acidic cysteine-free amino terminal half and a basic cysteine-rich carboxyl-terminal half. However, there are many differences from MAGP-1. The amino-terminal region of MAGP-2 is rich in serine and threonine residues, and contains a RGD integrin binding motif and a consensus sequence for N-glycosylation.[9] MAGP-2 lacks the proline, glutamine and tyrosine-rich sequences found in MAGP-1 including the putative elastin / collagen VI-binding region and the motifs for tyrosine sulfation and transglutamination.[36,38] The carboxyl-terminal region of MAGP-2 contains 8 cysteines in contrast to the 13 found in MAGP-1, and MAGP-2 lacks a hydrophobic region at the extreme carboxyl-terminus. Close sequence similarity to MAGP-1 is confined to a central 60 amino acid region of MAGP-2 where there is precise alignment between the first seven cysteines of MAGP-2 and seven of the first eight cysteines of MAGP-1, with C3 the unaligned cysteine in MAGP-1.[9] The spacing between the aligning cysteines ($CX_6CX_{12}CX_4CX_{14}CX_4CX_{10}C$) is conserved throughout the three species examined to date.[2,34,35]

The regions of similarity and dissimilarity in the two proteins are reflected in the exon structures of the MAGP-1 and MAGP-2 genes. The human and murine MAGP-1 and MAGP-2 genes exhibit exact size, sequence and junction alignment of the two penultimate exons (7 and 8 in MAGP-1 and 8 and 9 in MAGP-2) which encode the first six of the seven aligned cysteine residues. The seventh cysteine is encoded close to the 5' end of the final exon in both genes. Few other similarities are evident in the exon and promoter structures of the MAGP-1 and MAGP-2 genes.[40,41] The structural differences between MAGP-1 and MAGP-2 proteins and their genes are consistent with the concept that the two proteins are structurally and functionally diverse proteins with distinct patterns of tissue and developmental expression. However, MAGP-1 and MAGP-2 share one characteristic cysteine-rich motif which is likely to confer important functionalities common to both proteins. This region is a strong candidate to contain the microfibril-binding sequence in both proteins.[9,39,42]

Function of MAGPs

MAGP-1

The functions of MAGP-1 remain unclear although several roles have been proposed for MAGP-1 in the biology of fibrillin-containing microfibrils. Possible roles include an integral structural component, an elastin-binding protein on the surface of the microfibrils and a link molecule mediating the interaction of the microfibrils with structural elements of the surrounding matrix. The evidence for each possible function is summarized below.

The widespread tissue codistribution of MAGP-1 with fibrillin-1-containing microfibrils and its periodic localization to the beads along the microfibrils suggest that the protein may be an integral structural element.[4,7] This is supported by evidence that MAGP-1 is covalently attached to the microfibrils by disulfide bonding and that solubilization of the glycoprotein requires disruption of these structures with strong reducing agents.[1,8] It has been suggested that MAGP-1 may stabilize by disulfide bonding the head to tail interaction of overlapping fibrillin-1 molecules within the "bead" regions of the microfibrils.[7] MAGP-1 has been shown to interact with fibrillin in coimmunoprecipitation experiments. However, only one MAGP-1 binding site has so far been identified on fibrillin-1, close to the amino-terminus of the molecule.[43] In the absence of evidence for a carboxyl-terminal MAGP-1 binding site on fibrillin-1, it would seem unlikely that MAGP-1 fulfills the stabilization role as proposed by Henderson et al.[7] However, it remains possible that MAGP-1 may be involved in the stabilization of lateral interactions of fibrillin-1 molecules which appear to form parallel nonoverlapping dimers as an early step in microfibril assembly.[43-45]

MAGP-1 may be involved in the lateral interactions between adjacent microfibrils. The cross-banding pattern obtained when tissues are pre-embedding labeled with anti-MAGP-1 monoclonal antibodies indicates that MAGP-1 is present on the surface of each microfibril and that there is lateral alignment of the "bead" regions of individual microfibrils within each bundle of microfibrils.[7] Since MAGP-1 readily self-aggregates,[36] it is possible that MAGP-1 forms bridges between the beads of adjacent microfibrils.[7] MAGP-1 is a substrate for tissue transglutaminase and transglutamine cross-links may contribute to any structural role for MAGP-1, although evidence for such cross-links involving MAGP-1 is yet to be obtained.[36,37]

MAGP-1 has also been shown to bind to the elastin precursor tropoelastin in vitro[36,38,43] and it has been proposed that MAGP-1 may mediate the binding and alignment of tropoelastin onto a microfibril template during elastinogenesis.[2] Further evidence for a role for MAGP-1 in elastinogenesis has come from cell culture experiments.[46] In particular, an antibody raised to the amino-terminal region of MAGP-1 (amino acids 21-35), when added to the culture medium, was found to prevent elastic ear cartilage cells from organizing newly-synthesized tropoelastin into elastic fibers. The distribution of MAGP-1 in the matrix appeared to be normal when detected by an antibody raised to a different region of the molecule. These findings led the authors to suggest that the interaction between tropoelastin and fibrillin-containing microfibrils may be mediated by a domain involving the N-terminal half of the MAGP-1.[46] In another study a peptide corresponding to amino acids 29-38 of MAGP-1was found specifically to inhibit the interaction of MAGP-1 and tropoelastin in an in vitro binding assay. This finding indicated that the sequence, close to the N-terminus of MAGP-1, contains a major elastin-binding site.[38] The sequence(s) on the tropoelastin molecule important for binding MAGP-1 is less well characterized. Additional antibody inhibition studies have shown that the cysteine-containing, carboxyl terminal sequence of tropoelastin is important for incorporation into the extracellular matrix and binding to MAGP-1.[46] However, recent in vitro interaction studies, using a comprehensive range of tropoelastin fragments, failed to identify regions important for interaction with MAGP-1. This result suggests that the conformation of the intact tropoelastin molecule may be important for the binding of MAGP-1.[43] It should also be noted that tropoelastin can bind directly to fibrillin-1 suggesting that elastin deposition onto the microfibrils may occur independently from MAGP-1.[47]

MAGP-1 has been demonstrated to interact with the small dermatan sulfate proteoglycans, decorin[16] and biglycan[17] (see below), and with collagen VI via the pepsin-resistant domain of the alpha 3(VI) chain.[38] These findings support the concept that MAGP-1 plays a role in mediating the interaction of the microfibrils with constituents of the surrounding matrix. Interestingly, inhibition studies have shown that the binding site for collagen VI on MAGP-1 appears to be very close to that for tropoelastin, near the N-terminus of the MAGP-1 molecule. It has been suggested that the interaction of MAGP-1 with collagen VI may be important for the anchorage of the elastin-associated, fibrillin-containing microfibrils to the surrounding matrix during elastic fiber stretching in tissues such as ligament, lung and aorta. It is also possible that the interaction provides an indirect structural link between elastic fibers and collagen fibers via collagen VI microfibrils.[38]

MAGP-2

As with MAGP-1, the functions of MAGP-2 remain to be determined. Major differences in the composition of the amino- and carboxyl- terminal regions of the two MAGPs, and the more restricted expression patterns of MAGP-2 indicate that its functions are likely to be different from those of MAGP-1.[5,9] An early suggestion that MAGP-2 may be exclusively associated with fibrillin-2 containing microfibrils,[5] based on some correlation of expression patterns in tissues such as kidney, lung and zonule appears to be unfounded. MAGP-2 was later shown to be absent from several fibrillin-2-rich tissues including ear cartilage and the medial layer of elastic blood vessels.[5,31] More recently yeast two hybrid studies have shown that MAGP-2 interacts with both fibrillin-1 and fibrillin-2. Interestingly, the binding site on each fibrillin is located within the final 7 EGF-like repeats in the carboxyl-terminal region which is at the opposite end of the molecule to the MAGP-1 binding site on fibrillin-1. The reciprocal binding site on MAGP-2 was identified within the central cysteine-rich region conserved between MAGP-1 and MAGP-2. The authors suggested that the interactions may be important for assembly of both fibrillin-1 and fibrillin-2 containing microfibrils.[42] In nuchal ligament development, MAGP-2 expression does mimic the expression patterns for fibrillins -1 and -2, supporting a possible role for MAGP-2 in microfibril assembly and elastinogenesis in this tissue.[5,32] However, the absence of MAGP-2 from microfibrils in several tissues has been documented, indicating that a ubiquitous structural role for MAGP-2 in the microfibrils is unlikely.

The amino acid sequence of MAGP-2 provides a clue to a possible function of the molecule. MAGP-2 lacks the putative tropoelastin and collagen VI-binding region of MAGP-1.[38] However, the corresponding region of MAGP-2 contains an active RGD integrin recognition sequence. This sequence has been shown to mediate the attachment and spreading of a range of cell types on MAGP-2 substrate via specific interaction with alpha V beta 3 integrin.[48] Thus MAGP-2 may modulate microfibril-cell interactions at specific stages during development and differentiation in particular tissue environments.[48]

Further insight into the functions of MAGP-1 and MAGP-2 should be provided when the results of gene knockout experiments in mice are published.

Other Small Microfibril-Associated Proteins (MFAPs)

In addition to MAGP-1 and MAGP-2, several other small proteins have been identified as potential components of fibrillin-containing microfibrils. As the human form of each protein was cloned it was given the name MFAP and the series currently includes four members, one of which, MFAP2, is MAGP-1. It should be noted that there are no structural similarities between any of the four proteins. Since the nomenclature can be confusing the data has been presented in table form (Table 1).

MFAP1 was originally identified as 'associated microfibril protein' or AMP, following expression screening of a whole chick embryo cDNA library with antiserum raised to the microfibril-rich bovine ocular zonule.[11] Antiserum raised to a synthetic MFAP1 peptide localized specifically to fibrillin-containing microfibrils in several tissues, including the zonule fibers. MFAP1 was characterized as a 54 kDa protein which is processed to a 32 kDa protein, that had previously been identified in zonular extracts.[49] The human gene for MFAP1 has been characterized and mapped, close to the fibrillin-1 gene locus on chromosome 15.[12,50]

MFAP-3 is a 41 kDa serine-rich protein also identified by cDNA library screening with a polyclonal antibody to zonular fibers.[13] Antibodies raised to recombinant MFAP3 were found to localize to zonular microfibrils and identify the 41 kDa protein in extracts of developing nuchal ligament. Northern blotting indicated that the protein was also expressed in fetal aorta and lung. Genomic analysis revealed that the MAFP3 gene contained only two exons and mapped to chromosome 5q32-q33.2.

MFAP4 was identified from a novel cDNA mapping to human chromosome 17p11.2.[14] The clone encoded a 29 kDa protein which was named MFAP4 due to its high level of sequence identity with a partially sequenced 36 kDa protein extracted from porcine and bovine aortas. The 36 kDa protein showed ultrastructural localization to fibrillin-containing microfibrils surrounding elastic fibers in aorta, skin and spleen. Interestingly, the protein appeared not to be associated with the elastin-free microfibrils in ocular zonule and kidney.[51-53] It has been suggested that the 36 kDa protein plays a role in elastinogenesis and it has been named MAGP-36.[53] MFAP-4 also shares some sequence similarity with a 40 kDa protein identified using IgG from the aortic wall of patients with abdominal aortic aneurysms. The protein has been named aortic aneurysm associated protein (AAAP-40) and MAGP-3 due to its similarity to MAGP-36.[54,55] It should be noted that MFAP4, MAGP-36 and AAAP-40 have no structural resemblance to the MAGP family of proteins (MAGP-1 and MAGP-2) described in the previous section. The functions of MFAP1, MFAP2 and MFAP4 in the biology of fibrillin-containing microfibrils remain to be elucidated.

Table 1. Characteristics of small candidate microfibril-associated proteins

Human Protein	Alternative Name(s)	Chromosome Location of Human Gene	Characteristics	Evidence for Role in Biology of Fibrillin-Containing Microfibrils
AAAP-40[54]	MAGP-3[54]		40-kDa protein. No homology with MAGPs -1 and -2.[54]	Immunoreactive with IgG from aorta of patients with abdominal aortic aneurysms.[54] Some sequence similarity to MFAP4.
Associated microfibril protein[11,12] (AMP)	MFAP1[12]	15q15-q21[12]	57 kDa protein processed to 32 kDa. Lacks cysteine residues necessary for disulfide bond formation.[11]	Immunolocalized to microfibrils in aorta, nuchal ligament and zonule. Extracted from bovine and chick elastic tissues.[11]
Microfibril-associated glycoprotein-1 (MAGP-1)[1-4,7, 8,16,17,36,38,43]	MFAP2[34]	1p36.1-p35[34]	31 kDa acidic glycoprotein. Consists of two domains, an extended proline- and glutamine-rich N-terminal domain and a cysteine-rich C-terminal domain.[1,2]	Specific co-distribution with fibrillin-1 in most tissues.[4] Periodic labeling of the "bead" regions of the micro-fibrils.[7] Co-purification with fibrillin-1 in tissue extracts and with microfibrils in density gradients.[8,56,83] Binds in vitro to fibrillin-1 and tropoelastin.[36,38,43]
Microfibril-associated glycoprotein-2 (MAGP-2)[5,9,42]	MP25[9,56] SD25[8]	12p12.3-p13.1[9]	25 kDa glycoprotein Central cysteine-rich region of over 50% homology with MAGP-1.[9] Interacts with a variety of cell types via the alpha V beta 3 integrin.[48]	Specific co-distribution with, and periodic labeling of, microfibrils in some tissues. Co-purification with fibrillin-1 in tissue extracts and with microfibrils in density gradients.[5,8,9] (Hanssen E, Gibson MA, unpublished observations). Binds fibrillins -1 and -2 in two-hybrid assays.[42]
MFAP3[13]		5q32-q33.1[13]	40 kDa serine-rich acidic protein[13]	Immuno localization to zonular microfibrils. Immuno-blotting of nuchal ligament reductive extracts.[13]
MFAP4[14]	MAGP-36[53] (porcine homologue)	17p11.2[14]	cDNA encoding a 28.6 kDa protein with homology to MAGP-36.[14,53] Gene deleted in Smith-Magenis syndrome.[14]	MAGP-36 has immuno-ultrastructural localization to elastin-associated microfibrils in skin, aorta and spleen.[53]

Proteoglycans

A number of ultrastructural and histochemical studies have identified proteoglycans in close association with developing elastic fibers in tissues such as skin and aorta,[56-61] and with ocular zonule microfibrils.[62,63] In addition, immunohistochemical studies on human skin have suggested that two small dermatan sulfate proteoglycans, decorin and biglycan, are closely associated with elastic fibers, decorin with the fibrillin-containing microfibrils and biglycan with the elastin component.[60,61] More recent studies have supported these observations. Kielty et al[15] demonstrated that chondroitinase AC treatment disrupts the bead component and increases the inter-bead periodicity of fibrillin-containing microfibrils. In addition they showed that a small chondroitinase AC sensitive proteoglycan could be coprecipitated with fibrillin from smooth muscle cell medium.[15] The proteoglycan could be from either the chondroitin sulfate or dermatan sulfate family since chondroitinase AC specifically cleaves glycosaminoglycans at D-glucuronate residues which occur in the side-chains of chondroitin sulfate and also dermatan sulfate proteoglycans containing copolymeric side-chains.[64] The authors concluded that a small chondroitin sulfate proteoglycan associates with fibrillin and contributes to microfibril assembly, and suggested that proteoglycan that could be decorin, biglycan or fibromodulin.

In other studies decorin has been shown to form a ternary complex with fibrillin-1 and MAGP-1 in chondrocyte culture medium. Further experiments with recombinant fibrillin fragments indicated that the decorin binding site on fibrillin-1 was located in an amino-terminal region containing the proline-rich domain and the following five EGF-like repeats.[16] This region of fibrillin-1 is considered to be in or close to the 'beads' within the microfibril and thus decorin may be the chondroitinase AC sensitive proteoglycan identified by Kielty et al[15] to be associated with these structures. The corresponding fragment of fibrillin-2 did not bind decorin suggesting that the binding was specific for fibrillin-1. Interestingly, decorin and MAGP-1 were also found to coimmunoprecipitate from chondrocyte cultures which suggested that there was also a direct interaction between these two macromolecules. The authors speculated that decorin may be involved in regulating assembly of microfibrils or in coordinating individual microfibrils into bundles.[16] Decorin has also been shown to bind tropoelastin in vitro suggesting that the proteoglycan may be involved in elastic fiber assembly.[17]

Biglycan may also be involved in microfibril biology. In studies on glycosaminoglycan and proteoglycan content of developing nuchal ligament, expression of a specific glycoform of biglycan was shown to correlate with the elastinogenic phase of elastic fiber formation in this tissue. In contrast, decorin expression was shown to peak early and was relatively low during the periods of maximum microfibril and elastin deposition.[65] Subsequent in vitro binding studies showed that biglycan binds via its core protein to tropoelastin and to MAGP-1 and that the three macromolecules can form a ternary complex. Decorin was also found to bind to tropoelastin but less strongly than biglycan. However, in contrast to biglycan, decorin showed no binding affinity for MAGP-1 in both solid phase and immunoprecipitation assays.[17] The authors suggested that the differential binding to tropoelastin and MAGP-1 points to decorin and biglycan possessing distinct functions in elastic fiber biology. They also suggested that biglycan may be involved with MAGP-1 in the deposition of tropoelastin onto the surface of the microfibrils during elastinogenesis or in the stabilization of the mature elastic fiber perhaps by mediating interactions with elements of the surrounding matrix.

Although evidence is mounting that decorin and biglycan are involved in microfibril and elastic fiber biology, it is interesting that decorin and biglycan null mice show no obvious elastic fiber abnormalities.[66,67] These findings suggests that, in the absence of the other proteoglycan, decorin and biglycan may be able to substitute for each other in their elastic fiber-related functions.

While roles for decorin and biglycan remain uncertain, recent evidence has underlined the importance of proteoglycans for microfibril and elastic fiber assembly. Independent cell culture studies have clearly demonstrated that disruption of proteoglycan synthesis, by blocking sulfation or GAG side-chain assembly, prevents incorporation of fibrillin-1 into the extracellular matrix.[16,18] Other proteoglycan candidates for involvement in microfibril function include, the large chondroitin sulfate proteoglycan, versican,[19] and an as yet unidentified heparan sulfate proteoglycan.[18] A combined ultrastructural, immunolocalization and molecular binding study has shown that versican binds, apparently covalently, to microfibrils and to fibrillin-1 via its C-terminal domain. The binding site on fibrillin-1 is located centrally between calcium binding EGF-like domains 11 and 21, the region involved in the severe 'neonatal' form of Marfan syndrome. Immunolabeling along the microfibrils was found to be relatively sparse and to lack a discernible periodicity. This finding suggested to the authors that versican does not serve an integral structural function but connects microfibrils to hyaluronan in the surrounding matrix.[19] In other recent studies it has been reported that over-expression of a variant of versican in cultured arterial smooth muscle cells significantly increased tropoelastin expression and formation of elastic fibers in the extracellular matrix. Moreover, the seeding of these transfected cells into rat carotid arteries, damaged by balloon catheter, induced the formation of a compact neointima which was rich in elastic lamellae. These results clearly implicate versican in elastic fiber assembly.[68]

Tiedemann et al[18] have shown that heparin binds to fibrillin-1 at three sites along the molecule and that heparin and heparan sulfate inhibit, when added to the medium, the assembly of fibrillin-1 into the matrix of cultured skin fibroblasts. The authors suggest that the binding of proteoglycan-associated heparan sulfate chains to fibrillin-1 is an important step in microfibril assembly. The proteoglycan (s) involved remain to be identified. Evidently, research into proteoglycan involvement in the function of fibrillin-containing microfibrils is entering an exciting phase.

Fibulins

Recent evidence indicates that fibulins play an important role in elastic fiber biology. The fibulin family consists of five distinct rod-like proteins containing multiple EGF-like repeat motifs and a globular C-terminal domain exhibiting some sequence similarity with the C-terminus of fibrillins.[69-71] Fibulins have extensive but distinct tissue distributions. Fibulin-1 (100 kDa) is found in plasma and the extracellular matrix of a wide variety of tissues where it is associated with structures such as the elastin core of elastic fibers and several basement membranes.[20,72] Fibulin-2 (195 kDa) has been shown to be colocalized with fibrillin-1 in some but not all tissues.[21] Ultrastructurally, fibulin-2 was found to be located at the junction between the microfibrils and the amorphous elastin core in elastic fibers of the skin.[21] Fibulins 1 and 2 have been shown to bind an extensive range of matrix macromolecules including basement membrane and elastic fiber components. In particular, both fibulin-1 and 2 bind tropoelastin[73] and fibulin-2, but not fibulin-1, binds to the N-terminal region of fibrillin-1.[21] It has been proposed that fibulin-2 in particular may have a role in mediating the attachment of the microfibrils to elastin[73] and to basement membranes.[21]

More recently, targeted disruption of the fibulin-5 gene in mice has demonstrated that this fibulin, at least, is profoundly involved in elastic fiber development. The phenotype exhibits severe disorganization of the elastic fiber systems throughout the body leading to the development of marked elastinopathies including cutis laxa, emphysema and tortuosity of the aorta.[74,75] Fibulin-5 has been shown to interact with tropoelastin and several integrins suggesting it may have a critical role in the anchorage of elastic fibers to cells during elastic fiber development. Interestingly, fibulin 5 did not bind fibrillin-1 in in vitro binding experiments[75] and its relationship with fibrillin-containing microfibrils remains to be elucidated. Overall the evidence suggests that several fibulins may be important binding partners for elastic fiber components rather than functioning as structural components of the microfibrils themselves.

LTBPs

Latent transforming growth factor-beta binding proteins (LTBPs) are a family of four proteins which share structural characteristics with fibrillins in that they are rod-like extracellular matrix molecules consisting predominantly of tandem EGF-like 6-cysteine repeats interspersed with 8-cysteine motifs. Since the 8-cysteine motifs have only been found in fibrillins and LTBPs, the two groups of proteins are now considered to comprise a superfamily.[76-78] A major functional characteristic of LTBPs is considered to be the intracellular covalent binding of latent forms of TGF-beta and the facilitation of subsequent secretion and targeting of the growth factor to sites in the extracellular matrix. Latent TGF beta has been shown to bind covalently to LTBPs 1, 3 and 4 via a site in the third 8-cysteine motif of each protein. However, LTBP-2, like fibrillins, does not appear to bind to TGF-beta.[79] In addition, LTBP-1 and LTBP-2 can be secreted and deposited into extracellular matrix without attached TGF-beta.[22,80,81] Thus the evidence indicates that LTBPs may have additional functions as structural components of the matrix.

Ultrastructural immunolocalization studies have identified fibrillin-containing microfibrils as the location of matrix-associated LTBPs in a range of tissues. LTBP-1 has been immunolocalized to microfibrils in skin,[24,25] bone[26] and kidney.[27] In a very recent immunohistochemical study of human tissues, LTBP-1 was found to codistribute extensively with fibrillin-1 in tendon, perichondrium and blood vessels but to be absent or scarce in skeletal muscle and lung.[28] LTBP-2 has been located on microfibrils in elastic nuchal ligament and aorta.[22] In another study, both LTBP-1 and LTBP-2 showed partial coimmunolocalization with fibrillin-1 in the wall of coronary arteries. In addition, the intensity of staining increased following angioplasty-induced injury to the arteries.[82]

It is still unclear if LTBPs are integral structural components of the microfibrils or are associated with the surface of these structures. Both LTBP-1 and LTBP-2 have been demonstrated to bind strongly to the microfibrils[22,23,80] and LTBP-1 has been found to contain three potential matrix binding domains.[81] Very recently, LTBP-1 has been shown to bind to the N-terminal regions of fibrillins 1 and 2 via its carboxyl-terminal region.[28] LTBP-4 also appears to bind fibrillins in a similar manner. However, LTBP-1 was not detected in association with microfibrils extracted from tissues using crude bacterial collagenase, suggesting that the protein is not an integral component of these structures.[28] LTBP-1 and LTBP-2 can also be separated from fibrillin-containing microfibrils using density gradient centrifugation, indicating a lack of covalent association with these structural elements.[83] This finding is partially supported by studies on extraction of LTBP-2 from developing nuchal ligament. Most of the LTBP-2 was shown to be solubilized from the tissue, in monomeric form, using the chaotrope, 6M guanidinium chloride. Since the microfibrils remain morphologically intact after the treatment, it would appear that the majority of the LTBP-2 was not covalently bound to these structures. However, a significant proportion of the LTBP-2 could only be solubilized using a reducing agent, suggesting that some of the protein was covalently attached to the microfibrils, presumably by reducible disulfide bonding.[22]

Overall, the balance of evidence points to LTBPs being located on the surface of fibrillin-containing microfibrils rather than occurring as integral components of these structures. The functions of LTBPs in microfibril biology are yet to be determined. Some clues have been forthcoming from LTBP gene knock-out mice. LTBP-2 null mice have proved to be uninformative as they die in early embryogenesis.[84] The LTBP-3 null mice have bone abnormalities attributed to alteration of normal TGF-beta bioavailability.[85] However, the disruption of the gene for LTBP-4 in mice resulted in a homozygous phenotype exhibiting severe tissue specific disruption of elastic fiber structure in lung and colon.[86] Thus it appears that individual LTBPs may modulate microfibril function to suit particular tissue environments.

Emilins

Elastin-microfibril interface located proteins, or EMILINS, are a two-member family of matrix glycoproteins.[30] EMILIN-1 is a 115 kDa glycoprotein which, as its name suggests, has been ultrastructurally immunolocalized to the interface between the microfibrils and the elastin core of developing elastic fibers.[87] Molecular cloning has shown EMILIN-1 to contain several distinct domains including a short collagenous region and a C-terminal domain with close sequence similarity to C1q.[88] The molecule appears to form homotrimers and larger disulfide-bonded multimers. EMILIN-2 is similar in structure to EMILIN-1 and the two proteins have overlapping but distinct tissue expression patterns. EMILIN-1 has been shown to interact with EMILIN-2 via the C-terminal domains and to possess cell adhesive properties. However, binding of EMILINS to elastin and microfibrillar components has not yet been reported and EMILIN function in microfibril biology remains unknown. It has been suggested that EMILIN-1 may be involved in elastic fiber formation, in anchoring smooth muscle cells to elastic fibers and in regulation of blood vessel assembly.[30]

Other Proteins

Other proteins which may be involved with the function of fibrillin-containing microfibrils include the cross-linking enzyme, lysyl oxidase[89] and the 67 kDa cell surface elastin-binding protein.[90] Lysyl oxidase (32 kDa) has been localized by immuno-electron microscopy on, or very close to, the elastin-associated microfibrils in developing elastic tissues. This observation raises the possibility that an interaction of lysyl oxidase with fibrillin or other microfibrillar component may be important in the early stages of elastinogenesis when tropoelastin is deposited onto the microfibrils.[91]

The cell surface-associated, 67 kDa elastin binding protein appears to play a major role in tropoelastin secretion and assembly into elastic fibers.[90,92] Evidence suggests that the binding protein becomes bound to tropoelastin intracellularly and then accompanies the elastin precursor to the extracellular side of the plasma membrane. It is postulated that tropoelastin then remains bound to the 67 kDa protein until there is an interaction with a microfibril-associated galactoside sugar which induces the transfer of the tropoelastin onto an acceptor site on the microfibril.[93] The identity of the microfibril component(s) involved is unknown but it may be fibrillin-1 or MAGP-1 as both have been shown to bind tropoelastin and a proteoglycan component rich in galactosamine.

Involvement in Human Genetic Diseases

It has been postulated that mutations in the genes for non-fibrillin microfibrillar components may result in phenotypes with some similarities to the physical manifestations of Marfan syndrome. To date no mutations in human MAGP genes have been identified, although several alternatively spliced isoforms of MAGP-1 have been characterized including one isoform in MG-63 cells which lacks an alanine residue in the signal peptide.[94] A search for mutations in the MFAP-1 gene also proved unsuccessful.[50] The MFAP-4 gene has been found to be commonly deleted in Smith-Magenis syndrome, the characteristics of which include skeletal abnormalities of the head and mental retardation. However, the contribution of the lack of the MFAP-4 gene to this phenotype is unclear.[14]

Fibulin gene defects have been linked to several genetic disorders. Haploinsufficiency of the fibulin-1 gene has been linked to a case of synpolydactyly[95] and a mutation in the fibulin-3 gene has been shown to segregate with the autosomal dominant eye diseases, Malattia Leventinese and Doney honeycomb retinal dystrophy.[96] It is unclear if either of these mutations affects the function of fibrillin-containing microfibrils or elastic fibers. However, a missense mutation in the fibulin-5 gene has been linked to a severe recessive form of cutis laxa, exhibiting reduced, disorganized elastic fibers in the skin, thickened aortic valve, supravalvular aortic stenosis and the development of pulmonary emphysema.[97] These manifestations indicate disruption of normal elastin synthesis in these patients and are consistent with a proposed role for fibulin-5 in elastic fiber formation deduced from fibulin-5 gene knock out mice.[74,75] Interestingly, heterozygotes for the mutation were clinically normal, indicating that the mutant fibulin-5 molecules did not interfere with the function of normal fibulin-5 molecules. This finding suggested that fibulin-5 functions as a monomer molecule in its role in elastinogenesis and thus it is unlikely to be a structural component of the fibrillin-containing microfibrils.[97]

In preliminary studies, several distinct missense point mutations in the LTBP-2 gene have been reported in patients exhibiting a range of Marfan-like manifestations including mitral valve prolapse, aortic dilatation and spinal scoliosis. However, it needs to be determined if the mutations have a causative link to the phenotypes of these patients.[98] If such a link were to be established then it would point to a structural role for LTBP-2 in fibrillin-containing microfibrils in a variety of tissues.

It remains to be established if a mutation in the gene for any non-fibrillin microfibril-associated protein can result in physical manifestations attributable to the disruption of the normal function of fibrillin-containing microfibrils.

Concluding Remarks

In recent years a diverse range of macromolecules have been identified as candidates for nonfibrillin components of the 10 nm microfibrils. The evidence for an association can vary in each case from immunolocalization only, to binding interactions with fibrillins and biological effects of knock-out mice. Some of the macromolecules discussed above appear to be exclusively associated with the microfibrils suggesting that their functions relate specifically to the microfibrils. This group can be divided into macromolecules which have widespread codistribution with the microfibrils such as MAGP-1 and macromolecules which have limited codistribution with the microfibrils suggesting that they may modulate the function of these structures to suit particular tissue environments, for example MAGP-2 and some MFAPs. Another group of microfibril-associated macromolecules also have known association with other structural elements of extracellular matrices indicating that they also fulfill microfibril-independent functions. This group may include several proteoglycans, fibulins and certain LTBPs. As knowledge of each microfibril-associated protein grows, candidates are appearing for each category of function proposed in the introduction. Molecules providing structural and linkage functions within microfibrillar bundles would be expected to have widespread codistribution with the microfibrils and be difficult to extract without disruption of the microfibrils. From the evidence presented above, MAGP-1 would appear to be the chief candidate for such roles. In addition, there is evidence that an unidentified proteoglycan, possibly decorin, may have a structural role as a component of the bead substructures. Several nonfibrillin molecules, in close association with microfibrils, may be involved in the assembly of elastin onto these structures. Molecules in this category may include MAGP-1, emilin-1, fibulin-5, the 67 kDa elastin-binding protein, lysyl oxidase and biglycan. Molecules which may mediate interactions of microfibrils with the surrounding matrix include versican, fibulins 1 and 2, MAGP-1 and LTBP-2. MAGP-2 and fibulin-5 may be involved in specific microfibril-cell interactions. LTBPs may extend the function of the microfibrils by acting as stores for TGF-β on the surface of the microfibrils in particular tissue environments. However, despite the rapidly accumulating information about the above macromolecules, the role(s) of each protein in microfibril function remains unclear. Obviously, further research is needed to clarify which proteins are important in microfibril biology and to identify the full range of human genetic disorders linked to their genes.

Acknowledgments

The author would like to acknowledge the financial support the National Health and Medical Research Council of Australia for the contributions from his laboratory cited in this review. Dr. E.G. Cleary and Dr. E. Hanssen are also thanked for their advice and critical comment during the preparation of this article.

References

1. Gibson MA, Hughes JL, Fanning JC et al. The major antigen of the elastin-associated microfibrils is a 31kDa glycoprotein. J Biol Chem 1986; 261:11429-11436.
2. Gibson MA, Sandberg LB, Grosso LE et al. Complementary DNA cloning establishes microfibril-associated glycoprotein (MAGP) to be a discrete component of the elastin-associated microfibrils. J Biol Chem 1991; 266:7596-7601.
3. Gibson MA, Cleary EG The immunohistochemical localization of microfibril-associated glycoprotein (MAGP) in elastic and nonelastic tissues. Immunol Cell Biol 1987; 65:345-356.
4. Kumaratilake JS, Gibson MA, Fanning JC et al. The tissue distribution of microfibrils reacting with a monospecific antibody to MAGP, the major glycoprotein antigen of elastin-associated microfibrils. Eur J Cell Biol 1989; 50:117-127.
5. Gibson MA, Finnis ML, Kumaratilake JS et al. Microfibril-associated glycoprotein-2 (MAGP-2) is specifically associated with fibrillin-containing microfibrils but exhibits more restricted patterns of tissue localization and developmental expression than its structural relative MAGP-1. J Histochem Cytochem 1998; 46:871-885.
6. Kitahama S, Gibson MA, Hatzinikolas G et al. Expression of fibrillins and other microfibril-associated proteins in human bone and osteoblast-like cells. Bone 2000; 27:61-67.
7. Henderson M, Polewski R, Fanning JC et al. Microfibril-associated glycoprotein -1 (MAGP-1) is specifically located on the beads of the beaded-filament structure for fibrillin-containing microfibrils as visualized by the rotary shadowing technique. J Histochem Cytochem 1996; 44:1389-1397.
8. Gibson MA, Kumaratilake JS, Cleary EG. The protein components of the 12-nanometer microfibrils of elastic and nonelastic tissues. J Biol Chem 1989; 264:4590-4598.
9. Gibson MA, Hatzinikolas G, Kumaratilake JS et al. Further characterization of proteins associated with elastic fiber microfibrils including the molecular cloning of MAGP-2 (MP25). J Biol Chem 1996; 271:1096-1103.
10. Gibson MA, Kumaratilake JS, Cleary EG. Immunohistochemical and ultrastructural localization of MP78/70 (βig-h3) in extracellular matrix of developing and mature bovine tissues. J Histochem Cytochem 1997; 45:1683-1696.
11. Horrigan SK, Rich CB, Streeten BW et al. Characterization of an associated microfibril protein through recombinant DNA techniques. J Biol Chem 1992; 267:10087-10095.
12. Yeh H, Chow M, Abrams WR et al. Structure of the human gene encoding the associated microfibrillar protein (MFAP1) and localization to chromosome 15q15-q21. Genomics 1994; 23:443-449.
13. Abrams WR, Ma R-I, Kucich U et al. Molecular cloning of the microfibrillar protein MFAP3 and assignment of the gene to human chromosome 5q32-q33.2. Genomics 1995; 26:47-54.
14. Zhao Z, Lee C-C, Jiralerspong S et al. The gene for a human microfibril-associated glycoprotein is commonly deleted in Smith-Magenis syndrome patients. Hum Mol Genet 1995; 4:589-597.
15. Kielty CM, Whittaker SP, Shuttleworth CA. Fibrillin: Evidence that chondroitin sulphate proteoglycans are components of microfibrils and associate with newly synthesized monomers. FEBS Lett 1996; 386:169-173.
16. Trask BC, Trask TM, Broekelmann T et al. The microfibrillar proteins MAGP-1 and fibrillin-1 form a ternary complex with the chondroitin sulfate proteoglycan decorin. Mol Biol Cell 2000; 11:1499-1507.
17. Reinboth B, Hanssen E, Cleary EG et al. Molecular interactions of biglycan and decorin with elastic fiber components: Biglycan forms a ternary complex with tropoelastin and microfibril-associated glycoprotein-1. J Biol Chem 2002; 277:3950-3957.
18. Tiedemann K, Batge B, Muller PK et al. Interactions of fibrillin-1 with heparin/heparan sulfate, implications for microfibrillar assembly. J Biol Chem 2001; 276:36035-36042.
19. Isogai Z, Aspberg A, Keene DR et al. Versican interacts with fibrillin-1 and links extracellular microfibrils to other connective tissue networks. J Biol Chem 2002; 277:4565-4572.
20. Roark EF, Keene DR, Haudenschild CC et al. The association of human fibulin-1 with elastic fibers: An immunohistological, ultrastructural, and RNA study. J Histochem Cytochem 1995; 43:401-411.
21. Reinhardt DP, Sasaki T, Dzamba BJ et al. Fibrillin-1 and fibulin-2 interact and are colocalized in some tissues. J Biol Chem 1996; 271:19489-19496.
22. Gibson MA, Hatzinikolas G, Davis EC et al. Bovine latent transforming growth factor β1-binding protein-2: Molecular cloning, identification of tissue isoforms, and immunolocalization to elastin-associated microfibrils. Mol Cell Biol 1995; 15:6932-6942.
23. Taipale J, Saharinen J, Hedman K et al. Latent transforming growth factor-beta 1 and its binding protein are components of extracellular matrix microfibrils. J Histochem Cytochem 1996; 44:875-889.

24. Karonen T, Jeskanen L, Keski-Oja J. Transforming growth factor β1 and its latent form binding protein-1 associate with elastic fibers in human dermis: Accumulation in actinic damage and absence in anetoderma. Br J Dermatol 1997; 137:51-55.
25. Raghunath M, Unsold C, Kubitscheck U et al. The cutaneous microfibrillar apparatus contains latent transforming growth factor-beta binding protein-1(LTBP-1) and is a repository for latent TGF-beta 1. J Invest Dermatol 1998; 111:559-564.
26. Dallas SL, Keene DR, Bruder SP et al. Role of the latent transforming growth factor-beta binding protein 1 in fibrillin-containing microfibrils in bone cells in vitro and in vivo. J Bone Mineral Res 2000; 15:68-81.
27. Sterzel RB, Hartner A, Schlotzer-Schrehardt U et al. Elastic fiber proteins in the glomerular mesangium in vivo and in cell culture. Kidney Int 2000; 58:1588-1602.
28. Isogai Z, Ono RN, Ushiro S et al. LTBP-1 interacts with fibrillin and is a microfibril-associated protein. J Biol Chem 2003; 278:2750-2757.
29. Bressan GM, Daga-Gordini D, Colombatti A et al. EMILIN, a component of the elastic fibers preferentially located at the elastin-microfibrils interface. J Cell Biol 1993; 121:201-212.
30. Colombatti A, Doliana R, Bot S et al The EMILIN protein family. Matrix Biol 2000; 19:281-376.
31. Zhang H, Hu W, Ramirez F. Developmental expression of fibrillin genes suggests heterogeneity of extracellular microfibrils. J Cell Biol 1995; 129:1165-1176
32. Mariencheck MC, Davis EC, Zhang H et al. Fibrillin-1 and fibrillin-2 show temporal and tissue-specific regulation of expression in developing elastic tissues. Connect Tissue Res 1995; 31:87-97.
33. Bashir MM, Abrams WR, Rosenbloom J et al. Microfibril-associated glycoprotein: Characterization of the bovine gene and of the recombinantly expressed protein. Biochemistry 1994; 33:593-600.
34. Faraco J, Bashir M, Rosenbloom J et al. Characterization of the human gene for microfibril-associated glycoprotein (MFAP2) assignment to chromosome 1p36.1-p35, and linkage to D1S170. Genomics 1995; 25:630-637.
35. Chen Y, Faraco J, Yin W et al. Structure, chromosomal localization, and expression pattern of the murine Magp gene. J Biol Chem 1993; 268:27381-27389.
36. Brown-Augsburger P, Broekelmann T, Mecham L et al Microfibril associated glycoprotein binds to the carboxyl-terminal domain of tropoelastin and is a substrate for transglutaminase. J Biol Chem 1994; 269:28443-28449.
37. Trask BC, Broekelmann T, Ritty TM et al. Posttranslational modifications of microfibril-associated glycoprotein-1. Biochemistry 2001; 40:4372-4380.
38. Finnis ML, Gibson MA. Microfibril-associated glycoprotein-1 (MAGP-1) binds to the pepsin-resistant domain of the alpha 3 (VI) chain of type VI collagen. J Biol Chem 1997; 272:22817-22823.
39. Segade F, Trask BC, Broekelmann TJ et al. Identification of a matrix -binding domain in MAGP1 and MAGP2 and intracellular localization of alternative splice forms. J Biol Chem 2002; 277:11050-11057.
40. Frankfater C, Maus E, Gaal K et al. Organization of the mouse microfibril-associated glycoprotein -2 (MAGP-2) gene. Mamm Genome 2000; 11:191-195.
41. Hatzinikolas G, Gibson MA. The exon structure of the MAGP-2 gene. J Biol Chem 1998; 273:29309-29314.
42. Penner AS, Rock MJ, Kielty CM et al. Microfibril-associated glycoprotein-2 interacts with fibrillin-1 and fibrillin-2 suggesting a role for MAGP-2 in elastic fiber assembly. J Biol Chem 2002; 277:35044-35049.
43. Jenssen SA, Rienhardt DP, Gibson MA et al. Protein interaction studies of MAGP-1 with tropoelastin and fibrillin-1. J Biol Chem 2001; 276:39661-39666.
44. Trask TM, Ritty TM, Broekelmann T et al. N-terminal domains of fibrillin 1 and fibrillin2 direct the formation of homodimers: A possible first step in microfibril assembly. Biochem J 1999; 340:693-701.
45. Reinhardt DP, Gambee JE, Ono RN et al. Initial steps in the assembly of microfibrils. J Biol Chem 2000; 275:2205-2210.
46. Brown-Augsburger P, Broekelmann T, Rosenbloom J et al. Functional domains on elastin and microfibril-associated glycoprotein involved in elastic fiber assembly. Biochem J 1996; 318:149-155.
47. Trask TM, Trask BC, Ritty TM et al. Interaction of tropoelastin with the amino-terminal domains of fibrillin-1 and fibrillin-2 suggests a role for fibrillins in elastic fiber assembly. J Biol Chem 2000; 275:24400-24406.
48. Gibson MA, Leavesley DI, Ashman LK. Microfibril-associated glycoprotein-2 specifically interacts with a range of bovine and human cell types via alphaVbeta3 integrin. J Biol Chem 1999; 274:13060-13065.

49. Streeten BW, Gibson SA. Identification of extractable proteins from the bovine ocular zonule: Major zonular antigens of 32KD and 250KD. Curr Eye Res 1988; 7:139-146.
50. Lui W, Faraco J, Qian C et al. The gene for microfibril-associated protein-1 (MFAP1) is located several megabases centromeric to FBN1 and is not mutated in Marfan syndrome. Hum Genet 1997; 99:578-584.
51. Kobayashi R, Tashima Y, Masuda H et al. Isolation and characterisation of a new 36-kDa microfibril-associated glycoprotein from porcine aorta. J Biol Chem 1989; 264:17437-17444.
52. Kobayashi R, Mizutani A, Hidaka H. Isolation and characterization of a 36-kDa microfibril-associated glycoprotein by the newly synthesized isoquinolinesulfonate affinity chromatography. Biochem Biophys Res Commun 1994; 198:1262-1266.
53. Toyoshima T, Yamashita K, Furiucji H et al. Ultrastructural distribution of 36-kD microfibril-associated glycoprotein (MAGP-36) in human and bovine tissues. J Histochem Cytochem 1999; 47:1049-1056.
54. Xia S, Ozsvath K, Hirose H et al. Partial amino acid sequence of a novel 40-kDa human aortic protein, with vitronectin-like, fibrinogen-like and calcium binding domains: Aortic aneurysm-associated protein-40 (AAAP-40) Biochem Biophys Res Commun 1996; 219:36-39.
55. Hirose H, Kathleen J, Ozsvath KJ et al. Molecular cloning of the complementary DNA for an additional member of the family of aortic aneurysm antigenic proteins. J Vasc Surg 1997; 26:313-318.
56. Cleary EG, Gibson MA. Elastic tissue, elastin and elastin-associated microfibrils. In: Comper WD, ed. The structure and function of extracellular matrix. New York: Gordon and Breach Science Publisher, 1996:95-140.
57. Bartholomew JS, Anderson JC. Investigation of relationships between collagens, elastin and gly-cosaminoglycans in bovine thoracic aorta by immunofluorescence techniques. Histochem J 1983; 15:1177-1190.
58. Fornieri C, Baccarani-Contri M, Quaglino Jr D et al. Lysyl oxidase activity and elastin/glycosami-noglycan interactions in growing chick and rat aortas. J Cell Biol 1987; 105:1463-1469.
59. Baccarani-Contri M, Fornieri C, Pasquali-Ronchetti I. Elastin-proteoglycans association revealed by cytochemical methods. Connect Tissue Res 1985; 13:237-249.
60. Baccarani-Contri M, Vincenzi D, Cicchetti F et al. Immunocytochemical localization of proteoglycans within normal elastin fibers. Eur J Cell Biol 1990; 53:305-312.
61. Pasquali-Ronchetti I, Baccarani-Contri M. Elastic fiber during development and aging. Microsc Res Tech 1997; 38:428-435.
62. Chan FL, Choi HL. Proteoglycans associated with the ciliary zonule of the rat eye: A histochemi-cal and immunocytochemical study. Histochem Cell Biol 1995; 104:369-381.
63. Chan FL, Choi HL, Underhill CB. Hyaluronan and chondroitin sulfate proteoglycans are colocalized to the ciliary zonule of the rat eye: Histochemical and immunocytochemical study. Histochem Cell Biol 1997; 107:289-301.
64. Gu K, Liu J, Pervin A et al. Comparison of the activity of two chondroitin AC lyases on dermatan sulfate. Carbohydr Res 1993; 244:369-377.
65. Reinboth BJ, Finnis ML, Gibson MA et al. Developmental expression of dermatan sulfate proteoglycans in the elastic bovine nuchal ligament. Matrix Biol 2000; 19:149-162.
66. Danielson KG, Baribault H, Holmes DF et al. Targeted disruption of decorin leads to abnormal collagen fibril morphology and skin fragility. J Cell Biol 1997; 136:729-743.
67. Xu T, Bianco P, Fisher LW et al. Targeted disruption of the biglycan gene leads to an osteoporosis-like phenotype in mice. Nat Genet 1998; 20:78-82.
68. Merrilees MJ, Lemire JM, Fischer JW et al. Retrovirally mediated overexpression of versican v3 by arterial smooth muscle cells induces tropoelastin synthesis and elastic fiber formation in vitro and in neointima after vascular injury. Circ Res 2002; 890:481-487.
69. Argraves WS, Tan H, Burgess WH et al. Fibulin is an extracellular matrix and plasma glycoprotein with repeated domain structures. J Cell Biol 1990; 111:3155-3164.
70. Pan T-C, Sasaki T, Zhang R-Z et al. Structure and expression of fibulin-2, a novel extracellular matrix protein with multiple EGF-like repeats and consensus for calcium binding. J Cell Biol 1993; 123:1269-1277.
71. Giltay R, Timpl R, Kostka G. Sequence, recombinant expression and tissue localization of two novel extracellular matrix proteins, fibulin-3 and fibulin-4. Matrix Biol 1999; 18:469-480.
72. Kluge M, Mann K, Dziadek M et al. Characterisation of a novel calcium-binding 90-kDa glyco-protein (BM-90) shared by basement membranes and serum. Eur J Biochem 1990; 193:651-659.
73. Sasaki T, Gohring W, Miosge N et al. Tropoelastin binding to fibulins, nidogen-2 and other extracellular matrix proteins. FEBS Lett 1999; 460:280-284.
74. Nakamura T, Lozano PR, Ikeda Y et al. Fibulin-5/ DANCE is essential for elastinogenesis in vitro. Nature 2002; 415:171-175.

75. Yanagisawa H, Davis EC, Starcher BC et al. Fibulin-5 is an elastin-binding protein essential for elastic fiber development in vivo. Nature 2002; 415:168-171.
76. Sinha S, Nevett C, Shuttleworth A et al. Cellular and extracellular biology of the latent transforming growth factor-beta binding protein. Matrix Biol 1998; 17:529-545.
77. Saharinen J, Hyytiain M, Taipale J et al. Latent transforming growth factor -beta binding proteins (LTBPs)-structural extracellular matrix proteins for targeting TGF-beta action. Cytokine Growth Factor Rev 1999; 10:99-117.
78. Oklu R, Hesketh R. The latent transforming growth factor beta binding protein (LTBP) family. Biochem J 2000; 352:601-610.
79. Saharinen J, Keski-Oja J. Specific sequence motif of 8-cys repeats of TGF-beta binding proteins, LTBPs, creates a hydrophobic interaction surface for binding of small latent TGF-beta. Mol Biol Cell 2000; 11:2691-2704.
80. Hyytiainen M, Taipale J, Heldin C-H et al. Recombinant latent transforming growth factor beta-binding protein 2 assembles to fibroblast extracellular matrix and is susceptible to proteolytic processing and release. J Biol Chem 1998; 273:20669-20676.
81. Unsold C, Hyytiainen M, Bruckner-Tuderman L et al. Latent TGF-beta binding protein LTBP-1 contains three potential extracellular matrix interacting domains. J Cell Sci 2001; 114:187-197.
82. Sinha S, Heagerty AM, Shuttleworth CA et al. Expression of latent TGF-beta binding proteins and association with TGF-beta 1 and fibrillin-1 following arterial injury. Cardiovasc Res 2002; 53:971-983
83. Kielty CM, Hanssen E, Shuttleworth CA. Purification of fibrillin-containing microfibrils and collagen VI microfibrils by density gradient centrifugation. Anal Biochem 1998; 255:108-112.
84. Shipley JM, Mecham RP, Maus E et al. Developmental expression of latent transforming growth factor beta binding protein 2 and its requirement early in mouse development. Mol Cell Biol 2000; 20:4879-4887.
85. Dabovic B, Chen Y, Colarossi C et al. Bone abnormalities in latent TGF-beta binding protein (Ltbp)-3-null mice indicate a role for Ltbp-3 in modulating TGF-beta bioavailability. J Cell Biol 2002; 156:227-232.
86. Sterner-Kock A, Thorey IS, Koli K et al. Disruption of the gene encoding the latent transforming growth factor-beta binding protein 4 (LTBP-4) causes abnormal lung development, cardiomyopathy, and colorectal cancer. Genes Dev 2002; 16:2264-2273.
87. Colombatti A, Bonaldo P, Volpin D et al. The elastin associated glycoprotein gp115. J Biol Chem 1988; 263:17534-17540.
88. Doliana R, Mongiat M, Bucciotti F et al. EMILIN, a component of the elastic fiber and a new member of the C1q/TNF superfamily of proteins. J Biol Chem 1999; 274:16773-16781.
89. Smith-Mungo LI, Kagan HM. Lysyl oxidase: Properties, regulation and multiple functions in biology. Matrix Biol 1998; 16:387-398.
90. Hinek A. Biological roles of the nonintegrin elastin/laminin receptor. Biol Chem 1996; 377:471-480.
91. Kagan HM, Vaccaro CA, Bronson E et al. Ultrastructural immunolocalisation of lysyl oxidase in vascular connective tissue. J Cell Biol 1986; 103:1121-1128.
92. Mecham RP, Heuser JE. The elastic fiber. In: Hay ED ed. Cell Biology of Extracellular Matrix. New York: Plenum Press, 1991:79-109.
93. Mecham RP, Whitehouse L, Hay M et al. Ligand affinity of the 67 kDa elastin/laminin binding protein is modulated by the protein's lectin domain: Visualisation of elastin/laminin receptor complexes with gold tagged ligands. J Cell Biol 1991; 98:1804-1812.
94. Segade F, Broekelmann TJ, Pierce RA et al. Revised genomic structure of the human MAGP-1 gene and identification of alternate transcripts in human and mouse tissues. Matrix Biol 2000; 19:671-682.
95. Debeer P, Schoenmakers EF, Twal WO et al. The fibulin-1 gene (FBLN1) is disrupted in a t(12;22) associated with a complex type of synpolydactyly. J Med Genet 2002; 39:98-104.
96. Stone EM, Lotery AJ, Munier FL et al. A single EFEMP1 mutation associated with both Malattia Leventinese and Doyne honey comb retinal dystrophy. Nat Genet 1999; 22:199-2002.
97. Loeys B, Van Maldergem L, Mortier G et al. Homozygousity for a missense mutation in fibulin-5 (FBLN5) results in a severe form of cutis laxa. Hum Mol Genet 2002; 11:2113-2118.
98. Robinson PN, Godfrey M. The molecular genetics of Marfan syndrome and related microfibrillopathies. J Med Genet 2000; 37:9-25.

CHAPTER 14

The Fibrillins and Key Molecular Mechanisms that Initiate Disease Pathways

Lynn Y. Sakai

Introduction

In 1986, fibrillin was first identified as a large noncollagenous glycoprotein associated with a certain type of microfibril ubiquitous in the connective tissue space.[1] Because fibrillin monoclonal antibodies displayed a clear periodic labeling of microfibrils, we proposed that fibrillin was a major structural component of microfibrils, and not just a molecule bound to microfibrils. The implications of this hypothesis were that (a) fibrillin molecules should self-assemble or assemble together with equally essential microfibril components and (b) abnormalities in fibrillin molecules should affect the structural integrity of microfibrils. Subsequent investigations have been directed largely toward these areas.

After we showed that fibrillin was present in the microfibrils of the ciliary zonule, the suspensory ligament of the lens,[1] considerable interest was generated in fibrillin as a candidate gene for the Marfan syndrome. McKusick's suggestion (made some thirty five years before) that identification of protein components of the ciliary zonule common to the aorta would provide insight into the basic defect in the Marfan syndrome proved to be correct. This was fully demonstrated in 1991 when the first mutation in *FBN1* was described in an individual with the Marfan syndrome.[2]

Since the initial characterization of fibrillin in 1986, two additional fibrillins have been identified using cDNA cloning methods: fibrillin-2[3,4] and fibrillin-3.[5,6] Mutations in *FBN2* were found in congenital contractural arachnodactyly,[7] a genetic disorder related clinically to the Marfan syndrome. Recently, a gene locus for recessive Weill-Marchesani syndrome, another Marfan-related syndrome, was mapped to chromosome 19p13.3-13.2,[8] within a region that contains *FBN3*.[6] Mutations have not yet been reported in *FBN3* in Weill-Marchesani syndrome. However, it is interesting that a mutation in *FBN1* has been identified[9] in a family with dominant Weill-Marchesani syndrome, which had been previously linked to chromosome 15q21.1.[10] Thus, evidence from human genetics demonstrates that all three fibrillins perform related, although perhaps at times opposite, functions in skeletal tissues, regulating long bone growth and joint mobility, while fibrillin-1 performs an apparently exclusive function in cardiovascular tissue homeostasis.

In this chapter, fibrillin structure and biological function will be reviewed within the context of how mutations in fibrillin-1 may initiate disease pathogenesis in the Marfan syndrome. Two hypothetical models provide the current conceptual framework for understanding how mutated fibrillin-1 may result in the Marfan syndrome. These are depicted schematically in Figure 1. Assuming that mutated fibrillin-1 and wildtype fibrillin-1 are synthesized and secreted equally by cells and that both mutated and wildtype fibrillin-1 molecules are incorporated equally well into a microfibril, then disease may be initiated through proteolysis of sensitive sites caused by the mutation. This scenario is depicted in part A of Figure 1. However, if

Marfan Syndrome: A Primer for Clinicians and Scientists, edited by Peter N. Robinson and Maurice Godfrey. ©2004 Eurekah.com and Kluwer Academic / Plenum Publishers.

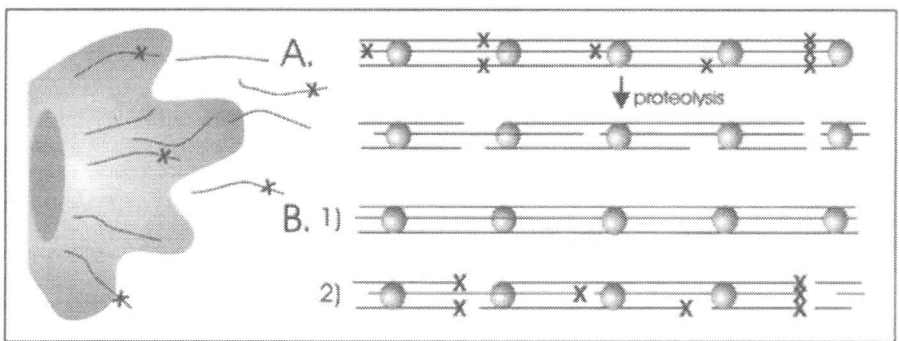

Figure 1. Model depicting two alternative mechanisms leading to fragmentation and loss of fibrillin-1 microfibrils. In (A), mutated molecules (marked with an X) are assembled along with wildtype molecules into microfibrils. Over time, the mutation creates sites susceptible to proteases, and microfibrils are cleaved into fragments. In (B), mutated molecules (marked with an X) cannot assemble properly into microfibrils, resulting in short fragmented microfibrils (B.2) and rare long, mostly wildtype microfibrils (B.1).

the mutated fibrillin-1 interferes with the assembly of the microfibril (part B of Fig. 1), then most microfibrils will consist of short fragments composed of both wildtype and mutated fibrillin-1 (part B2 of Fig. 1), while long microfibrils (composed of mostly all wildtype fibrillin-1) would be statistically rare (part B1 of Fig. 1). In both A and B, progression of disease occurs because long, intact microfibrils, particularly abundant in certain connective tissues like blood vessels, are fragmented or rare.

Molecular Structure of Fibrillin: How FBN Mutations Initiate Disease

Cloning and sequencing of fibrillin cDNAs[3,4,6,11-13] revealed similar primary structures for all three fibrillins. Each molecule is composed of domain modules whose overall organization is the same. The most abundant domain modules are the calcium-binding (cb) EGF-like modules (43 in each fibrillin). The remaining domain modules consist of 4 generic EGF-like modules, 7 modules that contain 8 conserved cysteines, 2 that appear to be "hybrid" modules, homologous carboxyl termini, and homologous amino termini in each fibrillin. A single region in the same location of each fibrillin is devoid of cysteines and is proline-rich (in fibrillin-1), glycine-rich (fibrillin-2) or proline- and glycine-rich (fibrillin-3). These modules and domains are shown schematically in Figure 2.

We first purified single fibrillin molecules from the medium of human dermal fibroblasts in culture and, using both rotary shadowing electron microscopy and velocity sedimentation, demonstrated that fibrillin molecules are linearly extended molecules with some flexibility.[14] Based upon NMR data[15] and the measured lengths of fibrillin monomers[14] and recombinant fibrillin polypeptides,[16] we can conclude that the cbEGF-like modules impart a linear shape to fibrillins. The influence of calcium upon the linear character of fibrillins was demonstrated experimentally by the finding that the measured lengths of fibrillin monomers or recombinant polypeptides are shorter in the presence of EDTA, a calcium-chelator.[16]

Around half of all of the identified mutations in *FBN1* occur in cbEGF domains. What effects might a mutation in one cbEGF domain have upon the normal biological functions of fibrillin-1? In addition to conferring a more extended linear shape to fibrillin, calcium binding helps to stabilize the structure of the fibrillin molecule. Removing calcium from fibrillin renders fibrillin more susceptible to proteolysis.[17] Mutations that are predicted to disrupt calcium binding in a cbEGF domain were specifically tested for increased sensitivity to various proteases, and results demonstrated increased proteolytic susceptibility both within the cbEGF domain containing the mutation as well as in adjacent domains.[18]

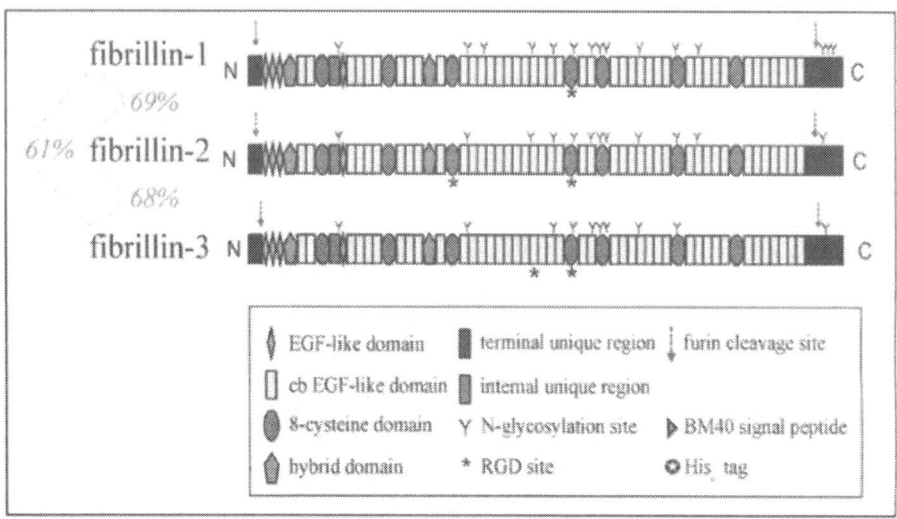

Figure 2. Schematic drawings of fibrillin-1, fibrillin-2, and fibrillin-3, demonstrating homologous domain structures and similar overall organization of domains. Identities of amino acid residues between the fibrillins is shown as percentages. Processing sites in the N- and C- terminal propeptides are marked with arrows.

Since even small perturbations of secondary structure could be easily recognized by proteases,[18] it is likely that any mutation in fibrillin-1 that results in an abnormal structure may also result in increased degradation by proteases. A cysteine substitution in a cbEGF domain produced a more dramatic increase in susceptibility to proteolysis than a different mutation in the same cbEGF domain,[19] suggesting that the large structural changes predicted to accompany cysteine substitutions in fibrillin-1 would significantly destablize the molecule.

Fibrillins are proteins rich in cysteine residues. 361 of the 2,871 amino acid residues present in fibrillin-1 are cysteines. Each of the major domain modules in fibrillins (EGF modules and 8-cys modules) is stabilized by the formation of intramolecular disulfide bonds: each EGF-like domain contains three intramolecular disulfide bonds; each 8-cys module contains four intramolecular disulfide bonds. There are nine cysteine residues in the first hybrid domain and eight in the second hybrid domain. The extra cysteine in the first hybrid domain is in the free sulfhydryl form, presumably available to form an intermolecular disulfide bond.[20] There are four cysteine residues in the amino terminus, and these are paired in two intramolecular disulfide bonds.[20] The status of the two cysteine residues in the carboxyl terminus is currently the subject of investigations.

Determination of the status of cysteine residues is important because a first step in the assembly of fibrillin into polymers (microfibrils) is formation of intermolecular disulfide crosslinks.[20] Therefore, identification of free cysteine residues can implicate the domains that are involved in the initial steps of microfibril assembly. At the present time, the first hybrid domain, which may mediate dimer formation,[20] and the carboxyl terminus, which may catalyze further multimerization, are candidate regions for these initial interactions.

Cysteine substitutions were discussed earlier in the context of mutations that can destabilize domain structure, rendering fibrillin-1 more susceptible to proteolysis. Alternatively, missense mutations involving either a cysteine substitution or the creation of a cysteine residue can lead to the formation of aberrant intermolecular disulfide bonds, due to an abnormal free cysteine residue. In addition, mutations of other residues that adversely affect domain folding may result in improperly free cysteine residues. The first mutation in *FBN1* identified in the Marfan syndrome (R1137P) was shown to prevent domain folding in vitro.[21] Improper domain folding could then interfere with microfibril assembly, which is dependent upon the correct formation of intermolecular disulfide bonds.

Consensus sequences for furin-type proteases are found at the amino and carboxyl termini of each fibrillin. If cleavage of the signal peptide occurs between G^{24} and $A,^{25}$ then the N-propeptide is composed of the subsequent twenty amino acid residues[20,22] In contrast, the C-propeptide contains 140 residues. The functions of the propeptides are not known. However, it is speculated that the propeptides prevent fibrillin-1 from assembling prematurely inside the cell, since cleavage appears to be required for deposition of fibrillin-1 into the extracellular matrix.[23]

In the Marfan syndrome, mutations occurring in exon 65 (W2756X; R2776X; 8236delGA; 8525del5) have been reported. Each of these mutations is predicted to truncate the C-propeptide of fibrillin-1. In addition, a mutation (124delG) affecting the N-propeptide has been identified in an individual with classic Marfan syndrome. It is currently a mystery why these mutations in *FBN1* result in the Marfan phenotype, because these mutations are not located within the functional mature fibrillin-1 molecule. Because they result in classic Marfan phenotypes, the same general mechanism initiating disease (i.e., a loss of fibrillin-1 molecules and/or microfibrils) is likely. Therefore, mutations in the propeptides of fibrillin-1 probably adversely affect intracellular folding and/or secretion of fibrillin-1 molecules. Accordingly, the function of the propeptides must be to facilitate intracellular folding and/or secretion, in addition to preventing intracellular assembly of fibrillin.

Fibrillin Microfibrils: Mutations Result in Poorly Assembled Microfibrils and/or Microfibril Instability

Fibrillin-containing microfibrils are uniform small diameter (10-12nm) fibrils that can be identified with the electron microscope. Fibrillin microfibrils are nonstriated with a characteristic hollow appearance in high resolution cross sections as well as in longitudinal sections (Fig. 3). They are found in bundles and also in association with elastic fibers. Monoclonal antibodies demonstrate that most, if not all, microfibrils in postnatal connective tissues contain fibrillin-1.[1] Fibrillin-2 is an abundant component of microfibrils in the developing fetus, but in postnatal tissues, fibrillin-2 appears to be limited in amounts and apparently restricted to certain connective tissues like peripheral nerve.[24] Fibrillin-3 has also been immunolocalized to microfibrils in developing fetal tissues.[6]

Fibrillin-1 microfibrils are ubiquitous in the connective tissue space, both in the developing fetus and in the mature adult. They are abundant in those tissues affected in the Marfan syndrome (ocular tissues, especially the ciliary zonule; skeletal tissues, especially the perichondrium/periosteum and joint capsules; cardiovascular tissues; dura; skin; lung). Fibrillin-1 microfibrils are also apparently plentiful in tissues that do not display a prominent clinical phenotype (e.g., kidney, liver, and peripheral nerve). Hence, perturbation of the intrinsic physicomechanical properties of fibrillin-1 microfibrils may not be sufficient to cause clinical phenotypes. Beyond the effects of physicomechanical defects in microfibrils, the tissue context (architectural, biomechanical, and molecular) appears to play a determining role in the phenotypic outcome.

Nevertheless, attention was first focused on assembly and architecture of fibrillin microfibrils as readouts to determine differential effects of mutations in *FBN1*. Original studies demonstrated that immunofluorescence of skin biopsies and dermal fibroblasts in culture could distinguish differences in fibrillin-1 fibril patterns in samples from individuals with the Marfan syndrome compared to unaffected controls.[25] These first studies suggested that in Marfan syndrome most mutations in *FBN1* would result in fragmentation or loss of fibrillin fibrils in skin and a failure to assemble de novo fibrillin fibrils in culture.

Pulse-chase analyses of fibrillin synthesis and deposition into the extracellular matrix over a twenty-hour period by fibroblasts demonstrated that in 20 of 25 samples from individuals with cysteine substitutions fibrillin synthesis was equivalent to control samples but matrix deposi-

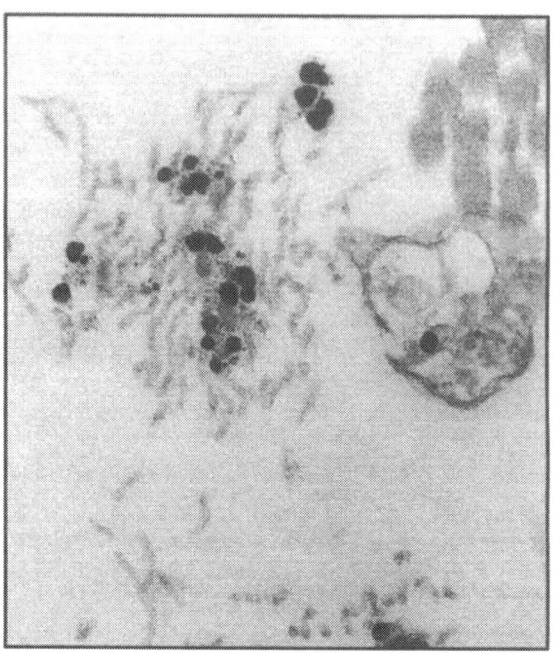

Figure 3. Fibrillin microfibrils display a characteristic uniform, small diameter and sometimes "hollow" appearance. These microfibrils are immunolabeled with a monoclonal antibody specific for one of the fibrillins.

tion was substantially lower.[26] In this study, 5 of the 25 samples showed reduced fibrillin synthesis and reduced or substantially lower matrix deposition.[26] These data, in addition to results from previous studies,[27] suggest that reduced synthesis, delayed secretion, and/or defective matrix deposition of fibrillin may underlie the failure of Marfan fibroblasts to assemble fibrillin fibrils in the immunofluorescence assay. Each of these factors could reduce the available fibrillin protein concentration to levels below a critical concentration required for fibril assembly in the immunofluorescence assay.

Neither of these assays directly addresses whether mutated fibrillin-1 is assembled into microfibrils in individuals with the Marfan syndrome. Each assay focuses on events occurring within a defined time period (20 hours in the pulse-chase assay and 3-4 days in the immunofluorescence assay). It is likely that both fibrillin protein concentration as well as required cellular factors will dictate microfibril assembly in a cell-specific manner.[28] Therefore, it is difficult to predict from these limited data whether in vivo and in which tissues mutated fibrillin-1 is incorporated into microfibrils (Fig. 1A) or interferes with microfibril assembly (Fig. 1B).

Studies have been performed to investigate whether structural differences can be observed in microfibrils extracted from Marfan fibroblast cultures compared to control fibroblast cultures.[29,30] These studies focused on finished products (extracted beaded string structures visualized by rotary shadowing and electron microscopy) rather than on the assembly process. Presumably, any morphological differences observed between Marfan microfibrils and control microfibrils could be the result of incorporation of mutated fibrillin-1. Only a few of these studies were performed, so visible differences in microfibril morphology were not systematically correlated with genotype. Moreover, these studies did not distinguish morphological differences derived during microfibril assembly from those produced after microfibril assembly by proteases. Nevertheless, visualization of rotary shadowed Marfan microfibrils, the end products of long term cultures, demonstrated that, even though almost all Marfan fibroblast cultures failed to elaborate a fibrillin fibril network during the standardardized 3-4 day immunofluorescence assay,[25] some Marfan fibroblast cultures were able to elaborate microfibrils over a longer time period.

Why is it important to determine whether mutations in fibrillin-1 result in defective microfibril assembly or in assembled microfibrils that are sensitive to proteases? It is generally thought that mutations that disrupt or impair microfibril assembly will lead to severe disease. Hence, it was proposed that the "neonatal" region of fibrillin-1, where mutations causing neonatal Marfan syndrome are clustered, is critical for microfibril assembly.[31] On the other hand, mutations that do not impair microfibril assembly but confer increased sensitivity to proteases may result in classic Marfan syndrome. These hypotheses have not been fully tested, even though the outcome of these investigations could contribute significantly to the development of biochemical diagnostics for the Marfan syndrome.

In order to test whether proteolysis contributes to the pathogenesis of the Marfan syndrome, we have developed a sandwich ELISA for quantitating fibrillin-1 fragments in blood. We have assessed different capture and detector antibody pairs (directed toward different epitopes in fibrillin-1) in order to optimize the assay and to compare values obtained from standard curves. Preliminary results indicate that low values of fibrillin-1 are found in nonMarfan blood samples, allowing for the potential utility of this assay. Our goal is to determine whether the presence of fibrillin-1 fragments in blood samples will correlate positively with the Marfan syndrome and disease severity and whether the amounts of fibrillin-1 fragments measured over time will correlate positively with disease progression. This blood test may then be an additional tool for the clinical management of the Marfan syndrome. In addition, these investigations may confirm biochemical studies indicating that mutations in fibrillin-1 predispose microfibrils to proteolytic fragmentation and would encourage research to move in the direction of identifying the relevant proteases and developing strategies to inhibit their destructive activities.

Microfibrils and Morphogenesis: Effects of Fibrillin Mutations on Growth Factors

The fibrillins are related to the latent TGFβ binding proteins (LTBPs), a family of four molecules that consist of similar types of domain modules as those found in the fibrillins (see Fig. 4). The LTBPs are smaller than fibrillins, and unlike fibrillins, they are variable in size. Each lacks a carboxyl terminus homologous to the fibrillins. The sequence of LTBP-1 was first published in 1990,[32] just as we were preparing the sequence of fibrillin-1. We first identified the homologies between fibrillin and LTBP-1[11,12] and distinguished the seven modules containing 8 cysteine residues as "8-cys modules" from the "hybrid" modules.[12] No other molecules containing 8-cys domain modules have been identified in the human genome, suggesting that fibrillins and LTBPs perform related functions. There are 33 8-cys domain modules contributed to the human genome by the three fibrillins and four LTBPs.

An important function of 8-cys modules is to mediate interactions between LTBPs/fibrillins and members of the TGFβ super family of growth factors. One of the three 8-cys modules in LTBP-1 has been shown to interact with the propeptide of TGFβ.[33,34] The homologous 8-cys module in LTBP-3 and LTBP-4 also interacts with TGFβ.[35] Since fibrillins each contain seven 8-cys modules, an hypothesis based upon molecular structure would predict that fibrillins interact with members of the TGFβ super family. It was found that fibrillins do not bind to TGFβ.[35] Therefore, other members of the TGFβ super family should be considered candidate ligands.

Inactivating the fbn2 gene in mice resulted in syndactyly and located a place for fibrillin-2 in the BMP signaling pathway that regulates digit formation in the developing autopod.[36] The role of fibrillin-2 in this signaling pathway has been determined biochemically. Current investigations in our laboratory demonstrate that at least BMP-7, and probably other BMPs, are targeted to the extracellular matrix through interactions with fibrillins.[37]

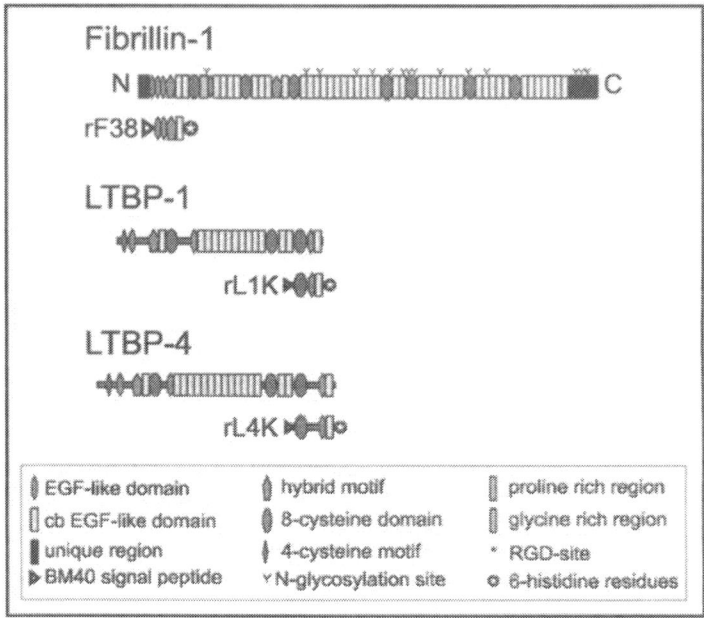

Figure 4. Schematic drawings of fibrillin-1, LTBP-1, and LTBP-4, depicting the presence of similar domain modules. Recombinant polypeptides (rF38, rL1-K, rL4-K) containing the fibrillin/LTBP binding sites are shown.

LTBP-1[38,39] and LTBP-2[40] were immunolocalized to fibrillin-containing microfibrils. Biochemical and immunochemical studies indicated that LTBP-1 is associated with fibrillin microfibrils but is not a principal component of extracted "beads-on-a-string" microfibrils.[41] Binding assays showed that both LTBP-1 and LTBP-4 interact with fibrillins with high affinity.[41] Fibrillin binding sites are present in the C-terminal ends of LTBP-1 (rL1-K) and LTBP-4 (rL4-K), and the LTBP binding site in fibrillin-1 is within rF38 (Fig. 4). Based upon these conclusions, we proposed a model wherein LTBPs are associated with fibrillin microfibrils through an interaction between the C-terminal end of LTBPs and a specific site close to the N-terminal end of fibrillin (Fig. 5). This model implies that interactions between LTBPs and fibrillins help to target and stabilize latent TGFβ complexes in the extracellular matrix. We predict that this model applies to the full spectrum of tissue-specific arrays of microfibrils (polymerized fibrillins, whether these are fibrillin-1, -2, or -3, with associated LTBPs).

Our investigations have led to the following overall working hypothesis: extracellular microfibrils, composed of fibrillins with associated LTBPs, form tissue-specific, temporally determined information highways along which growth factor signals are embedded. The three different fibrillins and four different LTBPs found in humans provide enough temporal and spatial diversity to accommodate a variety of different signals important to embryogenesis, growth and repair processes.

In vivo evidence that disease pathogenesis in the Marfan syndrome may involve dysregulation of growth factors was provided by analyses of fibrillin-1 deficient mice.[42] These studies demonstrated that activation of TGFβ resulted in a lung phenotype similar to an emphysema-like condition found in individuals with the Marfan syndrome. However, the lung phenotype in fibrillin-1 deficient mice was shown to be a developmental failure of distal alveolar septation, rather than a destructive process.[42] A plausible mechanism for the observed phenotype in the lung is that loss of fibrillin-1 (and therefore loss of the LTBP binding site) resulted in inappro-

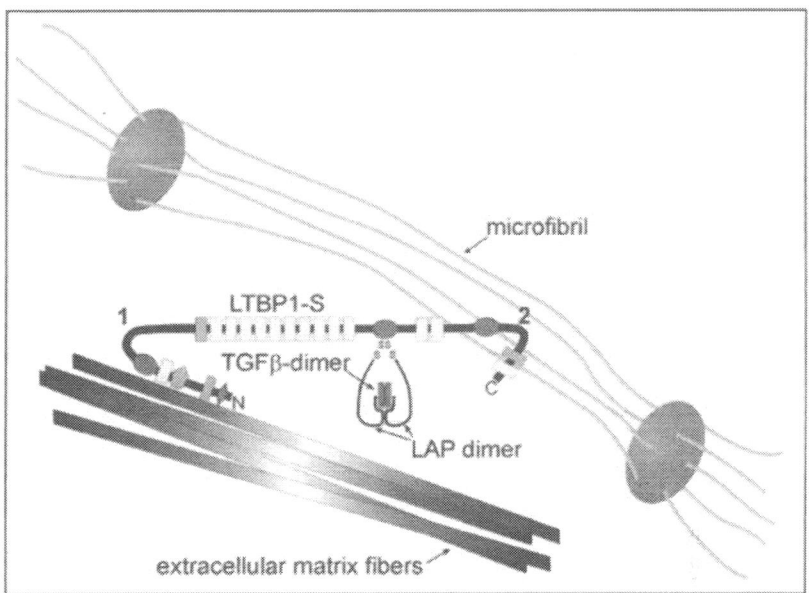

Figure 5. Model showing fibrillin microfibrils with associated LTBP. Interactions between LTBPs and fibrillins may sequester and stabilize latent TGFβ complexes.

priate sequestration of latent TGFβ complexes, which then led to abnormal activation of TGFβ. However, this potential mechanism must be tested.

Is abnormal activation of TGFβ responsible for some of the features in individuals with the Marfan syndrome? In general, mutations in *FBN1* result in fragmentation and loss of fibrillin-1 microfibrils, a condition modeled in fibrillin-1 deficient mice. However, among the 421 novel mutations in *FBN1* in the Marfan database,[43] there is one mutation that strongly implicates our proposed molecular mechanism for dysregulation of TGFβ signaling and for pathogenesis in the Marfan syndrome. A missense mutation (N164S) in *FBN1* was reported in a family with dominant ectopia lentis and skeletal features.[44] Unlike most other *FBN1* mutations, this mutation does not substitute a residue predicted to be important for domain folding, nor does it affect a domain thought to be critical for microfibril assembly. Therefore, N164S does not fit into our current understanding of how mutations in *FBN1* initiate disease pathogenesis. However, based on our unpublished data that further define the LTBP binding site in fibrillin-1, we hypothesize that this mutation affects the LTBP binding site, resulting in a lower affinity interaction between fibrillin-1 and LTBP. A decreased ability to bind to the large latent TGFβ complex may be sufficient in this case to cause disease. This hypothesis is under investigation. If this hypothesis is true, then there is already human genetic evidence supporting the general concept that dysregulation of growth factor signaling plays an important role in the pathogenesis of the Marfan syndrome.

New Research Opportunities Relevant to the Marfan Syndrome

Opportunities exist to translate basic research discoveries into treatment strategies and aids for improved clinical management for the Marfan syndrome. In this chapter, I have reviewed important progress that now makes translational research opportunities possible to consider. Two questions open for investigation are: (1) can specific protease inhibitors prevent disease progression in the Marfan syndrome? and (2) can growth factor agonists or antagonists modify disease progression in the Marfan syndrome? Important unanswered questions were also discussed, and progress in these areas may afford additional opportunities. Future research will be directed toward how fibrillins interact with and modulate cellular behavior, not only through growth factor regulation; which fibrillins perform overlapping or unique functions in various tissues; and how fibrillins are themselves regulated. As for the progress made over the last decade, research in these areas will allow us to develop more opportunities to better treat and care for individuals with the Marfan syndrome.

Acknowledgements

I thank the following individuals who have made fundamental contributions over many years: Doug Keene, Hans Peter Bächinger, Noe Charbonneau, Glen Corson, and Rob Ono. Several former postdocs (Dieter Reinhardt, Zenzo Isogai, and Kate Gregory) deserve thanks for their ideas and diligence, as does Helena Hessle, who has developed the sandwich ELISA described briefly in this chapter. I also thank Susan Hayflick who has helped both conceptually and materially with the development of the Marfan blood test. Work described in this chapter was supported by grants from the Shriners Hospitals for Children, the National Institutes of Health, and the National Marfan Foundation.

References

1. Sakai LY, Keene DR, Engvall E. Fibrillin, a new 350kD glycoprotein, is a component of extracellular microfibrils. J Cell Biol 1986; 103:2499-2509.
2. Dietz HC, Cutting GR, Pyeritz RE et al. Marfan syndrome caused by a recurrent de novo missense mutation in the fibrillin gene. Nature 1991; 352:337-339.
3. Lee B, Godfrey M, Vitale E et al. Linkage of Marfan syndrome and a phenotypically related disorder to two different fibrillin genes. Nature 1991; 52:330-334.
4. Zhang H, Apfelroth SD, Hu W et al. Structure and expression of fibrillin-2, a novel microfibrillar component preferentially located in elastic matrices. J Cell Biol 1994; 24:855-863.
5. Nagase T, Nakayama M, Nakajima D et al. Prediction of the coding sequences of unidentified human genes. XX. The complete sequences of 100 new cDNA clones from brain which code for large proteins in vitro. DNA Res 2001; 8:85-95.
6. Corson GM, Charbonneau NL, Keene DR et al Differential expression of fibrillin-3 adds to microfibril variety in human and avian, but not rodent, connective tissues. Genomics 2004; in press.
7. Putnam EA, Zhang H, Ramirez F et al. Fibrillin-2 (FBN2) mutations result in the Marfan-like disorder, congenital contractural arachnodactyly. Nat Genet 1995; 11:456-458.
8. Faivre L, Megarbane A, Alswaid A et al. Homozygosity mapping of a Weill-Marchesani syndrome locus to chromosome 19p13.3-p13.2. Hum Genet 2002; 110:366-370.
9. Faivre L, Gorlin RJ, Wirtz MK et al. In frame fibrillin-1 gene deletion in autosomal dominant Weill-Marchesani syndrome. J Med Genet 2003; 40:34-6.
10. Wirtz MK, Samples JR, Kramer PL et al. Weill-Marchesani syndrome—possible linkage of the autosomal dominant form to 15q21.1. Am J Med Genet 1996; 65:68-75.
11. Maslen CL, Corson GM, Maddox BK et al. Partial sequence of a candidate gene for the Marfan syndrome. Nature 1991; 352:334-337.
12. Corson GM, Chalberg SC, Dietz HC et al. Fibrillin binds calcium and is coded by cDNAs that reveal a multidomain structure and alternatively spliced exons at the 5' end. Genomics 1993; 17:476-484.
13. Pereira L, D'Alessio M, Ramirez F et al. Genomic organization of the sequence coding for fibrillin, the defective gene product in Marfan syndrome. Hum Mol Genet 1993; 2:961-8.
14. Sakai LY, Keene DR, Glanville RW et al. Purification and partial characterization of fibrillin, a cysteine-rich structural component of connective tissue microfibrils. J Biol Chem 1991; 266:14763-14770.
15. Downing AK, Knott V, Werner JM et al. Solution structure of a pair of calcium-binding epidermal growth factor-like domains: Implications for the Marfan syndrome and other genetic disorders. Cell 1996; 85:597-605.
16. Reinhardt DP, Mechling DE, Boswell BA et al. Calcium determines the shape of fibrillin. J Biol Chem 1997; 272:7368-73.
17. Reinhardt DP, Ono RN, Sakai LY. Calcium stabilizes fibrillin-1 against proteolytic degradation. J Biol Chem 1997; 272:1231-6.
18. Reinhardt DP, Ono RN, Notbohm H et al. Mutations in calcium-binding epidermal growth factor modules render fibrillin-1 susceptible to proteolysis. A potential disease-causing mechanism in Marfan syndrome. J Biol Chem 2000; 275:12339-45.

19. Booms P, Tiecke F, Rosenberg T et al. Differential effect of FBN1 mutations on in vitro proteolysis of recombinant fibrillin-1 fragments. Hum Genet 2000; 107:216-24.
20. Reinhardt DP, Gambee JE, Ono RN et al. Initial steps in assembly of microfibrils. Formation of disulfide-cross-linked multimers containing fibrillin-1. J Biol Chem 2000; 275:2205-10.
21. Wu YS, Bevilacqua VL, Berg JM. Fibrillin domain folding and calcium binding: Significance to Marfan syndrome. Chem Biol 1995; 2:91-7.
22. Ritty TM, Broekelmann T, Tisdale C et al. Processing of the fibrillin-1 carboxyl-terminal domain. J Biol Chem 1999; 274:8933-40.
23. Milewicz DM, Grossfield J, Cao SN et al. A mutation in FBN1 disrupts profibrillin processing and results in isolated skeletal features of the Marfan syndrome. J Clin Invest 1995; 95:2373-8.
24. Charbonneau NL, Dzamba BJ, Ono RN et al. Fibrillins can coassemble in fibrils, but fibrillin fibril composition displays cell-specific differences. J Biol Chem 2003; 278:2740-2749.
25. Hollister DW, Godfrey M, Sakai LY et al. Immunohistologic abnormalities of the microfibrillar fiber system in the Marfan syndrome. N Engl J Med 1990; 323:152-159.
26. Schrijver I, Liu W, Brenn T et al. Cysteine substitutions in epidermal growth factor-like domains of fibrillin-1: Distinct effects on biochemical and clinical phenotypes. Am J Hum Genet 1999; 65:1007-20.
27. Aoyama T, Francke U, Dietz HC et al. Quantitative differences in biosynthesis and extracellular deposition of fibrillin in cultured fibroblasts distinguish five groups of Marfan syndrome patients and suggest distinct pathogenetic mechanisms. J Clin Invest 1994; 94:130-7.
28. Dzamba BJ, Keene DR, Isogai Z et al. Assembly of epithelial cell fibrillins. J Invest Dermatol 2001; 117:1612-1620.
29. Kielty CM, Shuttleworth CA. Abnormal fibrillin assembly by dermal fibroblasts from two patients with Marfan syndrome. J Cell Biol 1994; 124:997-1004.
30. Kielty CM, Rantamaki T, Child AH et al. Cysteine-to-arginine point mutation in a 'hybrid' eight-cysteine domain of FBN1: Consequences for fibrillin aggregation and microfibril assembly. J Cell Sci 1995; 108:1317-23.
31. Liu W, Qian C, Comeau K et al. Mutant fibrillin-1 monomers lacking EGF-like domains disrupt microfibril assembly and cause severe marfan syndrome. Hum Mol Genet 1996; 5:1581-7.
32. Kanzaki T, Olofsson A, Moren A et al. TGF-beta 1 binding protein: A component of the large latent complex of TGF-beta 1 with multiple repeat sequences. Cell 1990; 61:1051-61.
33. Saharinen J, Taipale J, Keski-Oja J. Association of the small latent transforming growth factor-beta with an eight cysteine repeat of its binding protein LTBP-1. EMBO J 1996; 15:245-53.
34. Gleizes PE, Beavis RC, Mazzieri R et al. Identification and characterization of an eight-cysteine repeat of the latent transforming growth factor-beta binding protein-1 that mediates bonding to the latent transforming growth factor-beta1. J Biol Chem 1996; 271:29891-6.
35. Saharinen J, Keski-Oja J. Specific sequence motif of 8-Cys repeats of TGF-beta binding proteins, LTBPs, creates a hydrophobic interaction surface for binding of small latent TGF-beta. Mol Biol Cell 2000; 11:2691-704.
36. Arteaga-Solis E, Gayraud B, Lee SY et al. Regulation of limb patterning by extracellular microfibrils. J Cell Biol 2001; 154:275-81.
37. Isogai Z, Gregory K, Ono RN et al. Microfibrils and morphogenesis. In: Tamburro AM, Pepe A, eds. Elastin 2002 Potenza. Italy: EditricErmes, 2003:213-223.
38. Dallas SL, Miyazono K, Skerry TM et al. Dual role for the latent transforming growth factor-beta binding protein in storage of latent TGF-beta in the extracellular matrix and as a structural matrix protein. J Cell Biol 1995; 131:539-49.
39. Dallas SL, Keene DR, Bruder SP et al. Role of the latent transforming growth factor beta binding protein 1 in fibrillin-containing microfibrils in bone cells in vitro and in vivo. J Bone Miner Res 2000; 15:68-81.
40. Gibson MA, Hatzinikolas G, Davis EC et al. Bovine latent transforming growth factor beta 1-binding protein 2: Molecular cloning, identification of tissue isoforms, and immunolocalization to elastin-associated microfibrils. Mol Cell Biol 1995; 15:6932-42.
41. Isogai Z, Ono RN, Ushiro S et al. Latent transforming growth factor beta-binding protein 1 interacts with fibrillin and is a microfibril-associated protein. J Biol Chem 2003; 278(4):2750-7.
42. Neptune ER, Frischmeyer PA, Arking DE et al. Dysregulation of TGF-beta activation contributes to pathogenesis in Marfan syndrome. Nat Genet 2003; 33:407-11.
43. Collod-Beroud G, Le Bourdelles S, Ades L et al. Update of the UMD-FBN1 mutation database and creation of an FBN1 polymorphism database. Hum Mutat 2003; 22:199-208.
44. Comeglio P, Evans AL, Brice G et al. Identification of FBN1 gene mutations in patients with ectopia lentis and marfanoid habitus. Br J Ophthalmol 2002; 86:1359-62.

CHAPTER 15

Insights into Fibrillin-1 Structure and Function from Domain Studies

Pat Whiteman and Penny A. Handford

Structure of Fibrillin-1 Domains

Introduction

Structural information is required to understand the assembly of fibrillin-1 into 10-12 nm microfibrils and to gain insight into the consequences of Marfan syndrome (MFS)-causing mutations. Since fibrillin-1 is a modular protein (Fig. 1), a dissection approach has been used to generate structural information for individual or small numbers of domains, where an analysis of the complete protein is unlikely to be feasible due to its physicochemical properties (size, disulphide-rich, post-translational modifications, rapid macromolecular association). From these data, one can begin to produce a homology model of fibrillin-1.

The two predominant domain types in fibrillin-1 are the calcium-binding epidermal growth factor-like domain (cbEGF) and the transforming growth factor β binding protein-like (TB or 8-cysteine) domain. The properties of individual cbEGF domains (Fig. 2) have been well studied previously due to their widespread distribution amongst extracellular and transmembrane proteins. A common feature of these domains is their involvement in protein-protein interactions. They contain a calcium-binding consensus sequence which is defined as D/N-X-D/N-E/Q-X_m-D*/N*-X_n-Y/F where m and n can be a variable number of residues and * denotes a potentially β-hydroxylated residue.[1] These residues, located in the N-terminal region of the cbEGF domain, donate side-chain oxygen ligands to Ca^{2+} or stabilise the Ca^{2+}-binding pocket by hydrogen bonding or hydrophobic interactions.[2] In contrast to the cbEGF domain, the TB domain has a much more restricted distribution and is found only in members of the fibrillin/LTBP (latent transforming growth factor β binding protein) family of proteins. Since these two domain types comprise the majority of fibrillin-1 structure and most of MFS-causing missense mutations affect one or other domain type, studies have focussed on identifying their contribution to native fibrillin-1 structure.

A prokaryotic (E. coli) expression system has been employed to produce the large quantities of protein (~10 mg) required for high resolution structural studies.[3] Use of a 6xHis-tag sequence at the N-terminus of each domain construct has facilitated purification by Ni^{2+} affinity chromatography. The intracellular environment of E. coli is reducing, and since both the cbEGF and TB domains require oxidation to form their native structures which are stabilised by disulphide bridges, peptides purified from this prokaryotic source have been reduced and subsequently oxidised using an in vitro refolding procedure. In addition, because these fragments are expressed in E.coli they lack the post translational modifications associated with these domains in vivo (N- and O-linked glycosylation, β-hydroxylation).

High resolution structures of fibrillin-1 domain fragments have been solved using either X-ray crystallography or nuclear magnetic resonance (NMR), the latter identifying the solu-

Marfan Syndrome: A Primer for Clinicians and Scientists, edited by Peter N. Robinson and Maurice Godfrey. ©2004 Eurekah.com and Kluwer Academic / Plenum Publishers.

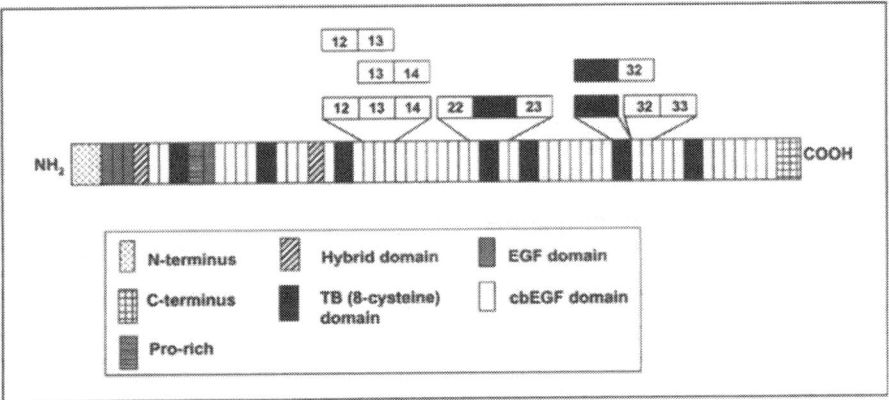

Figure 1. Schematic illustration of the module organisation of fibrillin-1 showing domain fragments referred to in the text.

tion structure. The solved structures comprise the cbEGF32-33 and cbEGF12-13 domain pairs, TB6, and most recently cbEGF22-TB4-cbEGF23 (Fig. 1). These structures provide insights into the three dimensional shape of human fibrillin-1.

cbEGF32-33 and cbEGF12-13

The NMR structure of the cbEGF 32-33 domain pair was the first high resolution structure of a fibrillin-1 domain fragment to be obtained.[4] Each cbEGF domain within the pair displayed the characteristic EGF fold, comprising a major and minor β-sheet stabilised by three disulphide bonds with a 1-3, 2-4, 5-6 arrangement. In the presence of Ca^{2+}, the two cbEGF domains are organised in a near-linear rod-like arrangement, stabilised by interdomain calcium binding and hydrophobic interactions between Y2157 in cbEGF32 packing against G2186 and its neighbour I2185 in cbEGF33 (Fig. 3a). The amino acid residues involved in stabilisation of the domain interface are highly conserved in other cbEGF pairs within fibrillin-1 suggesting that other tandem cbEGF repeats within the molecule adopt a similar orientation. The recent identification of the solution structure of cbEGF12-13, which adopts a similar calcium-dependent conformation to cbEGF32-33, has supported this hypothesis.[5] Homology modelling of multiple tandem repeats of cbEGFs suggests that, in the presence of Ca^{2+}, these form extended structures within the native protein. This is supported by the change in microfibril architecture that occurs on removal of Ca^{2+} by chemical chelation.[6,7]

TB6

The first TB structure to be determined was that of TB6.[8] The solution structure identified a novel globular fold comprising six antiparallel β-strands and two α-helices that are stabilised by hydrophobic interactions and four disulphide bonds (Fig. 3b). The cysteine residues are paired in a 1-3, 2-6, 4-7, 5-8 arrangement. An unusual Cys triplet is localised to the hydrophobic core, and two salt bridges formed from conserved residues within the TB domain consensus occur on the surface of the structure (D2055 - R2057 and E2097 - K2080). Like the cbEGF domain, it is clear that the TB domain can participate in protein-protein interactions since a functional integrin binding "RGD" motif is located within TB4 (see below)[9,10] and a TB domain from LTBP1 mediates a covalent interaction with the small latent complex TGFβ-LAP.[11,12]

cbEGF22-TB4-cbEGF23

The most recent structure obtained for fibrillin-1 is that of a cbEGF22-TB4-cbEGF23 triple domain fragment.[13] This was obtained by X-ray crystallography and identifies a

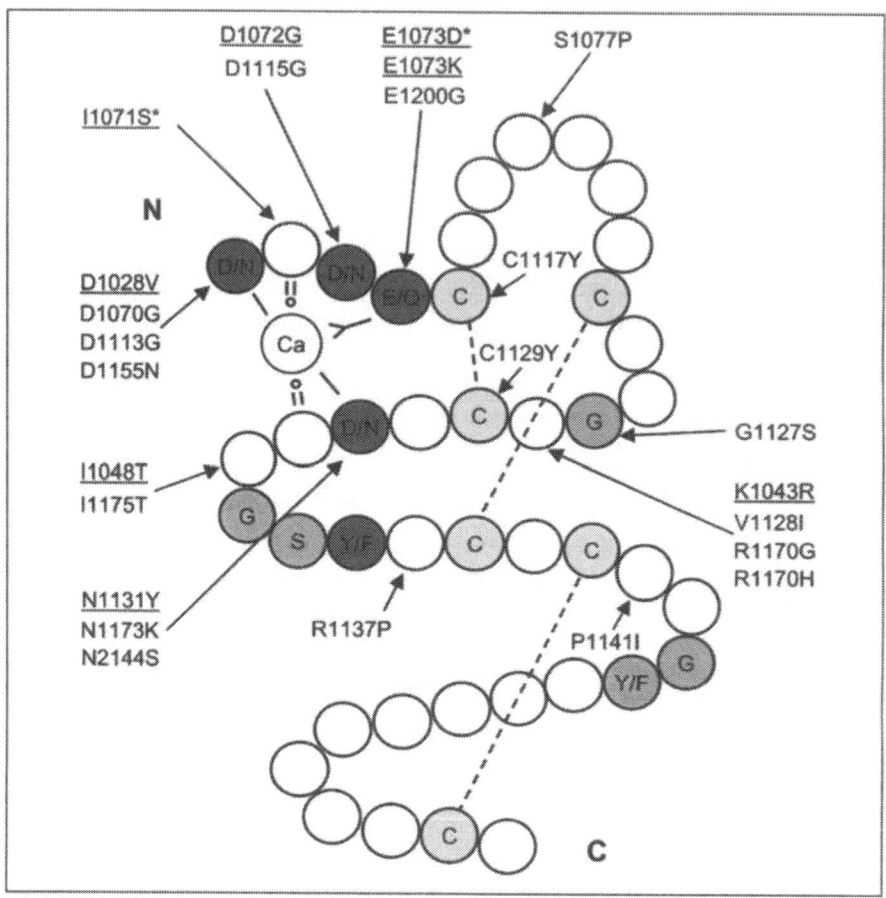

Figure 2. The secondary structure and consensus sequence of the cbEGF domain. Conserved cysteine residues are in yellow and the calcium-binding consensus residues in red. Ligands predicted to bind directly to calcium are indicated. Hydrophobic residues involved in stabilising inter-domain interactions are in blue and other conserved residues in green. The missense mutations discussed in the text and mapped onto Figure 5 are indicated. Those associated with neonatal MFS (nMFS) are underlined. A double mutation is indicated by asterisks.

calcium-stabilised tetragonal pyramidal conformation (Fig. 3c). The "RGD" integrin binding site localises, as predicted from the structure of TB6, to the tip of a β-sheet and is thus accessible to cell-surface integrins. Comparative sequence alignments of the linker regions from cbEGF-TB domains within fibrillin-1 suggest that the relative orientation of cbEGF22-TB4 is likely to be preserved at homologous sites within fibrillin-1. In contrast, the variation in amino acid number and composition of TB-cbEGF linker sequences suggests that these pairs will adopt different orientations with respect to one another within fibrillin-1, and may contribute to the biomechanical properties of microfibrils.

From these data, homology modelling has been used to generate a structural model of a large region of fibrillin-1 (cbEGF 11-TB5). The model suggests that although the protein is in an extended conformation, it is not simply linear. A significant bend is introduced by the packing of cbEGF22 against TB4.[13] Information such as this will allow more precise models of fibrillin organisation within microfibrils to be constructed.

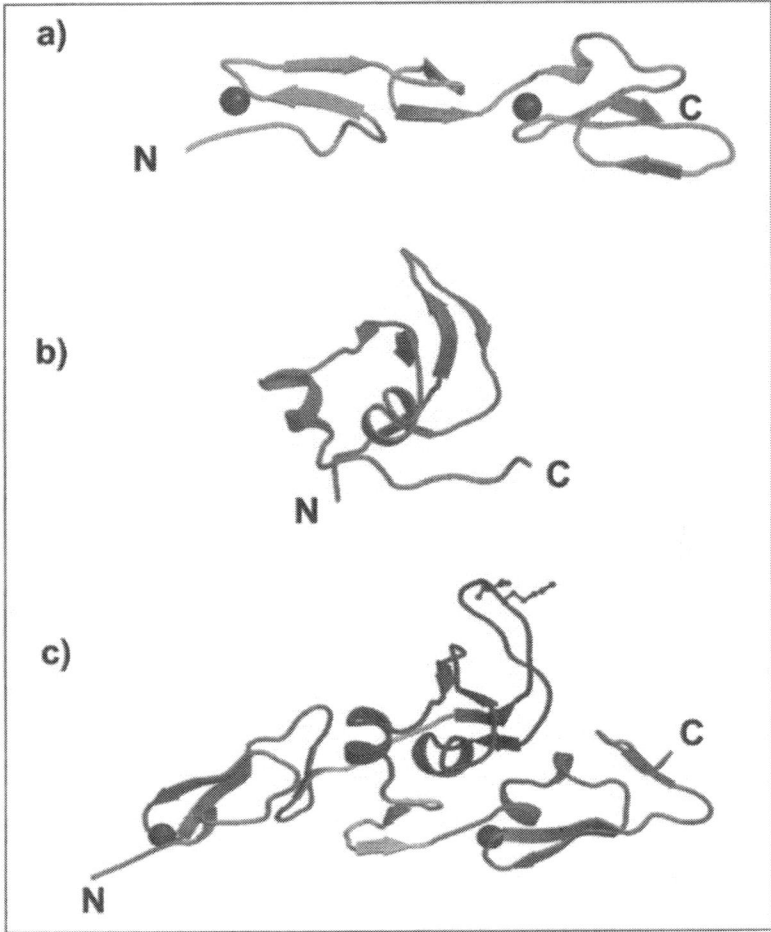

Figure 3. Structures of fibrillin-1 domain fragments: a) cbEGF32-33, b) TB6, c) cbEGF22-TB4-cbEGF23. The cbEGF domains are coloured green, TB domains blue and Ca^{2+} ions red. The position of the RGD motif in TB4 is indicated. Figures were produced using Bobscript[26] and rendered with Raster3D.[27]

Calcium Binding Properties of Fibrillin-1

Different methods have been used to measure the calcium-binding properties of cbEGF domain fragments from fibrillin-1. These include NMR, intrinsic protein fluorescence and equilibrium dialysis. NMR allows assignment of a calcium-binding site to a specific domain and has been used to analyse the effects of different domain linkages on calcium-binding affinity. It is known that single cbEGF domains expressed in isolation from fibrillin-1 and other proteins display low affinity binding in the mM range. However in fibrillin-1, and many other proteins, the cbEGF domains are often arranged as repeating tandem arrays. On covalent linkage of an N-terminal cbEGF, the affinity of the C-terminal cbEGF increases.[14] The bound calcium, together with the hydrophobic packing interaction, performs a key structural role in restricting interdomain flexibility[4,15] and therefore also protects the modules against proteolytic cleavage.[16,17] Dynamics studies show that the most stable region of a cbEGF pair is in the vicinity of the interdomain calcium-binding site.[15] Analysis of different cbEGF domain pairs has however identified a range of affinities associated with the C-terminal domain from 350 μM (cbEGF32-33) to < 30 μM (cbEGF12-13), suggesting that primary sequence variation, in addition to the pairwise domain interaction, must also influence affinity. Similar variation appears to exist within heterologous cbEGF domain pairs since TB6-cbEGF32 was observed to have an affinity of 1.6 mM,[18] while preliminary data indicate that the affinity of TB4-cbEGF23 is at least two orders of magnitude higher. These data suggest that under physiological conditions of I=0.15 and $[Ca^{2+}]_{free}$ ~1.5 mM, some of the fibrillin-1 cbEGF domains will not be fully saturated and may impart flexibility/extensibility to the native protein. These properties may be important for protein-protein interactions involved in higher order assembly or for biomechanical function within the specific tissues that contain fibrillin microfibrils. Further examination of other homologous and heterologous cbEGF domain pairs will be required to reveal the extent of variation in affinity.

Structural Consequences of *FBN1* Mutations

The availability of structural knowledge of the cbEGF and TB domain types has provided the basis on which to study the consequences of disease-causing *FBN1* mutations and understand their role in the pathogenesis of MFS and related disorders. Missense mutations represent the majority of identified *FBN1* mutations and most of these affect one of the 43 cbEGF modules of fibrillin-1. Consequently most studies have focussed on understanding their effects on cbEGF structure and function. A significant number of these mutations lead to a substitution of one of the six highly conserved cysteine residues hence disrupting one of the three disulphide bonds of the affected domain. These missense mutations are therefore predicted to cause domain misfolding, those that result in the introduction of extra cysteines are likely to have similar effects. The second most common type of missense mutation affects residues of the calcium-binding consensus sequence and thereby reduces the calcium-binding affinity of the affected domain.

Mutations which lead to substitution of glycine or proline residues adjacent to one of the six conserved cysteine residues have been reported. Glycine residues have greater conformational freedom than other amino acids while the structure of proline renders it the least flexible, consequently substitution of these residues would also be predicted to have effects on domain structure. These and other conformational mutations may compromise the fold of the cbEGF domain without disrupting the native arrangement of disulphide bonds and could therefore affect inter- or intra-domain interactions.

Other mutations causing amino acid substitutions with no obvious structural significance have been reported. These may be involved in intermolecular protein-protein interactions, either within the microfibril or with other components in the matrix.

Calcium Binding As a Marker of Structural Integrity—NMR Studies

NMR methodology is particularly suited to the study of the effects of disease-causing mutations in single and multidomain constructs. Following the determination of calcium-binding affinities for wild-type cbEGF domains, the effects of missense mutations that alter predicted ligands can be investigated. This provides information on the contribution of the ligand to calcium binding and identifies any effect that its substitution has on the global fold of the domain. Reduction of calcium binding caused by substitution of a calcium ligand would be predicted to destabilise the interdomain linkage and produce a less extended, more flexible structure within a region of fibrillin-1. This may result in increased proteolytic susceptibility and/or distort potential protein binding interfaces.

The heterologous domain pair TB6-cbEGF32 has been found by NMR analysis to have a calcium-binding site of low affinity (see above). This suggests that the linkage between TB6 and cbEGF32 is flexible and that pairwise interactions between TB6 and cbEGF32 are weak or absent. The structural effects of a calcium-binding substitution N2144S which is located on a β-strand in the N-terminal region of cbEGF32 have been determined. The substitution did not affect the structure of the mutant domain, or the adjacent N-terminal TB6 or C-terminal cbEGF33 when analysed in domain pairs. However, the affinity for calcium was decreased nine-fold in a TB6-cbEGF32 pair while calcium binding in the C-terminal cbEGF33 in a cbEGF 32-33 pair was unaffected.[18] The structural effect of the N2144S substitution thus appears to be localised.

Limited Proteolysis Studies of Domain Fragments

The observation that *FBN1* missense mutations often increase the susceptibility of fibrillin-1 peptides to proteolysis in the presence of Ca^{2+} has provided a useful tool with which to probe structural effects of both folding and calcium-binding substitutions. Analyses of the proteolytic degradation products obtained from digests of recombinant fibrillin-1 fragments have suggested that enzyme-specific cryptic cleavage sites are exposed when the calcium-binding properties are altered by the presence of a missense mutation. This is also observed in wild-type cbEGF domain fragments treated with a chelator such as EDTA/EGTA prior to protease digestion, and is presumably due to an increase in conformational flexibility. Identification of how far reaching the loss of calcium-dependent protection is, by mapping the sites of cleavage onto a three-dimensional model of the domain fragments, can be used to assess the short versus long range consequences of different mutations. The specific effect of a missense mutation on protease susceptibility of a cbEGF domain can be influenced by a number of factors such as the particular residue mutated and the position of the mutant domain within the fibrillin-1 peptide. Although a low resolution method, proteolysis can be used to probe structural effects of mutations in relatively large fragments which are not readily amenable to NMR analysis. Further, since it requires only small amounts of material it can be more easily applied to a number of different mutations.

Folding Substitutions

The structural consequences of a G1127S substitution in cbEGF13, which is associated with familial ascending aortic aneurysm, have been investigated by a combination of both NMR and limited proteolysis. NMR analysis of a peptide containing the G1127S substitution demonstrated that it disrupted the folding of cbEGF13. This is most likely due to the exchange of glycine for a less flexible residue at the start of the major β-hairpin.[19] Its localised consequences were demonstrated by lack of interference of the folding of the adjacent cbEGF12 or cbEGF14 in cbEGF12-13 and cbEGF13-14 domain pairs. In the mutant cbEGF12-13 pair, cbEGF13 retained the ability to bind calcium suggesting that the domain preserves a 'native-like' fold. This was supported by a more detailed NMR study of the mutant and wild-type cbEGF12-13 pair which showed only minor structural differences underscoring the subtlety of the substitution.[20]

The short-range effect was confirmed by protease digestion assays of cbEGF12-13 and also a cbEGF12-14 triple construct. In the mutant pair both cbEGF12 and 13 retained similar calcium-binding properties and thus tertiary structure to the wild type domain pair. In addition, all identified cleavage sites (determined by N-terminal sequence analysis of purified digestion products) showed calcium-dependent protection from proteolysis. These results were consistent with the NMR data for this domain pair described above and suggested that the mutant domain has suffered some disruption but is not severely misfolded. Additional cleavage sites identified in cbEGF12-14 G1127S triple construct indicated further subtle changes within the mutant domain but not the flanking domains. Calcium-dependent protection from proteolysis of all cleavage sites including those in mutant cbEGF13 was observed. The SDS-PAGE analysis (Fig. 4a) of the tryptic digests of the mutant cbEGF12-14 triple construct illustrates the protection against digestion conferred by Ca^{2+}. This correlation between the NMR analyses and protease digestion assays demonstrates the usefulness of proteolysis for probing the structural effects of mutations.

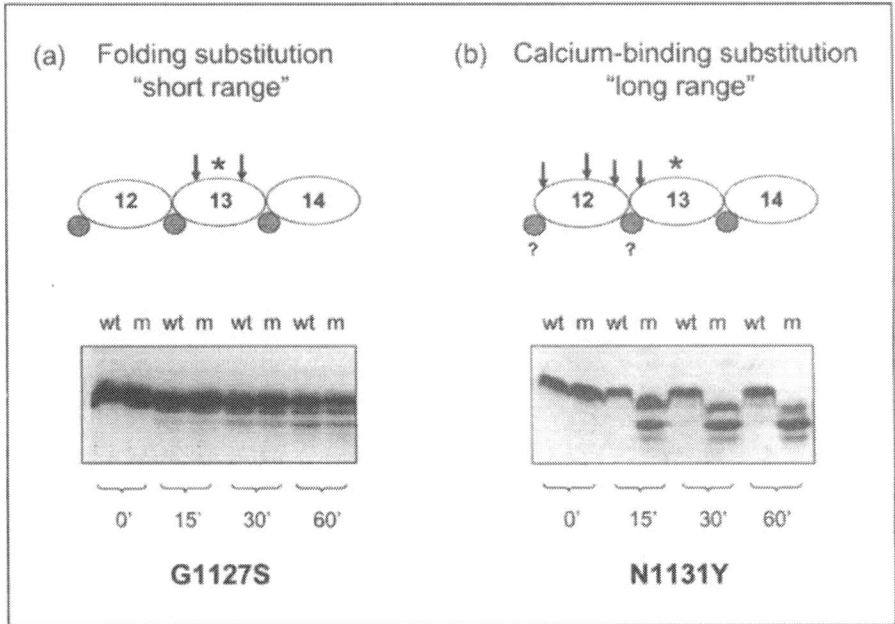

(a) Folding substitution
     "short range"

(b) Calcium-binding substitution
     "long range"

G1127S

N1131Y

Figure 4. Limited proteolysis as a tool for determining structural consequences of *FBN1* missense mutations. A comparison of wild-type (wt) and mutant (m) tryptic digests in the presence of 10 mM Ca^{2+} of the cbEGF12-14 triple domain fragment containing (a) a G1127S or (b) a N1131Y substitution in cbEGF13 (indicated with an asterisk). Cleavage sites identified by N-terminal sequence analysis are shown by arrows above a schematic representation of the fragment.

Calcium Binding Substitutions

The N2144S calcium-binding substitution in cbEGF32 described earlier did not alter the proteolytic susceptibility of a TB6-cbEGF32 pair. This is in contrast to the increased proteolytic susceptibility observed for a cbEGF32-33 pair containing an analogous N2183S substitution in cbEGF33.[17] This finding is consistent with the key role for interdomain calcium binding in stabilising tandem cbEGF domain linkages. The mechanism by which N2144S causes disease thus appears different from that of other calcium-binding substitutions which occur in the context of a cbEGF pair. Thus seemingly similar calcium-binding substitutions may cause variable intramolecular effects, dependent upon domain context.

In contrast to the localised effect of the G1127S substitution, significant structural changes can result from disruption of individual calcium-binding sites within fibrillin-1. Protease analysis of a recombinantly expressed cbEGF10-22 fragment containing a calcium-binding substitution E1073K in cbEGF12 demonstrated that the cleavage sites within the mutant domain showed enhanced susceptibility to proteolysis compared with the wild type. Furthermore, an additional cleavage site was revealed N-terminal to cbEGF11 in the mutant fragment indicative of a longer range structural effect of this mutation.[21]

A comparison of the SDS-PAGE analysis of tryptic digestion of an N1131Y calcium-binding substitution with the G1127S folding substitution, both in cbEGF13 of the cbEGF12-14 triple fragment, is shown in Figure 4. The more disruptive nature of N1131Y is evident and its longer range consequences are demonstrated by a lack of calcium protection in cbEGF12 as well as in cbEGF13 (manuscript in preparation).

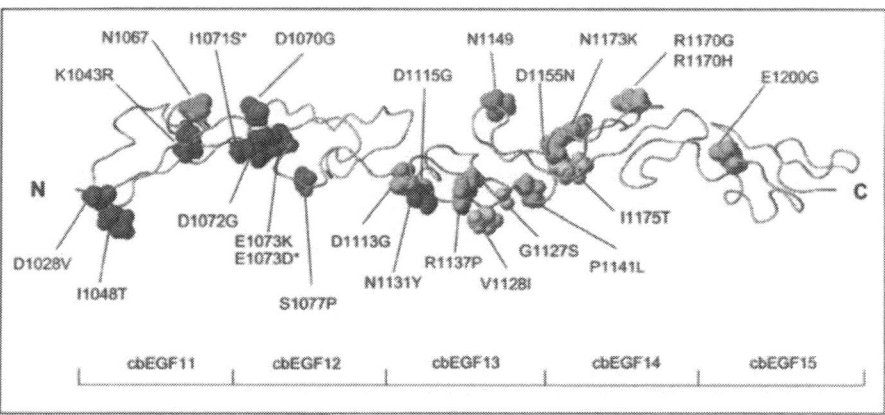

Figure 5. A model of cbEGF11-15 from human fibrillin-1 showing MFS-causing missense mutations in this region. Residues causing nMFS are shown in red, atypically severe MFS in magenta, MFS in cyan and related disorders in yellow. Potential N-linked glycosylation sites in cbEGF11 and cbEGF13 are shown in green. A double mutation is identified by asterisks. Cysteine substitutions are not considered here since they are predicted to affect protein folding rather than protein-protein interactions. The model was constructed based on the coordinates of cbEGF12-13 using Insight II (version 2000; MSI Inc.).

Modelling of FBN1 *Missense Mutations*

The molecular basis for the structural effects of different *FBN1* missense mutations is complex. In the neonatal region of fibrillin-1, for example, missense mutations which affect structurally analogous calcium ligands in different cbEGF domains, or cause substitution of different ligands coordinating the same Ca^{2+}, produce varying phenotypes. Modelling of regions of fibrillin-1 based on the solution structures of cbEGF domain pairs can provide insights into the effects of different substitutions.[5] The model of the cbEGF11-15 region of fibrillin-1 demonstrates the predicted rod-like structure for a contiguous set of cbEGF domains (Fig. 5). Substitution of the calcium binding residue N1131 by the bulkier tyrosine (associated with severe nMFS) would be predicted to result in a conformational change of the major β-hairpin of cbEGF13. It is interesting to speculate that the more moderate phenotype associated with D1113G results from the less disruptive nature of this change. While most of the MFS-causing mutations detected in this region have a predicted structural consequence, three missense mutations without a clear structural effect (K1043R, I1048T, V1128I) are found to cluster on one face of the model. This is located opposite to the potential N-glycosylation sites in cbEGF11 and cbEGF13 and may form part of a molecular interface. An unstructured, extended loop, present in cbEGF12 between cysteines 5 and 6, may also localise to this face of the model and be involved in intra or intermolecular contacts.[5] Analysis of the model shows that substitutions which may affect the calcium-binding properties of cbEGF12 give rise to severe phenotypes. An increase in the intrinsic flexibility of this region resulting from defective calcium binding could distort a potential binding interface which may be important for the microfibril assembly process and/or interactions with other microfibril components. Insights gained from the cbEGF11-15 model will provide a basis for future functional studies.

Correlation of Structural Studies with the Cellular Effects of Missense Mutations

The effects of missense mutations on fibrillin-1 biosynthesis, processing and matrix deposition have been studied by pulse-chase analyses of patient fibroblast cell cultures. The majority

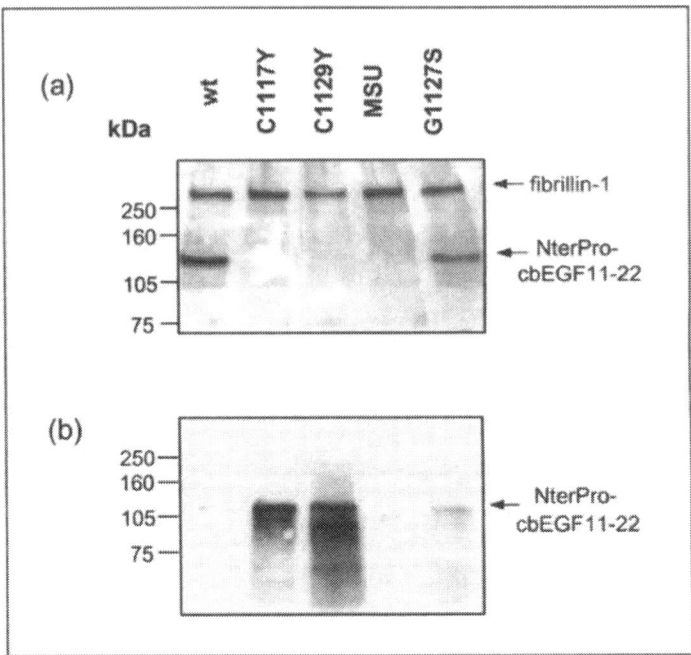

Figure 6. Fibroblast expression of a ~100 kDa wild-type (wt) recombinant fragment (NterPro-cbEGF11-22) from fibrillin-1 and comparison with a fragment containing a C1117Y, C1129Y or G1127S substitution in cbEGF13. MSU-1.1 (MSU) is the untransfected cell-line. The fragment was detected in a) conditioned media or b) cell lysates using an anti-Pro antibody specific for fibrillin-1 following Western blot analysis.

of studies reported in the literature have determined the effects of cysteine substitutions and have usually detected normal synthesis of fibrillin-1, but a delay in its secretion leading to a severe reduction of matrix deposition. However, in some cell lines containing cysteine substitutions a normal secretion profile has been characterised.[22-24] The interpretation of such pulse-chase studies however is complicated by the presence of normal fibrillin-1 produced from the wild-type allele which cannot be distinguished from the mutant product. Consequently it is difficult to identify if the mutant fibrillin-1 is retained in the cell or is delayed and eventually secreted into the extracellular space. To better understand the fate of mutant fibrillin-1 a recombinant system has been developed using a fibroblast host cell. This has been used to study the defects in intracellular trafficking and secretion associated with disease-causing mutations.[25]

In the recombinant system, fibrillin-1 fragments containing two cysteine substitutions associated with classic MFS, C1117Y and C1129Y in cbEGF13, were retained inside the cell. This suggests that the delay in secretion observed in the patient cells is due to selective retention of mutant protein in the cell. In contrast the G1127S folding substitution in the same domain was secreted into conditioned medium (Fig. 6a, b). This, together with the pulse-chase studies of patient fibroblasts containing G1127S, which showed normal synthesis and secretion of fibrillin-1, suggests that this substitution has an extracellular dominant negative effect. A greater disruption to cbEGF13 presumably results from the presence of an unpaired cysteine than from the localised structural effects of G1127S (see above) and consequently variable effects on cellular trafficking and secretion result. The recombinant system can thus be used to study functional effects of the structural changes introduced by missense mutations and to provide additional information on the pathogenic mechanisms leading to MFS.

Conclusions

An interdisciplinary approach has been used to gain insight into the structure of fibrillin-1 and the consequences of MFS-causing missense mutations. These studies have highlighted in particular the variable properties of cbEGF domains and the importance of domain context and type of amino acid substitution in determining the structural effects of mutations. Different pathogenic mechanisms appear to be associated with MFS-causing mutations and the future challenge will be to understand to what degree the complex phenotypes exhibited by patients are due to expression of a mutant protein or to other modulating factors.

Acknowledgements

We thank Stephen Lee and Kristy Downing for assistance with figures. We thank Jemima Cordle and Ji Young Suk for critical reading of this manuscript. This work was supported by the Medical Research Council, British Heart Foundation and the Wellcome Trust.

References

1. Handford PA, Mayhew M, Baron M et al. Key residues involved in calcium-binding motifs in EGF-like domains. Nature 1991; 351:164-167.
2. Rao Z, Handford P, Mayhew M et al. The structure of a Ca^{2+}-binding epidermal growth factor-like domain: Its role in protein-protein interactions. Cell 1995; 82:131-41.
3. Knott V, Downing AK, Cardy CM et al. Calcium binding properties of an epidermal growth factor-like domain pair from human fibrillin-1. J Mol Biol 1996; 255:22-7.
4. Downing AK, Knott V, Werner JM et al. Solution structure of a pair of calcium-binding epidermal growth factor- like domains: Implications for the Marfan syndrome and other genetic disorders. Cell 1996; 85:597-605.
5. Smallridge RS, Whiteman P, Werner JM et al. Solution structure and dynamics of a calcium binding epidermal growth factor-like domain pair from the neonatal region of human fibrillin-1. J Biol Chem 2003; 278:12199-206.
6. Kielty CM, Shuttleworth CA. The role of calcium in the organization of fibrillin microfibrils. FEBS Lett 1993; 336:323-6.
7. Cardy CM, Handford PA. Metal ion dependency of microfibrils supports a rod-like conformation for fibrillin-1 calcium-binding epidermal growth factor-like domains. J Mol Biol 1998; 276:855-60.
8. Yuan X, Downing AK, Knott V et al. Solution structure of the transforming growth factor beta-binding protein-like module, a domain associated with matrix fibrils. EMBO J 1997; 16:6659-66.
9. Pfaff M, Reinhardt DP, Sakai LY et al. Cell adhesion and integrin binding to recombinant human fibrillin-1. FEBS Lett 1996; 384:247-50.
10. Sakamoto H, Broekelmann T, Cheresh DA et al. Cell-type specific recognition of RGD- and nonRGD-containing cell binding domains in fibrillin-1. J Biol Chem 1996; 271:4916-22.
11. Gleizes PE, Beavis RC, Mazzieri R et al. Identification and characterization of an eight-cysteine repeat of the latent transforming growth factor-beta binding protein-1 that mediates bonding to the latent transforming growth factor-beta 1. J Biol Chem 1996; 271:29891-29896.
12. Saharinen J, Taipale J, Keski-Oja J. Association of the small latent transforming growth factor-beta with an eight cysteine repeat of its binding protein LTBP-1. EMBO J 1996; 15:245-53.
13. Lee SSJ, Knott V, Jovanovic J et al. Structure of the integrin-binding fragment from fibrillin-1 gives new insights into microfibril organisation. Structure 2004; in press.
14. Smallridge RS, Whiteman P, Doering K et al. EGF-like domain calcium affinity modulated by N-terminal domain linkage in human fibrillin-1. J Mol Biol 1999; 286:661-8.
15. Werner JM, Knott V, Handford PA et al. Backbone dynamics of a cbEGF domain pair in the presence of calcium. J Mol Biol 2000; 296:1065-78.
16. Reinhardt DP, Ono RN, Sakai LY. Calcium stabilizes fibrillin-1 against proteolytic degradation. J Biol Chem 1997; 272:1231-6.
17. McGettrick AJ, Knott V, Willis A et al. Molecular effects of calcium binding mutations in Marfan syndrome depend on domain context. Hum Mol Genet 2000; 9:1987-94.
18. Kettle S, Yuan X, Grundy G et al. Defective calcium binding to fibrillin-1: Consequence of an N2144S change for fibrillin-1 structure and function. J Mol Biol 1999; 285:1277-1287.
19. Whiteman P, Downing AK, Smallridge R et al. A Gly —> Ser change causes defective folding in vitro of calcium- binding epidermal growth factor-like domains from factor IX and fibrillin-1. J Biol Chem 1998; 273:7807-13.

20. Whiteman P, Smallridge RS, Knott V et al. A G1127S change in calcium-binding epidermal growth factor-like domain 13 of human fibrillin-1 causes short range conformational effects. J Biol Chem 2001; 276:17156-62.
21. Reinhardt DP, Ono RN, Notbohm H et al. Mutations in calcium-binding epidermal growth factor modules render fibrillin-1 susceptible to proteolysis. A potential disease-causing mechanism in Marfan syndrome. J Biol Chem 2000; 275:12339-45.
22. Aoyama T, Tynan K, Dietz HC et al. Missense mutations impair intracellular processing of fibrillin and microfibril assembly in Marfan syndrome. Hum Mol Genet 1993; 2:2135-40.
23. Aoyama T, Francke U, Dietz HC et al. Quantitative differences in biosynthesis and extracellular deposition of fibrillin in cultured fibroblasts distinguish five groups of Marfan syndrome patients and suggest distinct pathogenetic mechanisms. J Clin Invest 1994; 94:130-7.
24. Schrijver I, Liu W, Brenn T et al. Cysteine substitutions in epidermal growth factor-like domains of fibrillin-1: Distinct effects on biochemical and clinical phenotypes. Am J Hum Genet 1999; 65:1007-20.
25. Whiteman P, Handford PA. Defective secretion of recombinant fragments of fibrillin-1: Implications of protein misfolding for the pathogenesis of Marfan syndrome and related disorders. Hum Mol Genet 2003; 12:727-37.
26. Esnouf RM. Further additions to MolScript version 1.4, including reading and contouring of electron-density maps. Acta Crystallogr D Biol Crystallogr 1999; 55:938-40.
27. Merritt EA, Bacon DJ. Raster3D: Photorealistic molecular graphics. Methods Enzymol 1997; 277:505-524.

CHAPTER 16

Genetics of Marfan Syndrome in Mouse Models

Emilio Arteaga-Solis, Harry Dietz and Francesco Ramirez

Introduction

Tensile strength and resilience are critically important properties of the connective tissue that are afforded by the assembly of specialized collagenous and elastic fiber networks. In addition to conferring integrity to virtually every organ system, these extracellular macroaggregates are also involved in modulating several developmental programs and cellular activities through direct or indirect interactions with the surrounding tissues. To a large extent, our current understanding of the multiple functions of the extracellular matrix (ECM) is based on information acquired through comprehensive phenotypic characterization of human patients and mutant mice. Owing to the integrated nature of extracellular networks and cell-matrix interactions, these studies have underscored the need of placing linear genotype-phenotype correlations within the broader context of organismal function. Relevant to Marfan syndrome (MFS), a new paradigm has therefore emerged whereby causative mutations of fibrillin-1 have the capacity to interfere with microfibril assembly and stability, elastic fiber integrity, and cellular performance (or a combination thereof). This chapter reviews the evidence supporting this novel paradigm of MFS pathogenesis which has been gathered through the analysis of genetically engineered mice. It also discusses the implications of animal findings for the clinical management of the human condition. In order to provide a full account of elastic fiber contribution to tissue formation and homeostasis, the review includes a brief description of the components of the elastic fiber and their contribution to the assembly and function of the network.

Elastic Fiber Composition and Assembly

The elastic fiber is a complex network of secreted proteins that is assembled through a hierarchical and tightly regulated process. Postulated components of the elastic fibers include the structurally related fibrillins, latent TGFβ-binding proteins (LTBPs) and fibulins, as well as the emilin, microfibrillar associated glycoproteins (MAPGs) and microfibril-associated proteins (MFAPs) which each display unique primary sequences. Even though the variety of potential elastic fiber components theoretically correlates with tissue-specific arrangements of the extracellular network, pathological correlates have thus far been shown to be limited only to a few of them. They include MFS and fibrillin-1, congenital contractural arachnodactyly (CCA) and fibrillin-2, supravalvular aortic stenosis (SVAS) and elastin, and fibulin 5 and cutis laxa (CL). Replication of these pathological correlates in mice has provided invaluable information about the functions of these four extracellular proteins.

Early studies have suggested that the genesis of the elastic fiber involves deposition of tropoelastin on a pre-formed template of fibrillin-rich microfibrils.[1] More recent evidence had implicated fibulin-5 in directing early assembly of the elastic fiber by providing the molecular

Marfan Syndrome: A Primer for Clinicians and Scientists, edited by Peter N. Robinson and Maurice Godfrey. ©2004 Eurekah.com and Kluwer Academic / Plenum Publishers.

bridge between tropoelastin and cell surface receptors.[2,3] The evidence is based on the finding that fibulin-5 null mice exhibit loose skin, vascular abnormalities and emphysema resulting from disrupted elastogenesis in these tissues. The mouse phenotype closely resembles human CL, a condition subsequently shown to be associated with fibulin-5 mutations.[4] The role of fibulin-5 in guiding elastic fiber assembly is believed to be mediated by calcium-binding epidermal growth factor-like (cbEGF) motifs that bind elastin in a calcium-dependent manner, and by an arginine-glycine-aspartate (RGD) cell attachment sequence that interacts with $\alpha_v\beta_3$ and $\alpha_v\beta_5$ integrins.[2,3] It therefore appears that fibulin-5 coordinates integrin action by promoting and controlling the rate and orientation of elastic fiber growth over a specific cell surface area. Fibulin-5 action is however limited to early elastogenesis and does not extend to the maintenance of the extracellular network.

Similar genetic evidence has revealed the distinct function of elastin in elastic fiber assembly. Mice harboring a targeted deletion of the elastin gene show several phenotypic features of human SVAS.[5,6] They include abnormalities in the vascular wall and altered hemodynamics associated with changes in wall compliance. Elastin null mice survive gestation but die postnatally from subendothelial accumulation of proliferating smooth muscle cells (SMC) that ultimately occlude the vascular lumen. Although the precise underlying mechanism is unknown, the evidence indicates that elastin deficiency promotes abnormal SMC proliferation. The results also indicate that elastin haploinsufficiency is associated histologically with more elastic lamellae, suggesting that the deficiency affects normal vascular development. More generally, these findings can be accounted for by defect in normal cell-matrix interactions that lead to increased SMC proliferation or failure to exit the cell cycle. Hence, elastin's contribution appears to be restricted to elastic fiber growth rather than early organization, as is the case for fibulin-5.

Fibrillins are the building blocks of extracellular microfibrils that form macroaggregates by themselves or in association with elastin in the elastic fiber. There are three distinct fibrillins in humans (fibrillins 1, 2 and 3) and only two in mice (fibrillins 1 and 2). Fibrillins polymerize into a characteristic bead-on-a-string structure that gives rise to the microfibrillar lattice by lateral association and probable inclusion of other structural components. In vitro experiments have shown that fibrillins can co-assemble into fibrils, suggesting that they share common information for microfibril formation.[7] However, immunohistological localization studies have suggested unique functions for each fibrillin.[8-10] This postulate is further supported by the distinct clinical outcomes of fibrillins 1 and 2 mutations in both human patients and gene targeted mice.

Mouse Models of MFS

Three different mutations of the fibrillin 1 gene (*Fbn1*) have been thus far reported in the mouse which has provided new insights into microfibril assembly and MFS pathogenesis (Fig. 1). The first mutation (allele mgΔ) substituted exons 19-25 of *fbn1* with the neomycin resistance cassette (Neo[r]) resulting in an in-frame deletion that was predicted to have dominant-negative activity.[11] However, due to interference from Neo[r], only approximately 5-10% of the mutant allele is expressed compared to the wild-type counterpart. Heterozygous mgΔ/+ mice are born at the expected frequency. Homozygous mgΔ animals instead die suddenly prior to weaning following aortic dilation and dissection. Histologically, they display disruption of elastic laminae and loss of elastin content with the accumulation of amorphous matrix elements, as seen in mature human lesions. However, these abnormalities are focal in nature with the bulk of the aorta showing linear, uninterrupted and parallel elastic fibers. The phenotype of the mgΔ/mgΔ mice therefore demonstrate that minimal (if any) residual microfibril function is sufficient to support the deposition of extended elastic structures, in addition to highlighting the prominent role of fibrillin-1 in elastic fiber homeostasis.

The second mutant allele (mgR) resulted from an aberrant targeting event and produces structurally normal proteins at 15% of the normal level.[12] Heterozygous mgR/+ animals show no abnormalities throughout life, whereas homozygous mgR/mgR mice die between 3 and 6

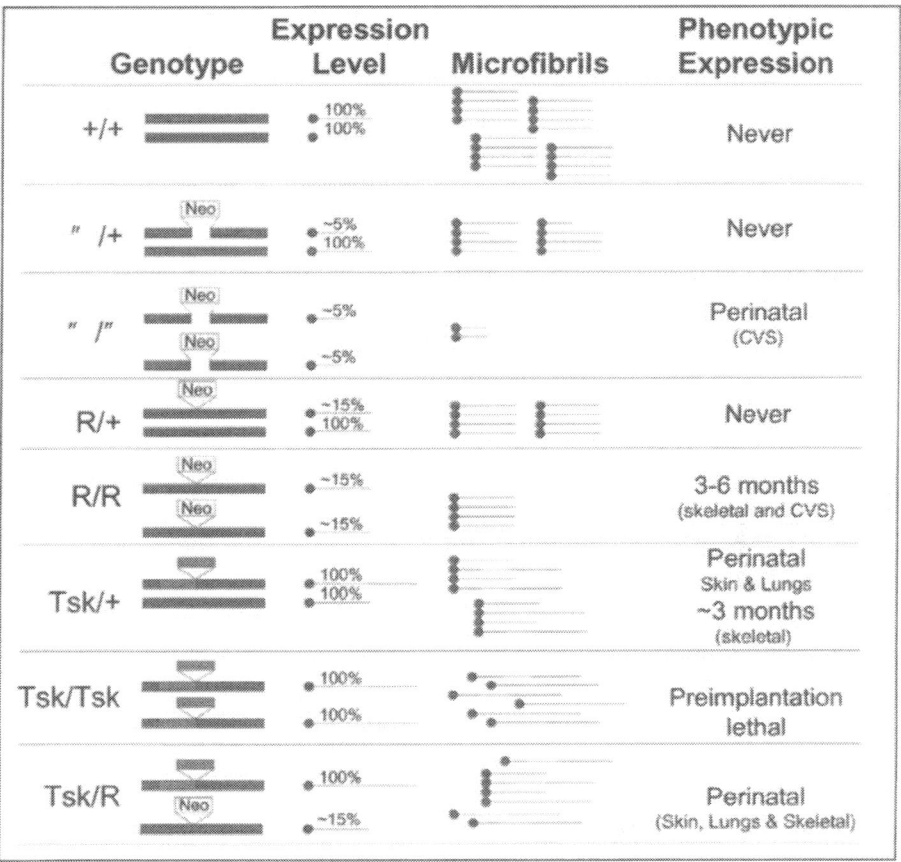

Figure 1. Genotype phenotype correlation in the various mouse models of MFS. Genotypes: wild type allele (+), mgR (R), mgΔ(Δ), tight skin allele (Tsk), neomycin resistance cassette (neo). Expression levels are relative to a wild type allele. Microfibrils, represented by a cluster of four fibrillin-1 monomers, indicate the abundance and character of the microfibrils in each mouse line. Phenotypic expression: Skeletal= kyphosis and bone overgrowth; Cardiovascular (CVS)= aortic dilatation and dissection; Skin= tight skin; Lungs= failure of distal alveolar septation.

months of age of dissecting aneurysm. They also display another MFS trait, bone overgrowth. The longer life span of mgR/mgR mice allowed demonstration of the pathogenic sequelae in these animals. The secondary events include regional elastic fiber calcification, intimal hyperplasia and adventitial inflammation; as a general rule, they were consistently found to be spatially coincident. More importantly, they were subsequently confirmed in MFS patients as well.[13] Focal calcification of intact elastic laminae is seen as the earliest event at about 6 weeks of age, and zones of calcification become more numerous and coalesced over time. It is important to note that calcification and intimal hyperplasia were observed in vessels of human patients that are not predisposed to dilatation or dissection in MFS.[13] This last finding suggests that calcification is not sufficient to commit to aneurysm formation. Intimal hyperplasia with excessive and disorganized deposition of matrix components is evident by 9 weeks of age in mutant mice. Finally, infiltration of mixed inflammatory cells into the media is associated with expression of matrix metalloproteinases (MMPs), intense elastolysis, and structural collapse of the vessel wall.

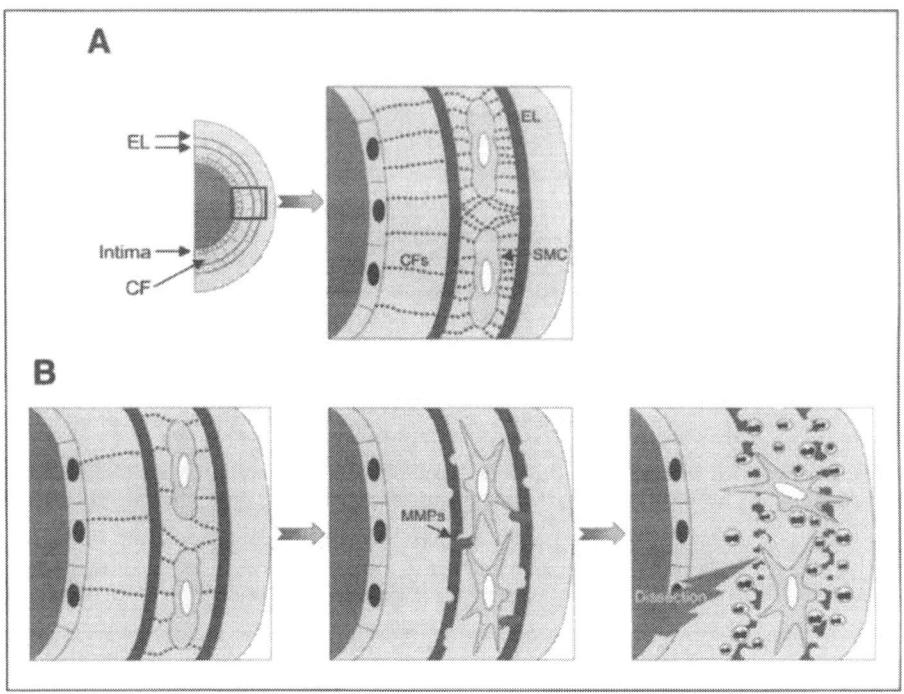

Figure 2. Model of the degradation of elastic fibers in MFS. A) In the normal vessel wall, the elastic lamella (EL) are anchored to the intima and smooth muscle cells (SMC) via connecting filaments (CF) composed by fibrillin-1 contributing to the structural integrity of the vessel wall. B) In MFS patients, there is a limited number of CFs that may decrease over time leading to dedifferentiation of the SMCs and over-production of metalloproteinases. The disruption of the integrity of the vessel allows the infiltration of the inflammatory cells, resulting in overt elastolysis, aneurysm formation and dissection.

Ultrastructural analysis defined events that initiate destructive changes in the aortic media of mgR/mgR mice (Fig. 2). Elastic laminae connect to adjacent endothelial cells and SMC through an intermediate structures composed of microfibrils termed connecting filaments[14-16] RGD sequences of fibrillin-1 support cellular adhesion in vitro via integrin $\alpha_v\beta_3$.[17-19] These interactions may contribute to the structural integrity of the vessel wall through cell anchorage and may coordinate contractile and elastic tensions.[15,16] Homozygous mgR mice show loss of these connections prior to overt elastolysis; as a result, the morphology and synthetic program of flanking vascular SMC change to a dedifferentiated state, perhaps in an abortive attempt to reconstitute a mature environment.[13] Abortive matrix remodeling is characterized by overproduction of multiple structural components and mediators of early elastolysis including MMPs 2 and 9. These observations are well in line with the reported high MMP immunoreactivity at the margins of mature vascular lesions in MFS patients.[20] Subsequent breach of the internal or external elastic laminae in fibrillin-1 deficient mice allows infiltration of inflammatory cells into the media, resulting in intense elastolysis that correlates temporally with aneurysm formation and dissection.

The third mutant allele of *Fbn1* is the so called *Tight skin* (*Tsk*) mutation, an internal gene duplication that results in a larger (418-kD) than normal (350-kD) protein.[21] *Tsk/+* mice display increased connective tissue, bone overgrowth, and lung emphysema; by contrast, *Tsk/Tsk* embryos die before implantation.[22,23] Lung emphysema, bone overgrowth and vascular complications are distinctive traits of MFS patients and mgR/mgR mice. However, *Tsk/+* mice do not exhibit vascular complications, in spite of producing equal amounts of the 418-kD and 350-kD proteins.[22] Analysis of compound heterozygotes for the *Tsk* and mgR mutation, displayed both *Tsk* and MFS traits.[24] This finding was interpreted to suggest that bone and lung abnormalities of *Tsk/+* mice are due to co-polymerization of mutant and wild-type molecules into functionally deficient microfibrils. It was also argued that vascular complications are absent in these animals because the level of functional microfibrils does not drop below a critical threshold. Indirect in vitro evidence suggests that a potential mechanism for the dominant negative effects of incorporating *Tsk* fibrillin-1 into microfibrils is increased proteolytic susceptibility conferred by the duplicated *Tsk* region.[24]

Instructive Roles of Microfibrils

The above studies have provided compelling evidence that multiple structural and cellular events participate in MFS pathogenic progression. More recent observations have also indicated that microfibrillar disorders are actually developmental abnormalities that affect multiple morphogenetic programs and become clinically manifest later in life. Multifunctional molecules (cytokines and growth factors) that are secreted into the extracellular space and that signal through membrane-bound, high-affinity receptors are the primary triggers and modulators of cellular activities underlying development and growth, and tissue homeostasis and repair. In contrast to the fairly well understood mechanisms that mediate the specificity of signal transduction within the confined and compartmentalized environment of the cell, significantly less is known about regulation of signaling molecules' activity in the extracellular space. Findings in fibrillin deficient mice have implicated the participation of extracellular microfibrils in signaling events that control patterning and morphogenesis.[25] In turn, these results have suggested functional coupling between tissue-specific organization of elastic fiber macroaggregates, and their ability to perform instructive as well as structural functions.

The phenotype of mice with a null mutation of the fibrillin 2 (*Fbn2*) gene provided direct evidence linking the organization of a particular matrix architecture with the execution of a specific developmental program.[26] Similar to human CCA, *Fbn2* null mice display joint contractures that improve as the animals age. They also exhibit a patterning defect, in the form of bilateral syndactyly with involvement of both soft and hard tissues (Fig. 3). Syndactyly results from defective commitment of the prospective interdigital mesenchyme, which normally undergoes BMP-induced apoptosis. The patterning defect correlates with altered microfibril assembly only and specifically in the developing autopod, and with inhibition of genes that lie downstream of the BMP-induced pathway of programmed cell death. Moreover, the disorganized matrix is also associated with loss of interdigital tissue competence to respond to apoptotic cues from exogenously administered BMP signals.

An epistatic test using BMP7[+/-] and Fn2[+/-] mice validated the causal relationship between the disorganized matrix and perturbed BMP signaling.[25] BMP7 null mice die soon after birth and display several developmental abnormalities, including polydactyly.[26,27] Like Fbn2[+/-] mice, BMP7[+/-] animals are morphologically normal and viable. However, double heterozygous Fbn2[+/-]; BMP7[+/-] mice exhibit the combined digit phenotype of each nullizygote as the sole phenotypic manifestation. Mechanistically, the ECM/BMP relationship translates into the requirement of a properly organized matrix for either sequestration, distribution, or subsequent signaling of effector molecules. The mechanism by which fibrillin regulates the activity of BMPs remains unclear. One possibility is that fibrillin modulates BMP bioavailability indirectly by organizing interactions with proteoglycans. Alternatively, BMPs and fibrillins may interact directly in a manner analogous to TGF-βs and LTBPs. It is also formally possible that an accessory molecule, such as noggin or chordin, mediates the interaction of growth factors with fibrillin.

Additional evidence for a direct role of the fibrillins in the regulation of cytokine activation and signaling derived from the study of lung disease in MFS. A distinct subset of individuals with MFS have clinically manifest lung disease, generally presenting as spontaneous pneumothorax.[28,29] A greater number show radiographic evidence of obstructive lung disease including upper lobe bullae and/or more diffuse airspace widening. Such findings have been

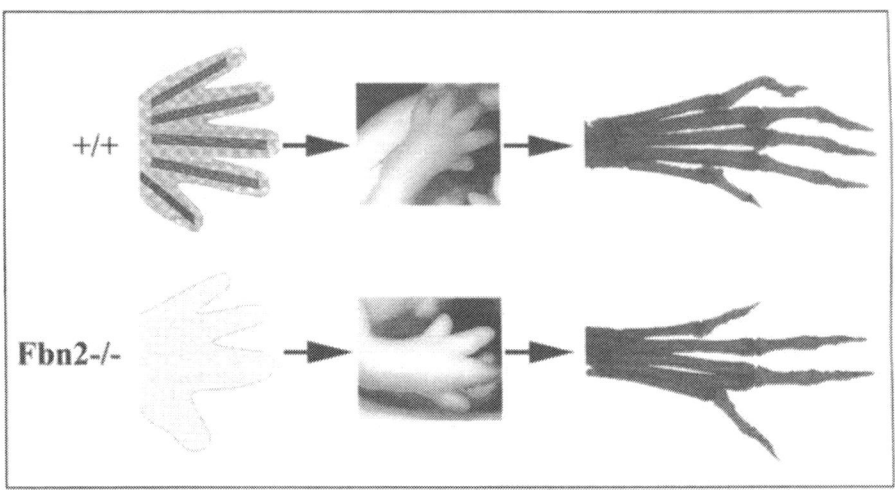

Figure 3. Lack of fibrillin-2 leads to a disorganization of the fibrillin-1 rich microfibrils in the developing mouse autopod causing synadactyly. In the wild-type, microfibrils form a parallel network in the detail region and a more complex network on the interdigital space. In the Fbn2 -/- mice, the fibrillin-1 rich microfibrils appear less abundant and more fragmented.

traditionally reconciled with models that invoke structural failure of the connective tissue with subsequent inflammation and destructive emphysematous changes. In contrast, phenotypic characterization of early postnatal mice deficient in fibrillin-1 revealed homogeneous widening of the distal alveolar saccules without any evidence of inflammation or tissue destruction.[30] Instead, a dramatic paucity of primordial alveolar septa was the predominant finding. These data are most consistent with primary developmental failure of distal alveolar septation rather than classic emphysema, as previously inferred.

TGFβ rapidly emerged as a potential mediator of this morphogenetic perturbation. TGFβs are multipotential cytokines that are synthesized as inactive precursor molecules containing an N-terminal prodomain termed latency associated peptide.[31] LAP remains associated with TGFβ (constituting the small latent complex) and ultimately becomes covalently linked to latent LTBPs 1, 3 or 4 to form the large latent complex.[31,32] It is currently believed that LTBP binding is a prerequisite for secretion. The LTBPs are structurally related to the fibrillins and are thought to participate in TGFβ regulation, in part by targeting TGFβ complexes to the ECM. Because LTBP1 has been immunolocalized to fibrillin-containing microfibrils[35-37] it seemed possible that morphogenetic abnormalities in the microfibril-deficient state (e.g., MFS) might manifest failure of latent complex sequestration and consequent excessive cytokine activation and signaling. Failure of lung septation in fibrillin-1-deficient mice correlates temporally and spatially with excess immunoreactive free TGFβ (Fig. 4).[30] A concomitant reduction in immunoreactive LAP suggests that this reflects increased local activation rather than production and secretion of TGFβ. Increased TGFβ signaling was also documented in the lungs of fibrillin-1-deficient mice.[30] Demonstration that LTBPs bind directly to fibrillin-1 and that antagonism with TGFβ neutralizing antibody rescues lung morphogenesis further substantiates the model.[30,38] Lung disease in aged *Fbn1*-targeted mice evolves to include tissue inflammation and destruction.[30] These data develop a paradigm whereby matrix sequestration of cytokines is critical to their regulated activation and that perturbation of this function can contribute to the pathogenesis of disease (including MFS). They also further the view that early developmental perturbations can predispose to late-onset and apparently acquired disease phenotypes.

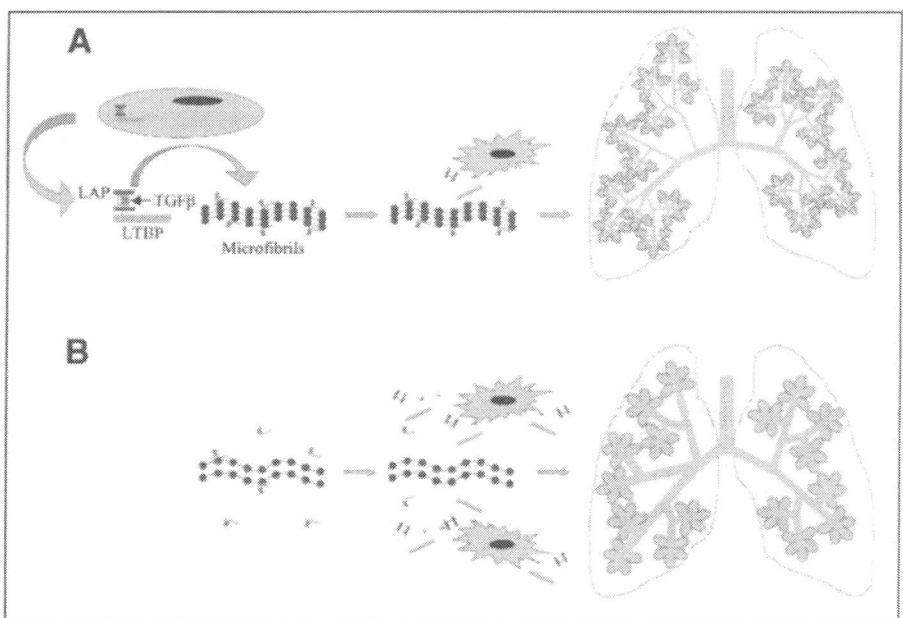

Figure 4. Model for TGFβ sequestration and activation in the developing lung. A) In the wild type lung, the large latent TGFβ complex localizes to the microfibrils regulating its activity leading to proper alveolar septation. B) In the fibrillin-1-deficient mice, decrease/deficient microfibrils lead to excess free TGFβ causing a deregulation of alveolar septation and the formation of an "emphysema like" lung.

Perspectives

The basis for dominant inheritance of MFS: Many lines of evidence have suggested a traditional dominant-negative mechanism for MFS and related fibrillinopathies. In this view, disease outcome is dependent on interference with matrix utilization and/or function of the normal protein imposed by the abnormal protein encoded by the mutated allele. Existing mouse models of MFS and CCA have added new complexity to this model. First, it is now clear that heterozygous hypomorphic alleles of *Fbn1* can be associated with disease manifestations including perturbation of distal alveolar septation.[30] While isolated haploinsufficiency for fibrillin-2 did not result in overt phenotypic manifestations, in the tested inbred backgrounds (an artificial state relative to the human population), the addition of other clinically silent genetic perturbations (e.g., Bmp7[+/-]) was sufficient to achieve relevant phenotypes.[25] While comprehensive characterization of newly created mouse models will be needed to fully scrutinize the dominant-negative hypothesis, existing data already document that haploinsufficiency for the fibrillins is a clinically relevant state. They also offer attractive hypotheses regarding the role of genetic modifiers in the modulation of disease severity.

The role of elastic fiber deficiency in aneurysm formation: While there is ample evidence that MMP-induced matrix degradation can contribute to the progression of aortic aneurysm, and can even independently initiate aneurysm formation under experimental conditions that yield overwhelming elastolysis, the contribution of matrix destruction to the etiology of genetically-predisposed aortic aneurysm remains to be fully elucidated. It is formally possible that elastic fiber destruction either serves as a marker of more critical pathologies, or that co-requisite pathogenetic events have yet to be discovered. Two independent lines of evidence, both derived from the study of mouse models of human disease, support this view. First, fibulin-5-deficient mice do not develop aneurysms despite global and overwhelming impairment of elastic fiber formation during embryogenesis.[2,3] Second, mice that completely lack fibrillin-1 die of aortic dissection within the first two weeks of life despite normal elastin content and largely preserved elastic fiber architecture in the aortic wall (unpublished results). It is also clear from experience with carotid endarterectomy in humans that the isolated normal adventitia has the ability to withstand hemodynamic force, at least in the short-term. Given the transmural expression of fibrillin-1, it seems clear that specific functional deficiencies in the intima and adventitia could contribute to the pathogenesis of MFS aneurysms. Furthermore, the full consequence of loss of matrix-cell attachments that are normally mediated by microfibrillar bundles devoid of elastin remains to be fully elucidated.

The fact that clinically manifest vascular disease in MFS is largely restricted to the root of the ascending aorta warrants further discussion. Prior rationalizations have centered on the unique hemodynamic environment in the ascending aorta. It should be noted, however, that enlargement of the proximal pulmonary artery is also common in MFS,[39] although progression to dissection is never observed. In view of the low-pressure state in the pulmonary circulation, the argument that the hemodynamic environment governs the initiation of aneurysm formation appears difficult to defend. Rather, this factor may play a restricted role in the progression of an established lesion.

As previously mentioned, abnormal performance of resident SMC prior to the onset of elastolysis appears to be the first abnormality seen in the vasculature of mouse models of MFS. These cells show abnormal proliferation and adopt a developmentally inappropriate synthetic repertoire that includes structural matrix elements and MMPs. It is interesting to note that the proximal aortic and pulmonary roots are uniquely populated by neural crest-derived SMC, as opposed to mesenchyme-derived cells in the remainder of the vascular tree.[40] Furthermore, it has been noted that SMC ontogeny dictates response to TGFβ, with neural crest derived cells uniquely adopting the fibroproliferative response characteristic of the MFS aortic root. TGFβ also has the capacity to stimulate MMP production.[41,42] Thus, the elastic fiber-deficient states imposed by the fibulin-5 and fibrillin-1 deficiency states could show discordance for aneurysm formation, because dysregulation of TGFβ is restricted to the latter and is coupled with the preservation of microfibrils in the former.

Abnormal cytokine regulation and other manifestations of MFS: Many manifestations of MFS are difficult to reconcile using pathogenetic models that singularly invoke structural failure of a biomechanically predisposed tissue. These include long bone overgrowth, myxomatous changes of the atrioventricular (AV) valve leaflets, skeletal muscle hypoplasia, and reduced adipogenesis. Furthermore, Shprintzen-Goldberg syndrome, an apparent fibrillinopathy, includes dramatic craniofacial abnormalities and craniosynostosis.[43] Strikingly, all of these findings appear to manifest abnormalities of cellular number and performance that could be adequately explained by excessive signaling of TGFβ superfamily members. Both BMPs and the TGFβs have been directly implicated as determinants of growth plate architecture and maturation. TGFβ plays a prominent role in the endocardial to mesenchymal transformation in the endocardial cushion, the progenitor of the AV valves and portions of the atrioventricular septum.[44] This may also be relevant to the observed increased incidence of AV septation defects in CCA patients deficient in fibrillin-2. Myostatin and other TGFβ superfamily members directly suppress skeletal muscle hypertrophy and adipogenesis. Finally, both TGFβ and selected LTBPs have been associated with critical events in craniofacial development and specifically in maturation of the cranial sutures.[45] Thus, it appears that perturbation of the instructive roles of the fibrillins could contribute to the multisystem pathogenesis of MFS and related disorders. This hypothesis can now be efficiently tested using existing mouse models and developed experimental strategies.

Treatment strategies: Despite the revolution in our understanding of the etiology and pathogenesis of MFS, none of the current treatment modalities derive from this knowledge. Indeed, with the exception of refinement in surgical and imaging techniques, all of the treatment protocols that are currently in place were conceived and initiated in the 1960s and 1970s. While these interventions have clearly improved life expectancy in MFS, significant morbidity and early mortality remain. At best, β-blocker therapy slows the rate of aortic root growth but does not preclude the need for reconstructive surgery that often mandates life-long use of anticoagulants. Virtually no prophylactic therapies are available to modulate the progression of pulmonary, orthopedic and ocular disease or myxomatous degeneration and malfunction of the AV valves.

The old view that fibrillin-1 is needed for basic elastogenesis suggested that individuals with MFS are born with an overwhelming and obligate structural predisposition for tissue failure. This boded poorly for the development of new treatment strategies. The refined understanding of disease pathogenesis afforded by the study of animal models offers a more optimistic view. It now seems plausible, even likely, that inhibitors of inflammation or MMPs, or antagonists of excessive cytokine activation and signaling will prove effective components of productive treatment strategies. Importantly, genetic proof-of-principle experiments can be performed efficiently in mouse models and will provide both rationale and conviction for the development of suitable pharmacologic regimens.

Acknowledgement

This work was supported by the Howard Hughes Medical Institute, National Institutes of Health (AR-41135 and AR-42044), the St Giles Foundation, and the Smilow Center for Marfan Syndrome Research.

References

1. Ross R, Fialkow RJ, Altman LK. The morphogenesis of elastic fibers. Adv Exp Med Biol 1977; 79:7-17.
2. Nakamura T, Lozano PR, Ikeda Y et al. Fibulin-5/DANCE is essential for elastogenesis in vivo. Nature 2002; 415:171-5.
3. Yanagisawa H, Davis EC, Starcher BC et al. Fibulin-5 is an elastin-binding protein essential for elastic fibre development in vivo. Nature 2002; 415:168-71.
4. Loeys B, Van Maldergem L, Mortier G et al. Homozygosity for a _antibáñ mutation in fibulin-5 (FBLN5) results in a severe form of cutis laxa. Hum Mol Genet 2002; 11:2113-8.
5. Li DY, Faury G, Taylor DG et al. Novel arterial pathology in mice and humans hemizygous for elastin. J Clin Invest 1998; 102:1783-7.
6. Li DY, Brooke B, Davis EC et al. Elastin is an essential determinant of arterial morphogenesis. Nature 1998; 393:276-80.
7. Ashworth JL, Kelly V, Rock MJ et al. Regulation of fibrillin carboxy-terminal furin processing by N-glycosylation, and association of amino- and carboxy-terminal sequences. J Cell Sci 1999; 112:4163-71.
8. Zhang H, Apfelroth SD, Hu W et al. Structure and expression of fibrillin-2, a novel microfibrillar component preferentially located in elastic matrices. J Cell Biol 1994; 124:855-63.
9. Zhang H, Hu W, Ramirez F. Developmental expression of fibrillin genes suggests heterogeneity of extracellular microfibrils. J Cell Biol 1995; 129:1165-76.
10. Quondamatteo F, Reinhardt DP, Charbonneau NL et al. Fibrillin-1 and fibrillin-2 in human embryonic and early fetal development. Matrix Biol 2002; 21:637-46.
11. Pereira L, Andrikopoulos K, Tian J et al. Targetting of the gene encoding fibrillin-1 recapitulates the vascular aspect of Marfan syndrome. Nat Genet 1997; 17:218-22.
12. Pereira L, Lee SY, Gayraud B et al. Pathogenetic sequence for aneurysm revealed in mice underexpressing fibrillin-1. Proc Natl Acad Sci USA. 1999; 96:3819-23.
13. Bunton TE, Biery NJ, Myers L et al. Phenotypic alteration of vascular smooth muscle cells precedes elastolysis in a mouse model of Marfan syndrome. Circ Res 2001; 88:37-43.
14. Davis EC. Immunolocalization of microfibril and microfibril-associated proteins in the subendothelial matrix of the developing mouse aorta. J Cell Sci 1994; 107:727-36.
15. Davis EC. Endothelial cell connecting filaments anchor endothelial cells to the subjacent elastic lamina in the developing aortic intima of the mouse. Cell Tissue Res 1993; 272:211-9.
16. Davis EC. Smooth muscle cell to elastic lamina connections in developing mouse aorta. Role in aortic medial organization. Lab Invest 1993; 68:89-99.
17. Pfaff M, Reinhardt DP, Sakai LY et al. Cell adhesion and integrin binding to recombinant human fibrillin-1. FEBS Lett 1996; 384:247-50.
18. Sakamoto H, Broekelmann T, Cheresh DA et al. Cell-type specific recognition of RGD- and non-RGD-containing cell binding domains in fibrillin-1. J Biol Chem 1996; 271:4916-22.

19. D'Arrigo C, Burl S, Withers AP et al. TGF-beta1 binding protein-like modules of fibrillin-1 and -2 mediate integrin-dependent cell adhesion. Connect Tissue Res 1998; 37:29-51.
20. Segura AM, Luna RE, Horiba K et al. Immunohistochemistry of matrix metalloproteinases and their inhibitors in thoracic aortic aneurysms and aortic valves of patients with Marfan's syndrome. Circulation 1998; 98:II331-7; discussion II337-8.
21. Siracusa LD, McGrath R, Ma Q et al. A tandem duplication within the fibrillin 1 gene is associated with the mouse tight skin mutation. Genome Res 1996; 6:300-13.
22. Green MC, Sweet HO, Bunker LE. Tight-skin, a new mutation of the mouse causing excessive growth of connective tissue and skeleton. Am J Pathol 1975; 82:493-512.
23. Bocchieri MH, Jimenez SA. Animal models of fibrosis. Rheum Dis Clin North Am 1990; 16:153-167.
24. Gayraud B, Keene DR, Sakai LY et al. New insights into the assembly of extracellular microfibrils from the analysis of the fibrillin 1 mutation in the tight skin mouse. J Cell Biol 2000; 150:667-80.
25. Arteaga-Solis E, Gayraud B, Lee SY et al. Regulation of limb patterning by extracellular microfibrils. J Cell Biol 2001; 154:275-81.
26. Dudley AT, Lyons KM, Robertson EJ. A requirement for bone morphogenetic protein-7 during development of the mammalian kidney and eye. Genes Dev 1995; 9:2795-807.
27. Luo G, Hofmann C, Bronckers AL et al. BMP-7 is an inducer of nephrogenesis, and is also required for eye development and skeletal patterning. Genes Dev 1995; 9:2808-20.
28. Rigante D, Segni G, Bush A. Persistent spontaneous pneumothorax in an adolescent with Marfan's syndrome and pulmonary bullous dysplasia. Respiration 2001; 68:621-4.
29. Wood JR, Bellamy D, Child AH et al. Pulmonary disease in patients with Marfan syndrome. Thorax 1984; 39:780-784.
30. Neptune ER, Frischmeyer PA, Arking DE et al. Dysregulation of TGF-, activation contributes to pathogenesis in Marfan syndrome. Nat Genet 2003; 33:407-411.
31. Munger JS, Harpel JG, Gleizes PE et al. Latent transforming growth factor-beta: structural features and mechanisms of activation. Kidney Int 1997; 51:1376-82.
32. Kanzaki T, Olofsson A, Moren A et al. TGF-beta 1 binding protein: a component of the large latent complex of TGF-beta1 with multiple repeat sequences. Cell 1990; 61:1051-61.
33. Munger JS, Huang X, Kawakatsu H et al. The integrin alpha v beta 6 binds and activates latent TGF beta 1: a mechanism for regulating pulmonary inflammation and fibrosis. Cell 1999; 96:319-28.
34. Munger JS, Harpel JG, Giancotti FG et al. Interactions between growth factors and integrins: latent forms of transforming growth factor-beta are ligands for the integrin alphavbeta1. Mol Biol Cell 1998; 9:2627-38.
35. Gibson MA, Hatzinikolas G, Davis EC et al. Bovine latent transforming growth factor beta 1-binding protein 2: molecular cloning, identification of tissue isoforms, and immunolocalization to elastin-associated microfibrils. Mol Cell Biol 1995; 15:6932-42.
36. Hyytiainen M, Taipale J, Heldin CH et al. Recombinant latent transforming growth factor beta-binding protein 2 assembles to fibroblast extracellular matrix and is susceptible to proteolytic processing and release. J Biol Chem 1998; 273:20669-76.
37. Raghunath M, Unsold C, Kubitscheck U et al. The cutaneous microfibrillar apparatus contains latent transforming growth factor-beta binding protein-1 (LTBP-1) and is a repository for latent TGF-beta1. J Invest Dermatol 1998; 111:559-64.
38. Isogai Z, Ono RN, Ushiro S et al. Latent transforming growth factor beta-binding protein 1 interacts with fibrillin and is a microfibril-associated protein. J Biol Chem 2003; 278:2750-7.
39. Nollen GJ, van Schijndel KE, Timmermans J et al. Pulmonary artery root dilatation in Marfan syndrome: quantitative assessment of an unknown criterion. Heart 2002; 87:470-1.
40. Bergwerff M, Verberne ME, DeRuiter MC et al. Neural crest cell contribution to the developing circulatory system: implications for vascular morphology? Circ Res 1998; 82:221-31.
41. Santibáñez JF, Guerrero J, Quintanilla M et al. Transforming growth factor-beta1 modulates matrix metalloproteinase-9 production through the Ras/MAPK signaling pathway in transformed keratinocytes. Biochem Biophys Res Commun 2002; 296:267-73.
42. Lin SW, Lee MT, Ke FC et al. TGFbeta1 stimulates the secretion of matrix metalloproteinase 2 (MMP2) and the invasive behavior in human ovarian cancer cells, which is suppressed by MMP inhibitor BB3103. Clin Exp Metastasis 2000; 18:493-9.
43. Sood S, Eldadah ZA, Krause WL et al. Mutation in fibrillin-1 and the Marfanoid-craniosynostosis (Shprintzen-Goldberg) syndrome. Nat Genet 1996; 12:209-11.
44. Nakajima Y, Yamagishi T, Hokari S et al. Mechanisms involved in valvuloseptal endocardial cushion formation in early cardiogenesis: roles of transforming growth factor (TGF)-beta and bone morphogenetic protein (BMP). Anat Rec 2000; 258:119-27.
45. Dabovic B, Chen Y, Colarossi C et al. Bone abnormalities in latent TGF-[beta] binding protein (Ltbp)-3-null mice indicate a role for Ltbp-3 in modulating TGF-[beta] bioavailability. J Cell Biol 2002; 156:227-32.

APPENDIX

Marfan Syndrome Patient Organizations

The First International Symposium was held in Baltimore, MD, U.S.A., in 1988. In 1992 the 2nd International Symposium on the Marfan Syndrome, sponsored by the National Marfan Foundation (U.S.A.), was held in San Francisco, U.S.A. It was during this 1992 Symposium that a number of Marfan Organizations from throughout the world gathered with the purpose to establish and International Federation of Marfan Syndrome Organizations (IFMSO). It was voted that each member organizations should retain its autonomy, while subscribing to the "IFMSO Statement of Purpose."

The purpose of this newly founded organization was threefold. To share current, accurate information about MFS worldwide and facilitate international communication among medical professionals and the general public; to establish standards for diagnosis and treatment of the MFS; and to support and foster research throughout the world and facilitate communication with research centers and researchers worldwide.

Following its founding in 1992, the IFMSO has fostered and supported six additional international Research Conferences: Berlin, Germany, 1994; Davos, Switzerland, 1996; Helsinki, Finland, 1998, Seattle, WA, U.S.A., 2001. The IFMSO has also supported the 1986 meeting that generated the diagnostic criteria for MFS, known as the "Berlin Nosology,"[1] and also, in 1996, the revised, "Ghent Nosology."[2]

In 2002 the IFMSO launched its website, www.marfanworld.org, to facilitate communication between MFS organizations and people with the MFS throughout the world. Through links to the various member organizations, visitors to the web site can access resources and publications that can help them in understanding and managing the MFS. The site averages 90 visitors a day, with a total of over 32,000 visitors in 2003. IFMSO member organizations are located in over 20 countries throughout the world. Although the IFMSO web site is in English, it provides links to the member organizations' web sites in their own languages. It also serves as a common source for research information and upcoming conferences and activities.

Priscilla Ciccariello
January, 2004

---

[1]1986: Diagnostic criteria for Marfan syndrome refined in the Berlin Nosology for Heritable Disorders of Connective Tissue. Beighton P, De Paepe A, Danks D et al. International nosology of heritable disorders of connective tissue, Berlin, 1986. Am J Med Genet 1988: 29:581-594. [2]1996: Diagnostic criteria for Marfan syndrome revised further in the Ghent Nosology. De Paepe A, Devereux RB, Dietz HC et al. Revised diagnostic criteria for the Marfan syndrome. Am J Med Genet 1996; 62: 417-426.

Marfan Syndrome: A Primer for Clinicians and Scientists, edited by Peter N. Robinson and Maurice Godfrey. ©2004 Eurekah.com and Kluwer Academic / Plenum Publishers.

The following list contains contact information for national Marfan Syndrome patient organizations. In addition, current information can always be obtained from the IFMSO website, www.marfanworld.org.

Argentina
Mario del Frade, Chairman
Marfan Argentina
Charcas 33913er piso, office 32
Buenos Aires, Republic Argentina
www.marfan.com.ar
E-mail: Marfan@argentina.com

Australia
Queensland
Susanne Hemphill
Marfan Association, Queensland Branch
P.O. Box 294
Sumner Park, Queensland 4074
Australia
Tel.: ++61 7 3376 6160
Fax: ++ 61 7 3279 1927
E-mail: marfanaustralia@hotmail.com
Victoria
Marfan Association Victoria Inc.
Sally Ferguson
P.O. Box 42
Sunbury
Victoria 3429
Australia
Tel./Fax: ++ 61- 3-5784 1357
E-mail: yvette1@bigpond.com
        jnix@ssc.net.au
New South Wales
Marfan Support Group
Trudy Whaite
Sydney
Tel.: ++61 2 8230 0424
E-mail: mail@marfan.net.au
www.marfan.net.au

Belgium
Yvonne Flemal - Jousten
Association Belge du Syndrome de Marfan
27A, rue Residence Air Pur
B 4623 - Magnee
Belgium
Tel.: ++32 4 355 13 59
Fax: ++32 4 355 39 27
E-mail: yvonne.jousten@easynet.be
www.marfan.be

Myriam Criel
Kontaktgroep Marfan
Van Hemelrijcklei 45
B2930-Brasschaat
Belgium
Tel./Fax: ++32 3 653 1950
E-mail: myriam.criel@eudoramail.com

Brazil
Adriano Rodrigues
Rua 01 Quadra A N.6-Ap 201
Residencial Filadelfia – Planalto
Sao Luis, MA, Brazil
CEP 85060 – 275
Tel.: 55 98 244 7120
E-mail: acrodrigues1977@yahoo.com.br

Canada
Canadian Marfan Association
Central Plaza Postal Outlet
128 Queen St. S., P.O. Box 42257
Mississauga Ontario L5M 4Z0
Canada
Tel.: (905) 826-3223
Fax: (905) 826-2125
E-mail: info@marfan.ca
www.marfan.ca

Cuba
Carolina Douglas
Cuban Association of Marfan Syndrome
Calle 34 No. 310, apto2
entre 3ra y 5ta, Miramar
Ciudad de La Habana 11300
Cuba
Fax: ++337 330145

Denmark
Ms. Bodil Davidsen
Danish Marfan Association
Frederikshave 34 st. th.
DK-3400 Hillerod
Denmark
Tel.: ++45 4826 3652
Fax: ++45 4826 3752
E-mail: msdk@get2net.dk
www.marfan.dk

Finland

Leila Raninen
Finnish Marfan Association
P.O. Box 1328 SF
00101 Helsinki,
Finland
Tel.: ++358 13 478 282
E-mail: hallitus@marfan.fi
www.marfan.fi

France

Annette Belloncle
Association Syndrome de Marfan Amis
d'Antoine (A.S.M.A.A.)
B.P. 63
F-95160 - Montmorency
France
Tel./Fax: ++33 139 64 56 12
E-mail: jtihierr@free.fr

Paulette Morin
Association Française du Syndrome
de Marfan (A.F.S.M.)
13, Allee des Terrasses
F-77200 Torcy
France
Tel.: ++33 164 620 375
Fax: ++33 160 057 910
E-mail: vivremarfan@caramail.com
www.geocities.com/vivremarfan

Germany

Marfan Hilfe (Deutschland) e.V.
Postfach 0145
D-23691 Eutin
Germany
Tel.: ++49 4521 8305035
Fax: ++49 4521 8305034
E-mail: kontakt@marfan.de
www.marfan.de

Greece

Greek Association of the Marfan
Syndrome
Xanthi Zafiris
Saliverou 5 N. Filothei
Marousi Athens Greece
Tel.: 0113-093-217-9451
Fax: 0113-010-685-4051
E-mail: marfansyndromegr@yahoo.com

Ireland

Mary O'Driscoll
Marfan Syndrome Support Group Ireland,
Ltd.
78 Whitehorn Drive
Palmerstown, Dublin 20
Ireland
E-mail: Marfan@eircom.net
www.marfan.ie

Italy

Silvia Salice
Associazione Vittorio
Strada alla Villa d'Aglié 23/3
I-10132 Torino
Italy
Tel./Fax: ++39 11898 0995
E-mail: michaela.tarantini@marfan.info
www.marfan.info

Japan

Yoko Shimozaki
Marfan Network Japan
Higashi-nakashinjuku2-1-23
Kashiwa, Chiba
Japan
Tel.: 81-4-7169-3484
Fax: 81-53-454-6764
E-mail: info@marfan.gr.jp
www.marfan.gr.jp/

Mexico

Marcos and Geraldine Duran
Sindrome de Marfan de Mexico A.C.
Paseo de Lilas #92
Loc. 58, Suite 143
Bosques de Las Lomas
Mexico D.F. C.P. 05120
Mexico
Tel/Fax: ++52-452260
E-mail: admin@marfan.org.mx
www.marfan.org.mx

Netherlands

Peter Alkemade
Contactgroep Marfan Nederland
Bultenkamp, 3
NL – 7429 – AV Colmschate
The Netherlands
Tel.: ++ 31 570 56 45 91
E-mail: info.marfan@shhv.nl
www.shhv.nl

New Zealand
Ian Edmunds
32 Marston Road
Timaru, South Canterbury
New Zealand
Tel.: ++ 6436883723
Fax: ++ 6436885198
E-mail: iannie@world-net.co.nz

Norway
Wenche Snekkevik
Marfanforeningen
5350 Brattholmen
Norway
Tel.: ++47 – 97 58 68 08
E-mail: info@marfan.no
   Wenche@marfan.no
www.marfan.no

Poland
Polish Marfan Association
Swietojanska 49/17
PI-81-391 Gdynia
Poland
Tel.: ++48 58 629 4425 (home)
   ++48 601 286 245 (mobile)
Fax: ++48 58 629 4425
E-mail: marfan@marfan.pl
   marfan@gdansk.sprint.pl
www.marfan.pl

Singapore
Willie Loh Chung Wei
Blk 471 Ang Mo Kio Ave 10 #11-774 SE
   560471
Republic of Singapore
Tel.: +65 91455708
E-mail: Lchungwe@yahoo.com
SingaporeMarfanSupportWebsite:
www.marfansingapore.org
E-mail: support@marfansingapore.org

Slovakia
Maria Dusinska
Slovak Marfan Association
Institute of Preventive and Clinical
   Medicine
Limbova' 14
833 01 Bratislava
Slovakia
Tel./Fax: ++42 12593 69270
E-mail: dusinska@upkm.sk
http://members.fortunecity.com/marfan

South Africa
South African Marfan Syndrome
   Organisation (Suid-Afrikaanse
   Marfansindroom
   Ondersteuningsgroep)
P.O. Box 7294
Pretoria 0001
South Africa
Tel.: ++012 333-9366

Spain
Nicolás Beltrán
Asociacion de Afectados Sindrome
   de Marfan (SIMA)
C/ San Agatangelo no 44 Bajo izda
E-03007-Alicante
Spain
Tel.: ++34 966 14 15 80
E-mail: sima@marfansima.org
   asociacion_sima@hotmail.com
www.marfansima.org

Sweden
Karin Olsson
Svenska Marfanföreningen
c/o Ulla Frick
St. Eriksgatan 50 A
112 34 Stockholm
Sweden
Tel.: +46 8 651 36 09
E-mail: bertil.frick@swipnet.se
   karin.olsson@vansterpartiet.se
www.marfanforeningen.se

Switzerland
Marfan Foundation (Switzerland)
Marktgasse 31
CH-3011 Bern
Switzerland
Tel.: ++41 31 312 11 22
Fax: ++41 31 312 11 20
E-mail: info@marfan.ch
www.marfan.ch

United Kingdom

Diane Rust
Marfan Association UK
Rochester House
5, Aldershot Road
Fleet, Hampshire GU51 3NG
United Kingdom
Tel.: ++44 1252 810472
Fax: ++44 1252 810473
E-mail: marfan@tinyonline.co.uk
www.marfan.org.uk/

United States of America

Carolyn Levering
National Marfan Foundation
22 Manhasset Avenue
Port Washington, New York 11050
U.S.A.
Tel.: ++ 516-883-8712
       ++800-862-7326
Fax: ++516-883-8040
E-mail: staff@marfan.org
www.marfan.org

World Organizations:

EMSN

Helga Straume
Chairman
European Marfan Support Network
    (EMSN)
Lottenzein, 16
N-2319-Hamar
Norway
Tel.: ++47 67 57 17 44
       ++47 984 26 920 (mobile)
E-mail: helga.straume@nrk.no
http://www.marfan.de/emsn/index.htm-

IFMSO

Priscilla Ciccariello
President
International Federation of Marfan
    Syndrome Organizations (IFMSO)
c/o National Marfan Foundation
22 Manhasset Avenue
Port Washington, NY 11050
E-mail: cilla71@aol.com
www.marfanworld.org
Revised: April 2, 2004

INDEX

Printed in Great Britain
by Amazon